A HISTORY OF EARLY TELEVISION

ROUTLEDGE LIBRARY OF MEDIA AND CULTURAL STUDIES

Other titles in this series

A HISTORY OF PRE-CINEMA
Edited with a new introduction by Stephen Herbert
3 volume set

A HISTORY OF EARLY FILM
Edited with a new introduction by Stephen Herbert
3 volume set

CAHIERS DU CINÉMA
Edited by Jim Hillier and Nick Browne
4 volume set

RACHAEL LOW'S HISTORY OF BRITISH FILM
Rachael Low
With a new introduction by Jeffrey Richards
7 volume set

A HISTORY OF EARLY TELEVISION

Volume I

Selected and with a new introduction by
Stephen Herbert

LONDON AND NEW YORK

First published 2004
by Routledge

2 Park Square, Milton Park, Abingdon, Oxfordshire OX14 4RN
52 Vanderbilt Avenue, New York, NY 10017

Routledge is an imprint of the Taylor & Francis Group, an informa business

First issued in hardback 2019

Typeset in Times by Keystroke, Jacaranda Lodge, Wolverhampton

British Library Cataloguing in Publication Data
A catalogue record for this book is available from the British Library

Library of Congress Cataloging in Publication Data
A catalog record for this book has been requested

ISBN 13: 978-0-415-32665-0 (Set)
ISBN 13: 978-0-415-32666-7 (Volume I) (hbk)

ISBN 13: 978-1-032-66035-6 (Volume 1 pbk)

Publisher's note
The publisher has gone to great lengths to ensure the quality of this reprint but points out that some imperfections in the original book may be apparent.

CONTENTS

VOLUME I: PART 1
Dreams and Experiments

VOLUME I: PART 2
Early Successes

VOLUME II: PART 4
A New Era (Britain and Europe)

ACKNOWLEDGEMENTS

The publishers would like to thank the following for permission to reprint their material:

The Hastings Museum and Malcolm Baird for permission to reprint **Letters, John Logie Baird/Will Day:**

Letter 1 [5 April 1924. Baird to Day]
Letter 2 [1 May 1924. Baird to Day]
Letter 3 [2 May 1924. Day to Baird]
Letter 4 [25 July 1924. Day to Baird]
Letter 5 [28 July 1924. Baird to Day]
Letter 6 [6 September 1924. Day to Baird]
'Television' (letter, Baird), 29 January 1927

Nature for permission to reprint 'Television', E. Taylor Jones, *Nature*, 18 June 1927.

Disclaimer

INTRODUCTION

A mighty maze of mystic, magic rays
Is all about us in the blue,
And in sight and sound they trace
Living pictures out of space
To bring a new wonder to you

(From 'Television', lyrics by James Dyrenforth.
Sung by Adele Dixon, accompanied by the BBC Television
Orchestra, at the Official Opening of the BBC Television
service in November 1936)

In the early twenty-first century, broadcast television is an established part of the lives of many millions all over the world, bringing information and entertainment directly into our homes, on many hundreds of channels. Webcams and mobile videophones add to our ability to 'see at a distance'. This ability has been, since ancient times, part of mankind's desire to extend the limits of its perceptive apparatus. The telescope fulfilled this ambition to some extent, and the camera obscura technique could bring live images of the outside world into a darkened room. But the challenge of sending images considerable distances, by day or night and with a range exceeding that of optical devices, had to wait until the manipulation of an electrical current became possible. Inventors then began to sketch out plans for television. The facsimile transmission of single static images (initially line drawings) was achieved in the mid-nineteenth century, but it would take many decades before the first fuzzy, flickering, but moving silhouettes of waving fingers, snapping scissors, and the head of a ventriloquist dummy would be sent by line, and then by radio, to a 'distant' receiver. Within a few years recognisable human faces were being broadcast to makeshift televisors put together by radio hams, and by the early 1930s a wider public was taking an interest in the entertainment possibilities of the new medium. The particular excitement of 'live' transmissions, transporting the viewer to a distant event as it happened, was of great importance in the early years, and this aspect was used to differentiate television from cinema. The broadcast of motion picture films was considered of secondary importance. By the summer of 1939 those pioneer 'lookers-in' with the financial means (and within range of limited broadcasting distances), in Britain, Germany and other European countries were enjoying regular reception of good quality images and varied content, and the first major public service was launched in the USA. With the outbreak of the Second World War, public television in Britain and Europe ceased, and in the USA was soon curtailed for the duration.

These volumes gather together a selection of books, articles and news items relating to this first developmental period of television, before it became the ubiquitous medium that we enjoy and suffer today. The selection has been limited to English-language material. When used as a source for historical research it should be remembered that national bias and other subjective attitudes are present in some pieces, and our understanding of the physiological/psychological mechanisms of moving-image perception has progressed. A full understanding of the reasons for the success or failure of particular inventions and ventures requires a reading of the recent accounts by today's researchers. I would suggest that readers use this collection in conjunction with at least one modern book on the subject; I would particularly recommend *Television, An International History of the Formative Years*, by R.W. Burns (Institute of Electrical Engineers, 1998), which covers the whole era in a comprehensive and informed way.

Volume I

Transmission of a complete image (or for 'real television' a complete *moving* image) simultaneously, is not possible. The scene to be sent must first be broken down into a sequence of pulses representing points from the image; these points are then reproduced in the correct order and brightness at the receiving end, to give a representation of the original scene. This was the task that faced the first experimenters. Eventually, television would become a complex field involving the multiple issues of aesthetics, economics and politics – as well as technical aspects – of programme production and distribution. Initially, however, the struggle to achieve the very difficult 'first step' of transmitting an image along a wire preoccupied the visionaries of the late nineteenth century, and their ideas and experiments peppered the pages of technical journals of the day. A small selection of these starts off our collection.

Part 1: Dreams and Experiments
Fantasies and predictions, and the first proposals for a practical system

The idea of an 'electric telescope' for distant viewing was suggested by several technical writers and experimenters in the late 1870s. A notice in the *English Mechanic and World of Science* (31 January 1879), of 'The Telectroscope', by Constantin Senlecq of France, repeated the earlier suggestion of Portuguese Professor Adriano de Paiva: the use of selenium as the material suitable for changing light fluctuations into the variable electrical current required. Senlecq's intended result was a pencil-drawn static picture, a form of facsimile ('fax') transmission and not television as we understand it.

Denis D. Redmond of Ireland apparently achieved some success transmitting 'built-up images of very simple luminous objects' with the use of platinum at the receiving end of a multiple-channel arrangement, but his report 'An Electric Telescope' – also in the *English Mechanic* (7 February 1879) – states that he failed in his attempts to produce a single-channel device with the picture formed sequentially, due to the sluggish response of selenium in the camera.

Following mention in the *Scientific American* in May 1879 concerning the work of George R. Carey, USA, 'Seeing by Electricity' (5 June 1880), reproduced here, described and illustrated his proposed multi-wire apparatus; but the result was a static image on paper. Mr William E. Sawyer wrote in, 'Seeing by Electricity' (12 June 1880), mentioning his own idea – a more sophisticated single-wire system using a spiral scanning technique producing a directly viewable image – but again lamenting the slow response of selenium.

The proposed purposes to which 'seeing at a distance' would be used, when achieved, were many and varied, from developments of facsimile transmission producing static paper copies at the receiver – but with 'live' scenes at the transmitting end rather than static paper originals – to television plays.

By the late 1870s there were rumours that Thomas Edison, and a little later Alexander Graham Bell, were working on television systems. These failed to appear but the popular press and satirists soon latched onto the idea, and most books covering television's origins reproduce one or two cartoons, such as 'Edison's Telephonoscope', which appeared in the English *Punch Almanack*, 1879; a father keeps an eye on his distant tennis-playing daughter by means of a videophone.

Two relevant cartoons, neither previously recorded in television histories, appeared in *Punch*'s rival, the satirical magazine *Fun*, in July 1889 and are reproduced here. In 'Professor Goaheadison's Latest' a gentleman wishes to consult his doctor, Sir Settemup Pilliboy. Not happy about the prospect of a journey to London, he is told of Professor Goaheadison's Far-Sight Machine: 'by means of which a person in Nyork can actually see another person in Shicaago, or Borston, or even "Frisco"'. The writer uses the concept to make a political comment about contemporary concerns. The text has an accompanying drawing of a videophone, by James Sullivan, best-known for his cartoons featuring the working man. The idea of medical diagnosis by television was still being used by British cartoonists in the early 1930s, in the *Daily Mirror* and *Punch*.

A subsequent issue of *Fun* (17 July 1889) has a three-panel cartoon strip by Sullivan: 'Goaheadison's Real Latest', featuring the Far-Touch Machine. I have included both the original cartoon, with manuscript captions on the mount, and the published version. The artist imagines a form of trans-Atlantic virtual-reality boxing. A Far-Touch Machine is seen as the logical follow-up to Edison's proposed Far-Sight Machine; the latter we are told is being used by the boxers to keep sight of each other as they trade punches. If Edison ever engaged in technical work on television it was soon abandoned, and instead his team started experiments in motion picture film production and exhibition.

While the possibilities of a Far-Touch Machine were not seriously considered by the press or scientific journals of the time, the period's visionary ideas of a machine that would allow 'seeing at a distance' – either one-way viewing for entertainment, as a method of domestic viewing of stage dramas, etc., or as a two-way videophone for personal communication – were not dismissed by all commentators. The telephone had enabled voices to be instantly heard at vast distances, so the idea of a visual equivalent was not a huge conceptual jump.

In 1884 German engineering student Paul Nipkow patented the scanning disc, which in the 1920s would be used to produce the first 'live' television images. (A scanning arrangement needed only one channel of communication between sending device and receiver, making a practical system possible.) Proposals for TV systems started to appear in popular science books, reaching a wider public. The engraving of 'Le Telephote' is reproduced from an 1891 French book of popular science, *Physique Populaire*, by Emile Desbeaux, and is based on an idea by Lazare Weiller, who proposed a system using a mirror scanner. Although this particular proposal was never achieved, facsimile scanning and transmission of line drawings, text, etc., had been accomplished as early as the 1840s, and many suggestions for early 'television' systems were in fact improved methods of transmission of static images.

In 'Seeing by Telegraph', H. Trueman Wood, M.A., *Science for All*, R. Brown, ed. (Cassell, 1894) considers a form of true television communication – what we would recognise as a videophone or webcam – suggesting 'There is a possibility, but, as yet, no great probability of it'. He outlines the basic requirements of a crude TV system, using selenium cells at the transmission end. But this theoretical arrangement required direct connection from each transmitted element to each received element: a bundle of 100 wires even for a very simple image; impractical for any useful purpose. Trueman Wood then reviews the progress of two experimenters, Ayrton and Perry, whose first suggestions for a television system date back to the late 1870s. They had recently demonstrated the principle of scanning, though not at real-time speed, as the selenium cells were not sensitive enough to act sufficiently fast. The writer then describes Shelford Bidwell's more easily achievable Telephotograph system for distance facsimile drawing.

The word 'Television' was first used as the title of a paper by Constantin Perskyi, read at the 1900 International Electricity Congress. In 1907, 'The Problem of Television', appeared in the *Scientific American Supplement* (15 June). This article is sometimes cited as being the first in English to use the term 'television', but using it to mean facsimile transmission. In fact, although it outlines various facsimile methods, by French experimenter Edouard Belin and others, it suggests that a successor to Belin's machine is 'expected some day to solve completely the problem of "television", when these images will be reproduced with a more rapid succession than that corresponding with the persistence of retina images'. So it actually uses the word 'television' to mean a future system that will reproduce scenes with images that change at an imperceptible rate; i.e., real live television. Also in 1907 Boris L. Rosing of Russia patented a system with a cathode-ray tube as a receiver.

The following year an important early text was published in *Nature*: 'Distant Electric Vision' (18 June 1908), a letter by A.A. Campbell Swinton, in which he proposes the use of 'kathode rays' as scanning beams in both transmitter (camera) *and* receiver, but recognises that no sufficiently sensitive device exists to change the electric current fast enough to produce a result.

Part 2: Early Successes
Television becomes a reality: the first successful experiments

In 1909, three experimenters had some success with demonstrations that incorporated certain aspects of television: Max Diekmann of Munich (but the 'images' sent were direct electrical contacts made by a metal stencil, not real pictures or objects); Ernst Ruhmer, who used a mosaic of twenty-five selenium cells; and Georges Rignoux, who achieved perhaps the most impressive result and would go on to further progress. Theoreticians continued to suggest television systems. Campbell Swinton gave an important address to the Rontgen Society in 1911 – a report from *The Times Engineering Supplement*, 'Distant Electric Vision' (15 November 1911) is reproduced here – with suggestions of how the sluggishness of the materials then in use may be overcome in the future.

In the following years important developments were made by Boris L. Rosing, Dionys von Mihaly of Hungary (of whom, more later), Charles Francis Jenkins, and others. Jenkins had been a pioneer of early motion picture technology in the USA. His first ideas relating to television were published in 1894, and in 1922 he patented a television system. By 1923 Jenkins was demonstrating silhouette transmission – the first by radio rather than direct wire – using a scanning disc, rather than the prismatic glass discs that would become a trademark of his system. It was several years before Jenkins was able to produce graduated-tone images. In December 1923, Russian Vladimir K. Zworykin, working in the USA for the Westinghouse electrical company, patented an all-electronic television system.

One of the first experimenters to achieve success in producing a television image was John Logie Baird, a Scot, who is often hailed as 'the inventor of television'. Certainly he was demonstrating a very simple image transmission system in the English south coast town of Hastings in 1924, and – most importantly – his images had tonal range soon afterwards.

An appeal by the inventor for help with his television experiments had appeared in *The Times* (London) in June 1923. Some small financial and practical help was received in the following months, and Baird continued his experiments. His first very basic results, using a display of torch bulbs as a receiver, were described in 'The Radio Kinema' (3 April 1924), published in *Kinematograph Weekly*, the organ of the cinema trade. This first-hand account of Baird's progress by F.H. Robinson, reported: 'I myself saw a cross, the letter "H," and the fingers of my own hand reproduced by this apparatus across the width of the laboratory', and predicted: ' . . . it is not too much to expect that in the course of time we shall be able to see on the screen the winner of the Derby actually racing home watched by hundreds of thousands of his worshippers at Epsom'. Through Robinson, Baird's work came to the attention of London-based cinema and radio equipment entrepreneur Wilfred E. (Will) Day. Day had been involved in the cinema trade, first as exhibitor and then equipment manufacturer, since the earliest days. He was also one of the cinema's first historian-collectors, gathering together in his shop in London's Lisle Street a mass of early equipment: magic lanterns, shadow devices and optical toys as well as pioneering film machines. Day believed that the modern motion picture was a culmination of centuries of evolving techniques of image projection and motion picture synthesis. He evidently saw television as the next development in this long history of optical communication and entertainment media, and was keen to be a part of the new phenomenon. With an involvement in radio technology too, he would have understood all aspects of the technical requirements. A fascinating cache of seventy-six letters between Will Day and John Logie Baird chronicles how Day supervised the specialist production of suitable electric motors, perforated discs and other technical items, using his extensive knowledge of audio-visual technology, and arranged the first private demonstrations to potential backers, while Baird was still in Hastings.

I am indebted to Hastings Museum and Malcolm Baird for allowing the reproduction of six of the 'Baird–Day letters (1–6)' in this collection, published in facsimile form for the first time. On 5 April 1924 Baird wrote to Will Day, 'Letter 1', after a suggestion that Day may be interested. Following a meeting, Day wrote to Baird recording their agreement that he (Day) would buy a one-third share in the patent for £200, and would assist in perfecting the patent. On 1 May 1924 Baird wrote to Day, 'Letter 2', with a drawing of a scanning disc, promising more drawings the following week, and acknowledging receipt of an amplifier. By now, Baird was testing new sensitive light cells obtained from various sources. This letter indicates the extent to which Day was by now assisting Baird, arranging for the production of important mechanical parts, and providing electronic equipment. Day replied the following day, 2 May 1924, 'Letter 3', with some thoughts concerning the physiological properties of colour perception and closing: 'assuring you at all times of my best help in the great achievement we both wish to see accomplished . . . '.

On 7 May an article by Baird, giving details of his experiments, appeared in *Wireless World and Radio Review*. The following day Baird sent a telegram informing Day that his mother had died. Day replied with his condolences, and mentions, 'We are very busy with your new apparatus'. The correspondence continues. On 27 May, Day wrote, ' . . . we are now waiting for delivery of the motor'; on 5 June, 'I notice that you have got your article mentioned in *Popular Wireless* . . . and hope you are not giving too much away'. Baird replied, 'They got no information or photographs from me . . . '. On 26 June, Day wrote that there was to be a Cinematograph Garden Party on 19 July, and wanted to know whether Baird would be able to give a demonstration. Baird was doubtful, but on 18 July he wrote to Day: 'I have the model running it runs splendidly . . . '.

On the 24th, Day wrote: 'I am anxious to hear that all is working well'. But there had been an accident. Day first became aware of this through a newspaper report the following morning, and wrote, 'Letter 4': 'I am sorry to see by the "Daily Chronicle" of today's date that you have again had a burn out'. Baird wrote to him the same day: 'I am pleased to say I managed to get a shadow of a strip of cardboard . . . ', then gave details of the accident: 'I got a shock from 1000 volts – it twisted me up and flung me across the laboratory . . . I must have come down with a big bump but don't remember much about it'. Baird's hands had been burnt. Day wrote again the next day, repeating his suggestion: 'you will be wise to obtain a rubber glove and not deal with these high voltages unless you have ample insulation'. On 28 July ('Letter 5'), Baird replied, 'I will box in the H.T. [high-tension current] and provide safety links [?] which will prevent a recurrence of the shocks', agreeing that '"Safety First" is perhaps the best policy'.

On 14 August John Logie Baird wrote an important letter to his co-patentee: 'I have now got the proper size of slots and am going ahead'. It must have been within days of this letter that he achieved the first demonstrable results with his new apparatus. Having got to this stage Will Day evidently felt that it was time Baird paid for some of the apparatus that had been supplied, and on 21 August his secretary wrote to Baird enclosing a statement of goods received, and asking for a remittance at an early date.

Early in September Day travelled to Hastings to see a demonstration, and on the 6th wrote to Baird, 'Letter 6', concerning a demonstration to a Mr Burney of the Sterling Telephone Company the following week, pressing the point that the demo must be by wireless rather than a closed-circuit, to ensure that the potential was not misunderstood. After the visit, Day wrote: 'I think Mr. Burney was impressed yesterday, although of course, being non-technical it is a job to make them understand'. The problem for the two collaborators, as Day evidently recognised, was that although Baird's apparatus clearly showed that the principle could be made to work, a leap of imagination was necessary to comprehend that the extremely simple images being shown at this time were only a glimpse of an enormous potential. The impoverished Baird had evidently ignored the bill of 21 August, and by late the following month a rift had developed between the two partners. On 26 September, Day wrote: 'I was rather surprised – and certainly extremely sorry – to hear from you regarding the matter of a loan'. Baird had suggested that if the loan was not forthcoming he would have to sell the apparatus. Understandably, Day (as part-owner in the project) was very unhappy with this. Angry, he wrote: 'I am sure I have helped you always, and so far have nothing tangible in return'. This last comment was unfair, as Baird had successfully demonstrated his apparatus. If by 'tangible' Day meant that he had received no monetary return on his investment, then he was being unreasonable. The reality is that Day could probably ill-afford to subsidise the project, and his enthusiasm for a fully-achieved television system made him impatient for fast progress. Baird's tactless dealings with regard to money had not helped the situation.

Shortly after his somewhat hair-raising experiments in Hastings, and despite the problems between them, Day arranged a room in London's Soho for Baird to use. But their relationship could not survive, and Day sold his interest in the invention. Baird went on to find other sponsors and co-workers, and by April 1925 was demonstrating his first flickering images in a prestigious department store: 'Selfridge's Present the First Public Demonstration of Television'. Some of his early demonstrations used a wire connection, but at other times he used wireless (radio) transmission.

During one of his demonstrations at the Radio Exhibition at Olympia in west London that same year, Baird televised a newspaper contents poster, while it was being read. 'The first advertisement to be sent by television in the world', claimed the paper, in its report, '"Daily Mail" Televised'. Baird was at this time improving his system so as to reproduce graduated tones – the first to achieve this essential step.

The science journal *Nature* resumed an interest in the subject. A short piece reporting Baird's 'series of interesting demonstrations' at Selfridge's was noted in the 4 April 1925 issue, and a letter from A.A. Campbell Swinton appeared on 23 October 1926. 'Electric Television' outlined the advantages of the proposed cathode-ray electronic method, with no moving parts. Cathode-ray experiments were continuing in America and Japan at this time.

'Television', *Nature* journal's leading item for 15 January 1927, was a somewhat dismissive comment on Baird's claims. The issue of 29 January published a response from Baird, also titled 'Television', strongly contesting some of the statements and conclusions in that editorial. An article in the 18 June issue by Professor E. Taylor Jones, again entitled 'Television', described a demonstration by Baird of a 'transmission' (by phone line) between London and Glasgow, with the eight-per-second scan providing a two-inch picture with 'light and shade . . . amply sufficient to secure recognisability of the person being "televised"'.

In the USA, meanwhile, experimental broadcasts were being made, and were being received by amateurs on home-made equipment and reported in technical magazines. *Science and Invention* (cover), edited by the influential *Amazing Stories* publisher Hugo Gernsback – the inventor of 'scientifiction', modern science fiction publishing – started a Television Department with the November 1928 issue. 'How to build the S & I television receiver' described the construction of a mechanical receiver built around an ordinary electric fan motor, and epitomises the home experimenter status of the medium at that time. In the same issue, a 'Television Timetable' listed thirteen experimental broadcasting stations and their technical requirements, 'Stereoscopic Television' described progress in 3-D TV, and 'Radio Movies Demonstrated' reviewed and illustrated technical developments by the Westinghouse company.

Part 3: Broadcasting Begins
Experimental transmissions

In 1928 experimental television broadcasting started in England, and books on the subject began to appear. *Practical Television,* by E.T. Larner (Ernest Benn Ltd, 1928) is reproduced here complete, and includes an introduction by Baird, in which he dwells on the benefits of increased general knowledge of optics and related disciplines that television will bring, rather than prophesying any uses of the medium. The book provides a basic international technical history of the subject to that date, with details of cathode-ray (all-electronic) television as well as the more widely used electro-mechanical systems.

The world's first magazine dedicated to the subject, *Television*, 'The Official Organ of the Television Society', appeared in Britain in March 1928, and I have selected for reproduction in this section a number of pieces from the first seven years of this important periodical (for this introduction grouping the articles by theme where more convenient, rather than the chronological order in which they were originally published and in which they appear reprinted in this work).

'Commercial Television – when may we expect it?' asked *Television* in March 1928. ('Commercial' in this context meant useful to a wide public, following the experimental transmissions then being made by Baird.) At the time of publication, there were still no commercially available television receivers, and the editor pleads for patience, citing the delay in other fields – powered flight, wireless – before the initial achievements had become commercially practicable. But the experiments were certainly moving on at some speed. A 'stop press' supplement in this first issue of the new periodical, entitled 'Seeing Across the Atlantic!' (March 1928), reports excitedly on the Baird Company's latest achievement: 'recognisable images of human beings seated in the heart of London were seen in New York, 3,500 miles away!'. The next issue of *Television* included the first of a series of cartoons: 'Clairvoyant' (April 1928), picturing one of the possible future uses of the new medium. 'Transatlantic Television' and 'Television in mid-Atlantic' (April 1928) printed more details of the latest technical feat, the latter piece an eye-witness account of images received on an ocean liner.

Television activity in the USA was the subject of several articles in *Television* that summer. 'Baird Televisors for America First' (June 1928) reports that New York's station WGY is to broadcast television three days a week, and announces that Baird Televisors will be on sale in that country before being made available to the British public. In 'America Leaves Us Behind Again' (June 1928), in the same issue, R.F. Tiltman (author of *Television for the Home*) despairs of the situation: 'How absurd for Britain – the birthplace of television – to sit calmly by while other countries reap all the benefits of a television system devised by British brains'. (After objections from the Radio Corporation of America, Baird's involvement in the American television world would cease a few years later.) With 'Moving Shadowgraph Experiments in America' (September 1928), *Television* gave full details of the system devised by Baird's early competitor, American inventor C. Francis Jenkins. 'Television in America' in the same issue, is a round-up of the accelerating pace of experiments in that country, with nightly broadcasts from WRNY in New York, the first one-act play broadcast from Schenectady, and experiments in broadcasting using natural light. 'Television and Broadcasting' (September 1928), suggests that Britain is still in the lead with TV development, but America would overtake if the British Broadcasting Corporation (BBC) failed to adopt a more proactive attitude.

Novelist and regular *Television* contributor Shaw Desmond anticipates videophones (or webcams), and shopping by TV, in 'Seeing Round the World'; in 'Television and the Films' (September 1928), the author reminds us that the 'talkies' were still very new at that time – quoting the 27-year-old Alfred Hitchcock concerning the possibilities of that medium – and makes the now unlikely-sounding proposal that shadow-shows projected live to a theatre audience by television would have a popular appeal. In 'Television as "Booster"' (October 1928), Desmond suggests that commercial advertising on television, by the BBC, is essential; but in fact it would be more than twenty-five years before commercial television (in the modern sense) would be launched in Britain, and then by independent companies, and not by the BBC. Technical expert J. Robinson, 'Television – an Appeal for Broadcasting Facilities' (October 1928), believes that the BBC is about to experiment with the transmission of 'still' pictures before any transmission of live motion is attempted, and warns that 'it is impossible to believe that the broadcasting of still pictures will produce anything but the bitterest form of disappointment'. He calls for the formation of a new corporation to control television broadcasting.

In 'The Entertainment Value of Television Today!' (November 1928), regular contributor R.F. Tiltman rails against the 'canting, carping, crabbing critics' who refuse to accept that television has reached a stage where it is indeed capable of providing true entertainment, as witnessed during demonstrations at the recent National Radio Exhibition, Olympia, London.

'My Impressions of Television' (November 1928) by science-writer Dr Frank Warschauer, compares the systems of the Telefunken Co. (two systems by Dr August Karolus), and Hungarian Dionys von Mihaly – all shown in Germany – with that of Baird. The German demonstrations comprised simple silhouette shadows and lantern slides. Warschauer reported, 'I therefore did not see television until I came to London', and praised the Baird system, predicting optimistically that 'The invention of television will bring all nations together in a way that could never previously be imagined'. Contributor A.W. Sanders, in 'Impressions and Opinions of a Layman' (December 1928), adds the thoughts of the man-in-the-street: 'the sooner television is broadcast the better'. 'Now *This Is* Television!' (December 1928), by Dr Alfred Gradenwitz, also anticipates a 'great future for this invention', but by also stressing the international 'mutual understanding and goodwill' that television would surely bring, weakens any claims to successful prophesy; and dashes them altogether by stating: 'Incidentally, . . . "Talkies," . . . will never supersede the silent film'. In 'Mihály's Tele-Cinema' (April 1929), Dr Gradenwitz describes a demonstration of Mihály's much improved system, but is unsure whether the result he saw was true live television, or the (easier to achieve) transmission of a cine film.

In 'The Future of Television' (February 1929), Dr C. Tierney, Chairman of the Television Society, is also critical of the BBC's apparent reluctance to take the medium seriously, and warns of the dangers of this 'garrulous ineptitude'. 'How the "War" ended' (March 1929), by 'A Student of Progress' gives a more balanced view of the BBC/Baird conflict, announcing: 'the almost traditional war between the Baird Companies and the B.B.C. has been brought to an end . . . '. In May, *Television* printed 'The Postmaster-General's Decision' (which had originally appeared in *The Times* on 28 March), that 'he would assent to a station of the British Broadcasting Corporation being utilised for this purpose [television broadcasting] outside [sound radio] broadcasting hours'. 'Television Broadcast by the B.B.C.' (November 1929) gave details of the inaugural thirty-line (Baird standard) programme which had been broadcast on 30 September, featuring the inevitable speeches – including one by Sir Ambrose Fleming, whose thermionic valve had been so important to broadcasting – a comic monologue and popular songs. This first broadcast was limited to one channel and it was not practicable to transmit sound and image together, so 'lookers-in' first heard each piece in audio only, and then saw the appropriate image, but mute. This rather unsatisfactory arrangement would become tiresome once the immediate novelty had worn off, and from 31 March 1930 this situation changed, as reported in 'Simultaneous Sight and Sound Broadcast' (May 1930); the participants of this programme included popular singer Gracie Fields.

Meanwhile there had been considerable activity in the USA, with public TV transmissions from several radio stations since 1928, and the world's first television play, *The Queen's Messenger*, broadcast by station W2XAD, Schenectady, on 11 September of that year. Back in Britain, 'The First B.B.C. Play to be Broadcast by Television' (July 1930), was to be Pirandello's *The Man with the Flower in his Mouth*, in rehearsal at the time this piece was written. An account of the actual broadcast, 'The Fourteenth of July, 1930' (August 1930), reflects that the date is 'celebrated in France in connection with a revolution. The 14th of July,

1930, had its revolution too . . . '. 'How the First Television Play was Received' (August 1930), is a selection of reports from 'telegazers'. One correspondent thought it wonderful but commented '"I should, however, have liked the lady to have screamed to make it more exciting".'

'Television in the Cinema' (August 1930) reveals plans to show television as part of the programme at the London Coliseum, in late July, with a display comprising 2,100 individual lamps. 'Baird Screen Television. The Coliseum Triumph' (September 1930) reports on the two-week run of the fifteen-minute presentations, which included the televised performers responding to spontaneous telephone questions from the live audience. 'The Big Screen in Germany' (November 1930), by Sydney A. Moseley (of the Baird Company), reports how that company's equipment was transported to Berlin for twenty-six performances at the Scala Theatre, featuring stars of stage and screen, politicians and other well-known figures, transmitted by wire from a studio seven kilometres away.

In 1895 the English Derby horse race had been one of the first subjects of the new 'animated photography', and (as anticipated by F.H. Robinson: see 'The Radio Kinema', 3 April 1924) it was a natural for early television transmission, accomplished by Baird and the BBC, for which a new mirror-drum camera was used. 'Seeing the Derby at a Distance' (July 1931) claimed this as the first outside broadcast; previously, television had been possible only with studio lighting. Advances were also being made in studio arrangements: 'Television as an Entertainment' (August 1931) describes how two or three characters could now be televised together, giving much greater subject flexibility.

'B.B.C. and Television, Official Statement of Policy' (June 1932) was something of a milestone announcement. Regular half-hour programmes were to start four nights a week, with responsibility for the content being taken over by the BBC, leaving the Baird Company to concentrate on technical development. 'The B.B.C. "First Night"' (September 1932) details the first TV transmission from Broadcasting House, London. The Heath-Robinson nature of much receiving apparatus in use at that time is indicated by the report of one viewer in Leeds: 'The "televisor" was being operated with the cover removed, and a moth chose this moment to commit suicide amongst the brush gear of the motor . . . synchronisation was lost for the time being'. A good indication of the range of programmes being broadcast at this time may be judged from the reviews selected here: 'Last Month's Programmes', from the November 1932 and March 1933 issues of *Television*.

'B.B.C. Television Policy, Rumours and Facts' (September 1933) suggests the likelihood of 120-line transmissions in the near future.

Despite *Television*'s inevitable focus on work in Britain, 'News from Abroad' (selections included here from September, October and November 1933, and January 1934) provided a round-up of developments elsewhere, with further details from two countries reported in 'Television in Japan' (September 1934), and 'A Television Service for Germany' (December 1934): 'Big events such as Hitler speeches will be handled by the special television van . . . '.

'The Baird Company's Great Achievement' (April 1934) was a demonstration of 180-line pictures broadcast on 'ultra short waves', with a range of thirty miles, using a cathode-ray tube in the receiver. 'Studio and Screen' (August 1934) contains further details of BBC studio technique and programme material. In 1934 a government committee was set up in Britain to investigate the situation, and make initial recommendations for future television broadcasting. A delegation visited the USA and Germany. Its report was published in January 1935.

At this time, most people in the USA and Europe had never seen television, but the subject – if not the reality – was becoming a part of everyday experience. The idea of television percolated from the world of technical experiment, through entrepreneurial promotion, to the wider public, becoming a general topic of interest. Even in those households without a TV set – which was most households –children could join in the fun by building and performing with 'The Television Machine', details of which appeared in *The Modern Handy Book for Boys* by Jack Bechdolt (London: Hutchinson & Co., 1934). But television was soon to become more than a toy for wireless experimenters and a tiny audience of pioneer 'lookers-in'. In Germany and in England, a 'high definition' television information and entertainment service was about to become a reality.

Stephen Herbert
Hastings, 2004

Part 1

DREAMS AND EXPERIMENTS

Fantasies and predictions, and the first proposals for a practical system

THE TELECTROSCOPE.—M. Senlecq, of Ardres, has recently submitted to the examination of MM. du Moncel and Hallez d'Arros a plan of an apparatus intended to reproduce telegraphically at a distance the images obtained in the camera obscura. This apparatus will be based on the property possessed by selenium of offering a variable and very sensitive electrical resistance according to the different gradations of light. The apparatus will consist of an ordinary camera obscura containing at the focus an unpolished glass and any systen of autographic telegraphic transmission; the tracing point of the transmitter intended to traverse the surface of the unpolished glass will be formed of a small piece of selenium held by two springs acting as pincers, insulated and connected, one with a pile, the other with a line. The point of selenium will form the circuit. In gliding over the surface, more or less lightened up, of the unpolished glass, this point will communicate, in different degrees and with great sensitiveness, the vibrations of the light. The receiver will also be a tracing point of blacklead or pencil for drawing very finely, connected with a very thin plate of soft iron, held almost as in the Bell telephone, and vibrating before an electro-magnet, governed by the irregular current emitted in the line. This pencil, supporting a sheet of paper arranged so as to receive the impression of the image produced in the camera obscura, will translate the vibrations of the metallic plate by a more or less pronounced pressure on that sheet of paper. Should the selenium tracing-point run over a light surface the current will increase in intensity, the electro-magnet of the receiver will attract to it with greater force the vibrating plate, and the pencil will exert less pressure on the paper. The line thus formed will be scarcely, if at all, visible; the contrary will be the case if the surface be obscure, for, the resistance of the current increasing, the attraction of the magnet will diminish, and the pencil, pressing more on the paper, will leave upon it a darker line. M. Senlecq thinks he will succeed in simplifying this apparatus by suppressing the electro-magnet and collecting directly on the paper by means of a particular composition the different gradations of tints proportional to the intensity of the electric current.

AN ELECTRIC TELESCOPE.

[15374.]—It may be of interest to your readers to know the details of some experiments on which I have been engaged during the last three months, with the object of transmitting a luminous image by electricity.

To transmit light alone all that is required is a battery circuit with a piece of selenium introduced at the transmitting end, the resistance of which falling as it is exposed to light increases the strength of the current, and renders a piece of platinum incandescent at the receiving end thus reproducing the light at the distant station.

By using a number of circuits, each containing selenium and platinum arranged at each end, just as the rods and cones are in the retina, the selenium end being exposed in a camera, I have succeeded in transmitting built-up images of very simple luminous objects.

An attempt to reproduce images with a single circuit failed through the selenium requiring some time to recover its resistance. The principle adopted was that of the copying telegraph, namely, giving both the platinum and selenium a rapid synchronous movement of a complicated nature, so that every portion of the image of the lens should act on the circuit ten times in a second, in which case the image would be formed just as a rapidly-whirled stick forms a circle of fire. Though unsuccessful in the latter experiment, I do not despair of yet accomplishing my object as I am at present on the track of a more suitable substance than selenium.

Denis D. Redmond.

Belmont Lodge, Sandford, Dublin.

SEEING BY ELECTRICITY.

The art of transmitting images by means of electric currents is now in about the same state of advancement that the art of transmitting speech by telephone had attained in 1876, and it remains to be seen whether it will develop as rapidly and successfully as the art of telephony. Professor Bell's announcement that he had filed at the Franklin Institute a sealed description of a method of "seeing by telegraph" brings to mind an invention for a similar purpose, submitted

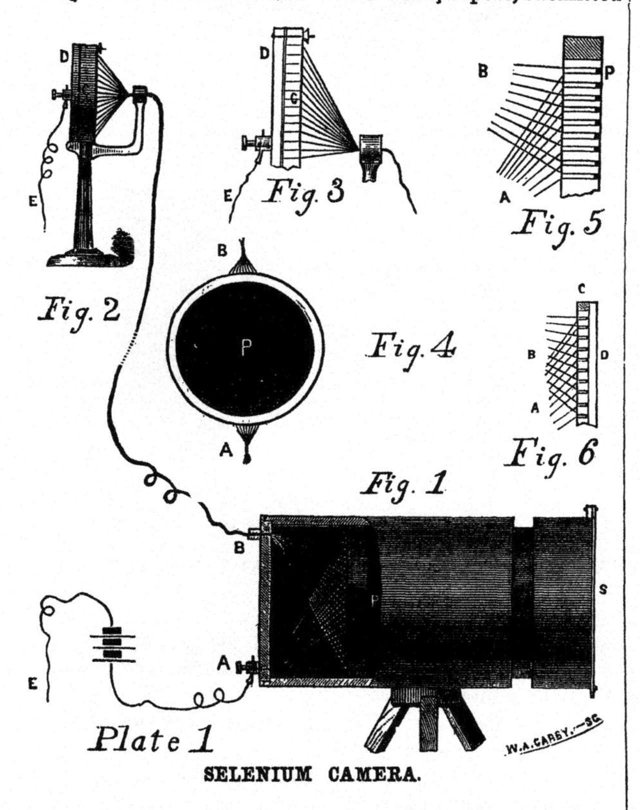

SELENIUM CAMERA.

to us some months since by the inventor, Mr. Geo. R. Carey, of the Surveyor's Office, City Hall, Boston, Mass. By consent of Mr. Carey we present herewith engravings and descriptions of his wonderful instruments.

Figs. 1 and 2, Plate 1, are instruments for transmitting and recording at long distances, permanently or otherwise, by means of electricity, the picture of any object that may be projected by the lens of camera, Fig. 1, upon its disk, P. The operation of this device depends upon the changes in electrical conductivity produced by the action of light in the metalloid selenium. The disk, P, is drilled through perpendicularly to its face, with numerous small holes, each of which is filled partly or entirely with selenium, the selenium forming part of an electrical circuit.

The wires from the disk, P, are insulated and are wound into a cable after leaving binding screw, B. These wires pass through disk, C (Fig. 2), in the receiving instrument at a distant point, and are arranged in the same relative position as in disk, P (Fig. 1).

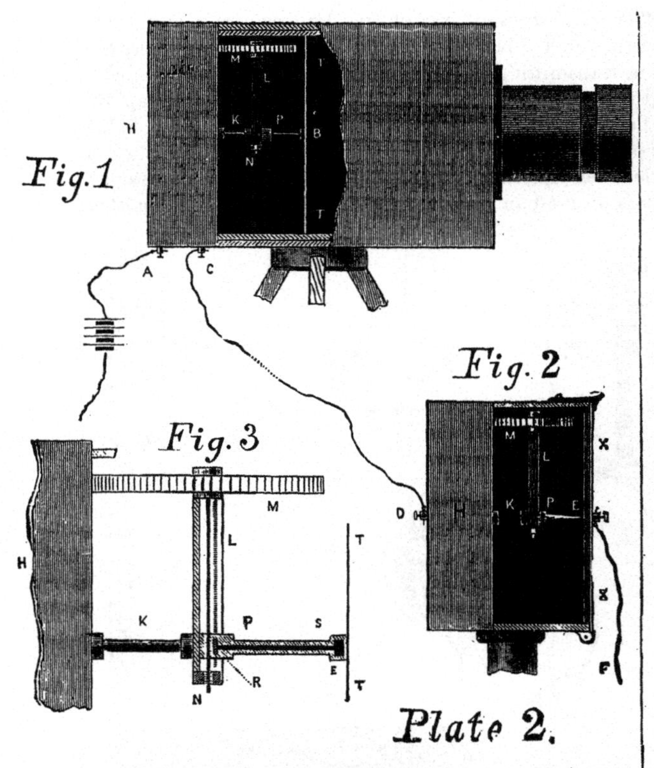

INSTRUMENT FOR TRANSMITTING AND RECORDING IMAGES.

A chemically prepared paper is placed between disks, C and D, for the image of any object projected upon disk, P (Fig. 1), to be printed upon.

Fig. 3 is a sectional view of Fig. 2, showing wires and the chemically prepared paper.

Fig. 5 is a sectional view of disk, P (Fig. 1), showing selen ium points and conducting wires.

Fig. 6 is a sectional view of another receiving instru ment with platinum or carbon points, covered with a glass cap, there being a vacuum between glass cap, D, and in sulating plate or disk, C.

These points are rendered incandescent by the passage of

the electrical current, thereby giving a luminous image instead of printing the same. These platinum or carbon points are arranged relatively the same as the selenium points in Plate P (Figs. 1 and 4); each platinum or carbon point is connected with one of the wires from selenium point in disk, P (Fig. 1), and forms part of an electrical circuit.

The operation of the apparatus is as follows: If a white letter, A, upon a black ground be projected upon disk, P (Fig. 1), all parts of disk will be *dark*, excepting where the letter, A, is, when it will be light; and the selenium points in the light will allow the electric current to pass, and if the wires leading from disk, P (Fig. 1), are arranged in the same relative position when passing through disk, C (Fig. 2), the electricity will print upon the chemically prepared paper between C and D (Fig. 2), a copy of the letter, A, as projected upon disk, P (Fig. 1). By this means any object so projected and so transmitted will be reproduced in a manner similar to that by which the letter, A, was reproduced.

Figs. 1 and 2, Plate 2, are instruments for transmitting and recording by means of electricity the picture of any object that may be projected upon the glass plate at T T (Fig. 1), by the camera lens. The operation of these instruments depends upon the changes in electrical conductivity pro duced by the action of light on the metalloid selenium.

The clock-work revolves the shaft, K, causing the arm, L, and wheel, M, to describe a circle of revolution. The screw, N, being fastened firmly to wheel, M, turns as wheel, M, revolves on its axis, thus drawing the sliding piece, P, and selenium point, disk, or ring, B, towards the wheel, M—see Fig. 3. These two motions cause the point, disk, or ring, B, to describe a spiral line upon the glass, T T, thus passing over every part of the picture projected upon glass, T T.

The selenium point, disk, or ring will allow the electrical current to flow through it in proportion to the intensity of the lights and shades of the picture projected upon glass plate, T T.

The electric currents enter camera at A, and pass directly to the selenium point, disk, or ring, B; thence through the sliding piece, P, and shaft, K, by an insulated wire to bind ing screw, C (Fig. 1); from this screw by wire to binding screw, D (Fig. 2), through shaft, K, and sliding piece, P, to point, E (Fig. 2); then through the chemically prepared paper placed against the inner surface of the metallic plate, X X, by wire, F, to the ground, thus completing the circuit and leaving upon the above mentioned chemically prepared paper an image or permanent impression of any object projected upon the glass plate, T T, by the camera lens.

Fig. 2 is the receiving instrument, which has a clock movement similar to that of Fig. 1, with the exception of the metallic point, E, in place of the selenium point, disk, or ring (Fig. 1), at B.

Fig. 3 is an enlarged view of clock-work and machinery shown in Figs. 1 and 2.

Seeing by Electricity.

To the Editor of the Scientific American :

Your article on "Seeing by Electricity" contained in the SCIENTIFIC of June 5, page 355, will prove of interest to many. Early in the fall of 1877, the principles and even the apparatus for rendering visible objects at a distance through a single telegraphic wire were described at No. 21 Cortlandt Street, in this city, to James G. Smith, Esq., formerly superintendent of the Atlantic and Pacific Telegraph Company, and now of the Continental Telegraph Company, I believe, and to Messrs. Shaw & Baldwin, telegraph constructors, also, I believe, now connected with the Continental. At that time I was engaged in perfecting an autographic telegraph by which maps and pictures were daily transmitted by telegraph over a single wire.

The recent announcements of this discovery in three different directions, each undoubtedly independent of my own experiments, show how the same idea often occurs in separate minds. There is no likelihood of any plan of this kind ever being reduced to practice, for some of the difficulties in the way of all of the plans are insuperable, as will be apparent from the following reasons:

1. The action of light upon selenium in changing its electric conductivity is slow; although new discoveries may remedy this feature.

2. To convey with any accuracy an image, one even so small as to be projected upon a square inch of surface (I am speaking now of the apparatus you describe), would necessitate that this surface should be composed of at least 10,000 insulated selenium points, connected with as many insulated wires leading to the receiving instrument; for the variation of the one-hundredth of an inch either way will "throw a line out of joint."

3. The most delicate apparatus would not indicate a change in resistance by the projection of light upon merely a selenium point.

4. Isochronism is unattainable, as required. The method I proposed involved the isochronous movement of the separate instruments. The transmitter consisted of a coil of fine selenium wire in a darkened case, having a diameter of say three inches. Light from the image to be transmitted was to be let into the chamber and upon the selenium coil by a fine tube which, starting at the periphery of the circle, would draw concentric imaginary spiral lines until reaching the center of the circle. Thus light emitted or reflected from the image to be transmitted would affect the selenium just in proportion to the brightness of the image at the different points within the compass of the circle traversed by the imaginary lines drawn by the opening in the tube. The speed of motion of the tube was to be such that in describing all the spiral lines from the periphery to the center of the circle, the impression made upon the retina while at the periphery of the circle would not have ceased until the light ray should have reached the center of the circle.

The receiver consisted of a darkened tube, having an inside diameter of three inches (corresponding to the transmitting circle), with its sides and bottom absolutely black. In this tube, describing imaginary lines just as the tube in the transmitter, was a blackened index carrying two fine insulated platinum points very close together connected with the secondary wire of a peculiar induction coil, the primary wire of which constituted a part of the main wire leading to the transmitter.

The transmitting ray of light and the invisible index in the darkened receiving tube were to start at the periphery and describe their spiral motions in exact unison until the center should be reached, and the speed being sufficiently great it is obvious that as the first spark between the receiving platinum points would not have ceased to affect the retina until the last spark, with the index at center, would have been produced, an exact image of the object before the transmitter would be reproduced before the eye of the observer placed at the darkened chamber of the receiver.

But the trouble is to make the selenium sufficiently active, *and to get the isochronous motion.* Perhaps some of your readers may like to try their hands at rapid synchronism.

W. E. SAWYER.

New York, June, 1880.

CONVERSATIONS FOR THE TIMES.

PROFESSOR GOAHEADISON'S LATEST.

A. I am much distressed in my mind. I am told that I cannot be cured unless I consult Sir Settemup Pilliboy, the eminent physician, and it's such a tremendous journey to London from here——

B. London? What the deuce do you want to go to London for? Haven't you heard of Professor Goaheadison's Far-Sight Machine? Bless my soul—you *are* behind the times! Oh dear, yes. Wonderful man is Goaheadison! Just invented a machine by means of which a person in Nyork can actually see another person in Shicaago, or Borston, or even 'Frisco. You just stand at a street corner in Nyork, and drop a dollar in the slot, and pull out a drawer, and there you see the party you want to have a look at, with a white choker on, and his hair all beautifully curled and oiled. If you keep on dropping dollars in the slot, bless your soul! you can keep your eye on him all day long—follow him in and out of all the saloons, and everything. Look here, these machines are not exactly laid on as yet in England, but that Goaheadison is such a smart chap that there's no doubt if you just drop a twenty-pound note in the slot—any slot—and pull out a drawer—any drawer—you'll find one of the machines in it.

A. Dear me! Well, this beats all! Here is the machine as you predicted! Lorks! Now, do I just lay it on to Doctor Pilliboy in Harley Street? Like so? * * * By Jingo! there he is, standing in his consulting room and smiling at me. He's motioning me to do something—but I can't quite grasp—what a pity I can't hear what——

B. My dear boy, simplest thing in the world—just drop a jubilee four-shilling piece (be careful that it *is* a four-shilling piece, as Mr. Boehm's coinage is carefully kept free from any means of identifying its value; that's its chief artistic merit—its problematical value. If you want to find out the value of a British coin you take it to a chemist to weigh, and then you go home and carefully work out how much such a weight of silver is worth)—just drop in a jubilee problem, and pull out a drawer, and you have a telephone laid on. There, now!—you can consult your London physician perfectly. * * Can he see your tongue?

A. Oh, yes, beautifully; but there's a difficulty about sounding my chest.

B. Difficulty? Pooh! Just lay on a microphone, and connect it with the telephone, and there you are.

A. Well, I *am* glad you turned up just in time to tell me about Goaheadison's inventions! I do feel so much better already! What a wonderful man Fullspeedaheadison must be!

B. Rather! and I see in the papers that it's all owing to his method of keeping up the nervous tension while at work. He does all his thinking seated on a powerful dynamo from which wires are connected with his double teeth. He sleeps head downwards, with his toes tightened up in a vice, and a thorough draught blowing through him. He must be a most uncomfortable person, but it's a tremendous gain to science.

* * * * * *

A. Let's have a general look round with this machine. Dear me! the Eiffel Tower doesn't look nearly so high as I imagined it would—it's not nearly so high as that other object that seems to be trying to hide behind it.

B. Let's have a look—oh, that other object is the Paris prices. Of course there's no comparison. Dear me! here's somebody struggling to get out of a bush. He *does* look tired, doesn't he? Why, it's Stanley! Ah, just make a memorandum to send him one of Plentyofbrainsinhisheadison's machines to find Emin with, or or the way out, or whatever he may happen to be looking for at the present moment. No, wait a bit—turn the machine all over Africa, and tell Stanley where everything is. * * Hullo! come and look!—I can see right down to the bottom of the truth in the Parnell Commission. Well, by Jingo! *Isn't* this a revelation? Whoever *would* have imagined it from the evidence? That just shows how much evidence is worth! Good gracious! Oh, I say—better present this machine to the judges when we've done with it. * * Why,—I *say*! *Do* look! Just fancy! I can actually see the intelligence of the London County Council! *That's* a thing nobody has ever succeeded in detecting before. What a wonderfully powerful lens this machine must have! It certainly *is* a triumph of science!

GOAHEADISON'S REAL LATEST.

EH? Goaheadison's Far-Sight Machine? Bless your soul, that's quite an antiquity now. Quite eclipsed by the Far-Touch Machine!

Haven't read the account of that mill between Dan Dotter of Doncaster and the Mc Flattener, the Boston Bumper? Oh, yes – all carried on by cable; one end of it at Doncaster, and the other at Boston.

Dan faced the machine at the Doncaster end in good form at 11.32. Some very pretty play. Dan dodged a cleverly-tried jaw-compresser from Mac's right, and got the electric current onto the ropes in the fifth round.

Then the machine planted several on Dan's ribs; & Dan came up groggy for the 17th round; but supplied the cable with a neat lifter under the right ear at 12.13¼.

On the call for the 24th round Mac forgot to come up; & Dan got the ~~belt~~.

It was a pretty sight throughout; the American champion being distinctly visible through a Far-Sight machine.

During the affair an amusing incident occurred. Just as Mac was ushering in a superior lightning dislocater with his left, a heedless visitor happened to pass in front of the transmitter.

Curable in six weeks, with reserve.

JAS. F SULLIVAN

GOAHEADISON'S REAL LATEST.

Eh? Goaheadison's Far-Sight Machine? Bless your soul, that's quite an antiquity now. Quite eclipsed by his latest—the Far-Touch Machine.

Haven't read the account of that mill the other day between Dan Dotter, of Doncaster, and the McFlattener, the Boston Bumper? Oh, yes, all carried on by cable, one end of it at Doncaster, and the other at Boston. Dan faced the machine at the Doncaster end in good form at 11.32. Some very pretty play. Dan dodged a cleverly tried jaw compresser from Mc's right, and got the electric current on to the ropes in the fifth round.

Then the machine planted several on Dan's ribs, and Dan came up groggy for the seventeenth round, but supplied the cable with a neat lifter under the right ear at 12 13¼ On the call for the twenty-fourth round Mc forgot to come up, and Dan got the belt. It was a pretty sight throughout, the American champion being distinctly visible through a Far-Sight machine.

During the affair an amusing incident occurred. Just as Mc was ushering in a superior lightning dislocator with his left, a heedless visitor happened to pass in front of the transmitter. Curable in six weeks, with reserve.

LE TÉLÉPHOTE.

Fig. 167. — Poste transmetteur du Téléphote : envoi d'une image.

LE TÉLÉPHOTE.

Fig. 168. — Poste récepteur du Téléphote : arrivée d'une image.

SEEING BY TELEGRAPH.

By H. Trueman Wood, M.A.

Secretary of the Society of Arts, London.

SINCE the telephone has descended from the rank of a scientific marvel to that of a commonplace and useful piece of apparatus, there has been a demand on the part of the insatiable public for some device which will enable it to see what its friends are doing, as well as to hear what they are saying, at a distance beyond the range of the unaided eye or ear. Is there any chance of this being effected? and if so, what chance? We can only answer, there is a possibility, but, as yet, no great probability of it. Any day some one of our many searchers into nature's secrets may announce that he has found the key to the problem; but in all likelihood it will be by the use of some means not yet imagined or discovered, rather than by the development of any system now in use. Bell found that a plate of iron could reproduce every vibration of the human voice, and the transmission of speech was effected. If anybody will discover a means of reproducing at a distant station the variations in the light vibrations by which we are enabled to see, the transmission of pictures, or rather *reflections*, by telegraph will become possible.

Failing, however, such definite successes to record, it may be interesting to consider what is being done in this direction by several energetic workers who are striving in various ways thus to extend the limits of human vision.

The devices which have been employed are two. One of them is the invention of Mr. Shelford Bidwell, the other of Messrs. Ayrton and Perry Before, however, saying anything about the apparatus, let us consider the problem to be dealt with. In the telephone we have a transmitter, into which the sender of the message speaks. This transmitter is connected by wires with a receiving instrument, by which the sounds spoken into the transmitter are reproduced. Various devices are now used in the transmitter, but in all, the vibrations of the air caused by speaking are made to vary the electrical condition of the line wire. These alterations in the condition of the line affect the receiver in such fashion as to produce vibrations therein, which, by throwing the air into motion, cause sounds corresponding to those which first set the whole system at work.* Now, it is not difficult to imagine a similar apparatus applied to sight instead of sound: a sensitive plate or mirror at one end, a connecting wire, a second mirror at the other end, capable of being so affected as to absorb and reflect light precisely as the light is absorbed or reflected from the surface of the first mirror. The result would be that the image of an object thrown on the first mirror would be seen in the second, it might be in black and white, as in a photograph, or in all its proper

* For our present purpose it is not necessary to refer to the means by which the result above stated is effected, but the reader may be referred to previous papers on the Telephone and its allies (Vol. I., pp. 124, 180; Vol. IV., p. 307; and Vol. V., p. 147) for a full explanation of them.

colours. Unfortunately, this is as yet only a philosopher's dream. Nothing approaching it has yet been done, or is likely to be done. Perhaps it would not be far from the truth if it were said that the great difficulty lies in the fact that the impression of sound results from a series of successive impulses, whereas the eye, in seeing any object, receives a vast number of undulations impinging simultaneously upon it. In the telephone the whole plate receives and transmits one vibration after another, however rapidly they may succeed. In our imaginary "Teleoptical" apparatus, the plate would receive a great number at once, and each on a different part of its surface. We cannot well conceive a single wire transmitting all these different impulses simultaneously, and we must therefore suppose our imaginary plates to be made up of a great number of small pieces, each piece of one plate in correspondence with the corresponding portion of the other plate. We should then get a sort of mosaic which would represent, with greater or less accuracy, the original image, according to the minuteness of the pieces composing it. An illustration may make this clearer. Suppose Fig. 1 to be a plate made up of a number of cells, sensitive to light, and capable of affecting a current of electricity passing through them. Suppose Fig. 2 to be a plate made up of cells—shall we say—capable of emitting more or less light, according to the strength of a current passing through them. Each cell of Fig. 1 is joined up electrically with the corresponding cell of Fig. 2—1 with 1, 2 with 2, 3 with 3, and so on. We will now throw a dense shadow of a capital letter E on the first plate, the part of the plate not shadowed being brilliantly illuminated. The cells which are lighted will cause the corresponding cells of Fig. 2 to emit light, while the cells of Fig. 2 corresponding to the shadowed cells of Fig. 1 will remain dark. Thus we shall get our E in mosaic. Now, one-half of our supposition is possible,* the other is not. The reader will remember that in a selenium cell we have precisely

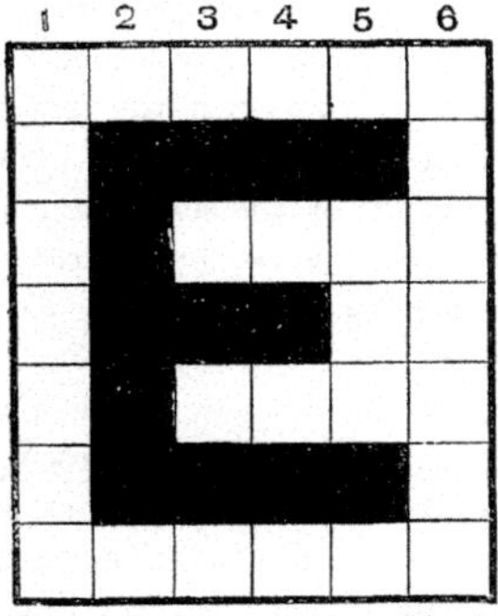

Fig. 1.—Diagram showing Mosaic of Sensitive Cells for Transmitting a Picture.

* "The Photophone:" "Science for All," Vol. IV., p. 307.

what is wanted for the transmitter, a device which is affected by light in such a way as to offer more or less resistance to an electrical current. Unfortunately we have not as yet any material which will act conversely, will emit, or reflect, light, when more or less excited by electricity; and our plan, as above suggested, must await realisation until some such material is discovered.

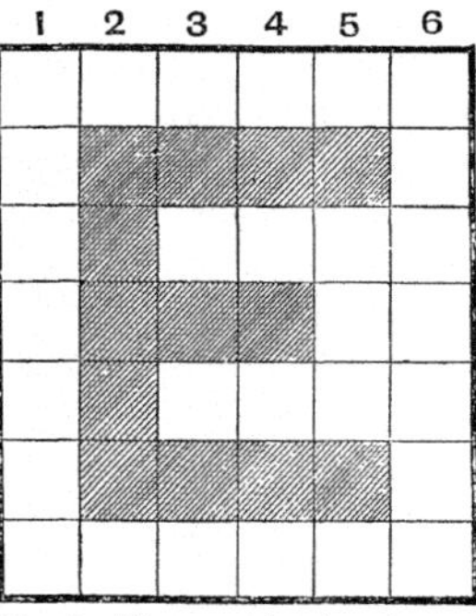

Fig. 2.—Diagram showing how a Mosaic of Sensitive Cells might reproduce a Picture.

Pending, however, this discovery, Messrs. Ayrton and Perry have devised a very ingenious method of exhibiting at one station the effect of light falling on a system of selenium cells at another—possibly a distant—station. As yet their work has not passed beyond the experimental stage, and the distance over which they have telegraphed has not exceeded the length of a lecture-room table. They have, however, demonstrated the possibility of sending—shall we say shadow pictures?—by telegraph; and this is alone a remarkable feat. It ought also to be stated that the notion of thus reproducing in mosaic the image of a distant object, seems to have been quite original with them.

The simplest way, perhaps, of getting a correct notion of the somewhat complicated apparatus which these inventors suggest should be employed, will be to consider the action of one unit of it, to see how the amount of light falling on a small square surface at Brighton can cause a similar surface at London to be illuminated with a corresponding amount of light. At Brighton we have a selenium cell—an arrangement of two wires laid as close as possible without touching, and the intervening space filled with selenium. An electrical current flowing through the system traverses the selenium more easily when a light is shining upon it than when it is in the dark.

Hence we get variations in the current corresponding with the amount of light falling on the selenium. Such variations can, of course, be detected by various means. Mr. Bell used a telephone, and hence his photophone. In previous experiments a galvanometer had been employed. Messrs. Ayrton and Perry cause the current to open and close a little shutter in a tube through which light is admitted. In the end of the tube is a lens, arranged to throw an image of a square hole on a screen. When the shutter is open all the light passes through; when it is closed no light passes; in the intermediate positions more or less light passes. For our present purpose it is not necessary to describe precisely the arrangement employed. It may be sufficient to say that the shutter is attached to a small magnet, arranged like the magnet of a galvanometer, so that it is moved by the action of the electric current which passes through a coil of wire surrounding the tube in which the shutter and magnet are mounted. Fig. 3 shows the arrangement. With a mosaic of selenium cells at one station, and a mosaic of receivers such as Fig. 3 at the other, there seems no reason to doubt that we might get a reproduction, at all events, of the shadow of an object thrown upon the receiving screen. The next step would be to reproduce a picture such as could be thrown on the screen by a magic lantern, and the ideal would be to reproduce an image such as is formed on the table of a camera obscura, or on the ground glass of the photographic camera. But when an attempt is made to convert theory into practice, difficulties multiply. It is evident that even for experimental purposes a mosaic with ten cells in a row would offer but a limited field. Only very simple images could be thus transmitted. Such a square screen would require a hundred cells and a hundred wires. Now, the manufacture of selenium cells has not yet arrived at such a pitch of perfection that a dozen, let alone a hundred, similar cells could be readily turned out, while the notion of a telegraph line containing a hundred wires is quite out of the question. Messrs. Ayrton and Perry therefore propose to make a few rapidly moving cells do the work of a number of stationary cells, and they rely on the permanence of the impression on the retina of the human eye for the production of a picture. To do this the sending and receiving apparatus would have to move in precise unison, but it is believed this might be effected. The main idea of the proposal may be drawn from the following diagram (Fig. 4) of a piece of apparatus used by Mr. Perry to illustrate a lecture at the Society of Arts. D is a selenium cell, which is drawn across the dark and illuminated spaces shown upon the screen. E is a receiver, similar to Fig. 3. The light from E falls on a mirror, F, and is by it reflected on a curved screen, G. D and E are connected in an electric circuit with a battery. The string which moves D also gives motion to the arc H, at the centre of which F is fixed. If the light from E be uniform, motion of F on its axis will obviously cause the spot of light on G to move to the right or left. If the motion be rapid, a line of light will be seen. If the light from E be interrupted, a broken line of light will be seen on the screen when F is rotated. Now, as D passes through light or dark spaces its resistance varies; the result of this is the opening or closing of the shutter in E, and the consequent appearance on G of a broken line, corresponding with the spaces of light and dark in B. As constructed, the apparatus marked the passage of the selenium cell through the light and dark spaces, but it could not be worked at sufficient speed to give a continuous visual impression.

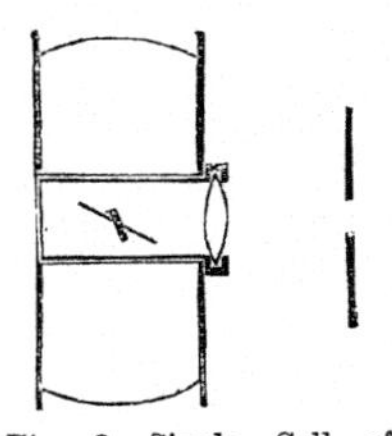

Fig. 3.—Single Cell of Ayrton and Perry's Receiving Apparatus.

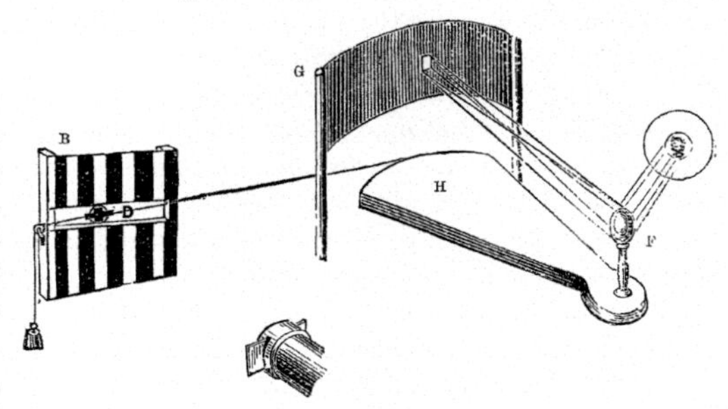

Fig. 4.—Model of Ayrton and Perry's Apparatus.

Mr. Shelford Bidwell's "Telephotograph" works in a totally different, but no less interesting fashion. The object of this ingenious apparatus is not to show you, as in a mirror, a representation of an object at any distance, but to produce, at a distance, a drawing of any object presented in front of the receiver, and held stationary there.

Fig. 5 is a diagram showing the principle on which Mr. Bidwell works. M is a metal plate on which is laid a piece of paper soaked in iodide of potassium—a salt which is easily decomposed by electricity. If a current be passed from a platinum style, P, through the moist paper, to the plate M, the paper is marked with a brown stain resulting from this decomposition. By drawing the style along while the current is flowing steadily, a line is marked on the paper. If the current be interrupted the line is broken, and thus a row of dots or dashes of any required length may be produced. The effect is the same if the paper be drawn along under the style. In the diagram, B represents a battery in circuit through a galvanometer, G, with P and M. B′ is another similar battery, arranged in the same manner, but including also in its circuit a selenium cell, S. The current in this circuit flows in the contrary direction to that in the first circuit. The effect of this is that if the currents in the two circuits are previously equal, they will counterbalance each other, and no effect will be produced at M; but if the current in either circuit is stronger than that in the other, then a current equal to the difference

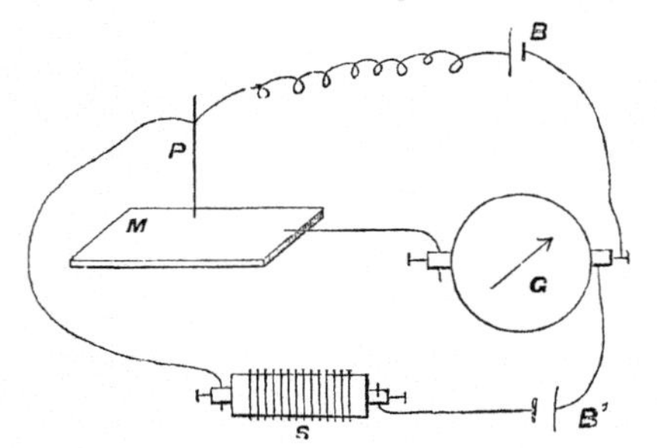

Fig. 5.—Diagram showing Action of the Telephotograph.

between the two will flow between P and M. We can make the two currents equal by inserting in the first circuit a "resistance," R, equal to the resistance of S, the selenium cell, in the dark. Then, if a light be thrown on S, the resistance in that circuit is caused to be less than the resistance of the other circuit, a current flows across the paper, and a mark is produced. If, then, the selenium is lighted and shaded at intervals, while the style is drawn steadily over the paper, we shall get a series of short lines, each line representing an illuminated interval, and the break between every two lines representing a shaded interval. It is not difficult to perceive that by a suitable arrangement of broken lines any figure can be drawn.

Fig. 6.—Form to be Reproduced by Telephotograph.

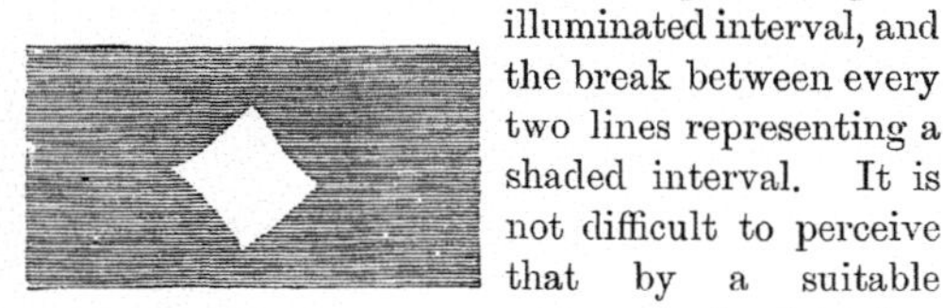

Fig. 7.—Reproduction by Telephotograph of form shown in Fig. 6.

Such a simple form as Fig. 6, for instance, would be represented by Fig. 7. which is indeed a reproduction of a bit of work actually done by the instrument. Instead of adopting any complicated mechanism to draw the marking style across the

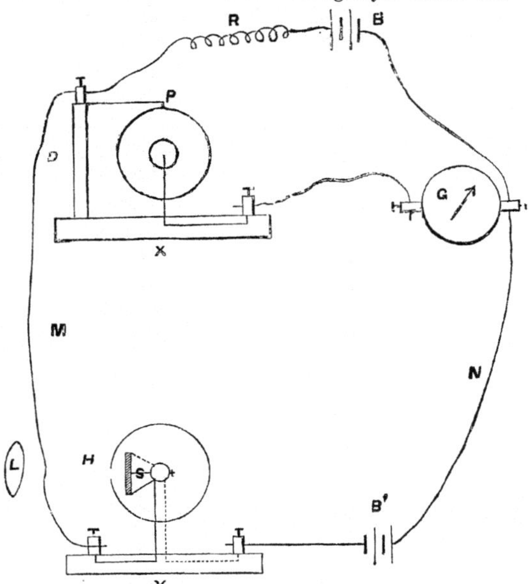

Fig. 8.—Diagram showing action of Telephotograph (Modification of Fig. 5).

paper in parallel lines, it is obviously simpler to put the paper on a cylinder the axis of which is cut with a fine screw, so that as it is turned it also travels along, and causes a point held steadily against it to describe a spiral line on the paper. When the paper is taken off the cylinder, the lines drawn spirally upon it appear practically straight and parallel, like the lines on the phonographic tinfoil, figured on p. 151. Such an arrangement is shown at X, in the upper part of Fig. 8. The arrangements here precisely correspond with those of Fig. 5, except that, instead of a simple selenium cell, we have the transmitter, Y. A platinum point, P, presses gently on the cylinder on which the prepared paper is placed. M and N in this figure represent the wires connecting the transmitting and receiving instruments; and the other letters (except H and L) represent corresponding parts with those of Fig. 5.

And now for the transmitter. It is evident that to produce the lines of Fig. 7 the selenium has to be lighted and shaded at intervals represented by the breaks in each line. No mere throwing of the shadow on the selenium will effect this. Let us see how Mr. Bidwell solves the problem. The selenium cell, S, is placed on a stand within a cylinder capable of revolving, and having on its axis a screw precisely like that of the receiver. In the cylinder is a pin-hole, H. Now, while this pin-hole is opposite the face of the selenium, light shines through it on the selenium; when the pin-hole is at the back of the cell, the light passing through it is non-effective. By means of the lens, L, an image of the figure to be reproduced is focussed on the surface of the cylinder. When the pin-hole is in the shaded part of the focussed picture, little or no light passes through it; when it is in the bright part a good deal passes. In its spiral path the pin-hole covers successively every part of the picture, and thus the selenium is lighted up and shaded at intervals, which, if the receiving cylinder be rotated in precise correspondence with that of the transmitter, will be represented by discontinuous lines drawn upon the moistened paper by the marking style.

In the experimental apparatus both the receiving and transmitting cylinders are mounted on one shaft, so as to secure absolute synchronism. This uniformity of motion would have to be obtained by special means, if the apparatus were really set to work over any appreciable distance. This is a difficulty, but it need not be regarded as an insuperable one.

THE PROBLEM OF TELEVISION.

Now that the photo-telegraph invented by Prof. Korn is on the eve of being introduced into general practice, we are informed of some similar inventions in the same field, all of which tend to achieve some step toward the solution of the problem of television.

Among the most promising schemes of this kind are doubtless the telegraphoscope and telestereograph invented by Belin, of Nancy. These two apparatus, although strictly different in principle and application, are destined to supplement each other in the same field.

The problem solved by the telegraphoscope is the production of an unalterable image of any illuminated object (person, monument, or landscape) at any desired distance, by a purely physical method, without the aid of photography. While this is actually achieved by the apparatus now constructed, the device is expected some day to solve completely the problem of "television," when these images will be produced with a more rapid succession than that corresponding with the persistence of retina images.

The problem solved by the telestereograph is identical with that solved by Prof. Korn's apparatus, viz., the reproduction at a distance of a photograph located at the transmitting station, by means of another photograph produced at the receiving station. However, the solution suggested by Belin is entirely different in principle from Prof. Korn's scheme.

The telegraphoscope comprises a number of novel processes intended for reproducing the real aerial image of a camera obscura by a purely physical method without the aid of any chemical manipulation. Sensitive plates or papers, as well as developing or fixing processes, are thus dispensed with, the image being immediately recorded on any kind of paper either in the vicinity of the object or at considerable distance, according to the length of line connecting the two stations.

The general principles underlying the construction of the telegraphoscope are as follows:

At the transmitting station there is installed a camera obscura similar to a photographic camera, an objective and a mirror reflecting the optical image on a row of eighty exceedingly small selenium cells. A collector is provided for successively inserting these cells in the same circuit.

The receiving station comprises a highly sensitive galvanometer with a very rapidly oscillating needle (Blondel's oscillograph) and three local circuits, viz.:

1. A circuit containing the most important organ of Belin's scheme, viz., the equilibrator, and a telegraph relay.

2. The primary circuit of a small induction coil, containing in turn the second part of the equilibrator.

3. The secondary circuit of the induction coil, comprising a spark gap, the spark of which, controlled by the distributing collector referred to, perforates a paper with a number of exceedingly small holes, all of which are situated at the same distance from the centers while their diameters are directly or inversely proportional to the corresponding points of the image. These points, situated side by side without covering each other, impart to the reproduction the aspect of an autotypy.

The ratio above referred to can be made at will either direct or inverse (the images of the telegraphoscope becoming positive or negative respectively), simply by handling a current-reversing switch.

It should be understood that the selenium cells used in this apparatus are constructed by the inventor himself, M. Belin having nowhere found a set of eighty identical cells of sufficiently small dimensions (1.5 millimeter). It was only after extensive researches continued for many years that the inventor developed his actual process, according to which the eighty cells are prepared simultaneously on the same support. As these cells show a considerable resistance (5 megohms), the resistance of the circuit becomes negligible in comparison. On the other hand, they are entirely free from inertia, so as to necessitate no compensating device.

Experimenters who have hitherto dealt with the same problem have been unable to obtain at the receiving station a satisfactory effect, either with the slight current intensities traversing the selenium or with the feeble differences in intensity corresponding with the difference in the resistance of the selenium between total darkness and full light in the camera. This difficulty (to which the failure of all previous attempts may be ascribed) is entirely eliminated by the equilibrator. In fact, quite independently of the current intensity at the receiving station, the local current always has a considerable intensity, which is always strictly proportional to the intensity traversing the line in spite of the unceasing variations in the latter, due to the reproduction of half-shades.

This equilibrator was tested as far back as two years ago, when some preliminary experiments were carried out between Paris-Havre-Paris (480 kilometers), the perforations obtained at the receiving station always reproducing faithfully any variations in the luminous intensity of the sending station.

The final apparatus has just been completed at Paris by M. Ducretet, the well-known constructor. After performing some additional regulations, a trial service will be commenced.

As regards the other invention, the telestereograph, this as above mentioned, is intended for telegraphically transmitting a photograph to be reproduced at the receiving station by another photograph. However, the solution suggested by Belin is entirely different from Prof. Korn's scheme.

In opposition to Korn's as well as to all earlier devices of a similar kind, this apparatus in fact dispenses with selenium both at the sending and receiving stations. The apparatus thus is entirely mechanical in construction and independent of such capricious organs as selenium resistances, while electricity as in telegraphy plays no other part than that of controlling the process. At the receiving station the image is reproduced by lines 1/6 millimeter in thickness and in distance, so that the pictures obtained show continuous shades which are very rich in detail. By virtue of the very principle underlying its construction, the drawing is necessarily of good definition.

While the reproductions are normally of the same dimensions as the original photograph, they can at will be made greater or smaller simply by controlling a nut, the same original photograph can be made to give a positive or a negative reproduction with other shades similar to those of the original photograph or with stronger or slighter gradations.

After a slight alteration the same apparatus can be used as a high speed facsimile telegraph without using any special paper for writing the telegram, the only condition being the use of a special ink.

The apparatus is now being constructed by M. J. Richard.

Distant Electric Vision.

REFERRING to Mr. Shelford Bidwell's illuminating communication on this subject published in NATURE of June 4, may I point out that though, as stated by Mr. Bidwell, it is wildly impracticable to effect even 160,000 synchronised operations per second by ordinary mechanical means, this part of the problem of obtaining distant electric vision can probably be solved by the employment of two beams of kathode rays (one at the transmitting and one at the receiving station) synchronously deflected by the varying fields of two electromagnets placed at right angles to one another and energised by two alternating electric currents of widely different frequencies, so that the moving extremities of the two beams are caused to sweep synchronously over the whole of the required surfaces within the one-tenth of a second necessary to take advantage of visual persistence.

Indeed, so far as the receiving apparatus is concerned, the moving kathode beam has only to be arranged to impinge on a sufficiently sensitive fluorescent screen, and given suitable variations in its intensity, to obtain the desired result.

The real difficulties lie in devising an efficient transmitter which, under the influence of light and shade, shall sufficiently vary the transmitted electric current so as to produce the necessary alterations in the intensity of the kathode beam of the receiver, and further in making this transmitter sufficiently rapid in its action to respond to the 160,000 variations per second that are necessary as a minimum.

Possibly no photoelectric phenomenon at present known will provide what is required in this respect, but should something suitable be discovered, distant electric vision will, I think, come within the region of possibility.

A. A. CAMPBELL SWINTON.

66 Victoria Street, London, S.W., June 12.

Part 2

EARLY SUCCESSES

Television becomes a reality: the first successful experiments

THE TIMES ENGINEERING SUPPLEMENT,

WEDNESDAY, NOVEMBER 15, 1911.

RONTGEN SOCIETY.

DISTANT ELECTRIC VISION.

In the course of his presidential address to the Röntgen Society, Mr. A. A. CAMPBELL SWINTON outlined an arrangement by which with the aid of cathode rays it might be possible to realize distant electrical vision, or in other words, do for the sense of sight what the telephone has done for the sense of hearing. He wished it to be distinctly understood, however, that his plan was an idea only, and that the apparatus had never been constructed; furthermore, he did not for a moment suppose it could be got to work without a great deal of experiment and probably much modification. But he believed it to be the first suggested solution of the problem of distant electric vision in which the difficulty of securing the required extreme rapidity and accuracy of motion of the parts was got over by employing for those parts things of the extreme tenuity and weightlessness of cathode rays. Indeed, apart from the revolving armatures of the alternators employed for synchronization, which presented no difficulty, there was no more material moving part in the suggested apparatus than almost immaterial streams of negative electrons. Furthermore, only four wires, or three wires and earth connexions at each end, were required.

CONSTRUCTION OF THE APPARATUS.

The transmitter consisted of a Crookes tube fitted with a cathode which sent a cathode ray discharge through a small aperture in the anode, the cathode rays being produced by a battery or other source of continuous electric current giving some 100,000 volts. Two electro-magnets were placed at right angles to one another, in such a way that when energized by alternating current, they would deflect the cathode rays in a vertical and in a horizontal direction respectively.

The receiving apparatus consists similarly of a Crookes tube fitted with a cathode, which transmitted cathode rays through an aperture in the anode. Again there were two electro-magnets placed at right angles, similar to those in the transmitter; the two magnets (in the transmitter and receiver) which controlled the vertical motions of the cathode ray beam were energized from the same alternating dynamo, which had a frequency, say, of ten complete alternations a second, while the other two magnets which controlled the horizontal movements of the cathode ray beam, were energized by a second alternating dynamo having a frequency of, say, 1,000 complete alternations a second.

In the receiver was a fluorescent screen upon which the cathode rays impinged, and the whole surface of which they searched out every tenth of a second under the combined deflecting influence of the two magnets, with the result that the screen fluoresced with what appeared to the eye as a uniform brilliancy. Similarly, in the transmitting apparatus, the cathode rays fall on a screen, the whole surface of which they searched out every tenth of a second under the influence of the magnets. Further, as the two pairs of magnets were energized by the same currents, the movements of the two beams of cathode rays would be exactly synchronous and the cathode rays would always fall on the two screens and on each corresponding spot simultaneously.

In the transmitter, the screen, which was gas-tight, was formed of a number of small metallic cubes insulated from one another, but presenting a clean metallic surface to the cathode rays on the one side, and to a suitable gas or vapour, say sodium vapour, on the other. The metallic cubes which composed this screen were made of some metal, such as rubidium, which was strongly active photo-electrically in readily discharging negative electricity under the influence of light, while the receptacle behind it was filled with a gas or vapour, such as sodium vapour, which conducted negative electricity more readily under the influence of light than in the dark. Parallel to the screen was another screen of metallic gauze, and the image of the object which was to be transmitted was projected by lens through the gauze screen upon the photo-electric screen through the vapour contained in the receptacle. The gauze screen was connected through the line wire to a metallic plate in the receiver, past which the cathode rays had to pass. There was in the receiver a diaphragm fitted with an aperture in such a position as, having regard to the inclined position of the cathode, to cut off the cathode rays coming from the latter and prevent them from reaching the fluorescent screen unless they were slightly repelled from the metallic plate, when they were able to pass through the aperture.

Action of the Device.

The whole apparatus was designed to function as follows:—Assuming a uniform beam of cathode rays to be passing in the two Crookes tubes in the transmitter and receiver, and the two pairs of magnets to be energized with alternating current, as mentioned, and assuming that the image to be transmitted was strongly projected through the lens and gauze screen upon the photo-electric screen; then, as the cathode rays in the sending Crookes tube oscillated and searched out the surface of that screen, they would impart a negative charge in turn to all the metallic cubes of which it was composed. In the case of cubes on which no light was projected, nothing further would happen, the charge dissipating itself in the tube; but in the case of such of the cubes as were brightly illuminated by the projected image, the negative charge imparted to them by the cathode rays would pass away through the ionized gas along the line of the illuminating beam of light until it reached the gauze screen, whence the charge would travel by means of the line wire to the metallic plate of the receiver. This plate would thereby be charged; would slightly repel the cathode rays in the receiver; would enable these rays to pass through the diaphragm, and, impinging on the fluorescent screen, would make a spot of light. This would occur in the case of each metallic cube of the transmitting screen which was illuminated, while each bright spot on the fluorescent screen would have relatively exactly the same position as that of each illuminated cube. Consequently, as the cathode ray beam in the transmitter passed over in turn each of the metallic cubes of the screen, it would indicate by a corresponding bright spot on the fluorescent screen whether any particular cube was or was not illuminated, with the result that the fluorescent screen, within one-tenth of a second, would be covered with a number of luminous spots exactly corresponding to the luminous image thrown on the transmitting screen, to the extent that this image could be reconstructed in a mosaic fashion. By making the beams of cathode rays very thin and employing a very large number of very small metallic cubes, and a high rate of alternation in the dynamo, it was obvious that the size of the luminous spots by which the image was constituted could be made very small and numerous, with the result that the more these conditions were observed the more distinct and accurate would be the received image.

It was also obvious that, by employing for the fluorescent material something that had some degree of persistency in its fluorescence, it would be possible to reduce the rate at which the synchronized motions and impulses need take place, though this would only be attained at the expense of being able to follow rapid movements in the image that was being transmitted. Further, as each of the metallic cubes acted as an independent photo-electric cell, and was called upon to act only once in a tenth of a second, the arrangement had obvious advantages over the other arrangements that had been suggested, in which a single photo-electric cell was called upon to produce the many thousands of separate impulses required to be transmitted through the line wire each second, a condition which no known form of photo-electric cell would admit of. Nor did sluggishness on the part of the metallic cubes or of the vapour in the receptacle in acting photo-electrically interfere with the correct transmission and reproduction of the image, provided all portions of the image were at rest; and it was only to the extent that portions of the image might be in motion that such sluggishness could have any prejudicial effect. In fact, sluggishness would only cause changes in the image to appear gradually instead of instantaneously.

THE RADIO KINEMA

by F. H. ROBINSON

The Receiver Disc at Rest.

In Action. Note X on right.

FEW people have seriously taken the view that the kinema has reached a state of finality. Mechanical developments in problems of projection are few and far between now, it is true, and on the production of pictures the tendency has been to follow the broad path of least resistance rather than to adventure into byways that may lead to a real steep forward. Coloured photography and talking pictures are, as far as the practical showman is concerned, not much more than the subject of experiment. But the rapid strides made in the world of broadcasting suggest a new line of thought.

There is a close association between the kinema and radio, obvious when one realises that light waves and wireless waves travel through the same medium—ether.

A definite link has now been established between these two types of waves. It is called television, and far from being the "dream of the future" is an established fact.

Not so very long ago I visited one John Logie Baird at his laboratory at Hastings, and saw a demonstration which proved that he has proceeded so far along the road to radio vision as to make it almost a commercial proposition, for the whole of the apparatus used in the experiment about to be described could have been purchased for £40.

The apparatus used can be applied to wire or wireless transmitters with ease and without the alteration of anything further than the microphone, in which circuit the "Radio Vision" machine is connected.

The Mechanism.

This roughly is the method employed by Mr. Baird. A powerful light, either a gas-filled electric or arc lamp, is used as the prime mover. The image to be transmitted is cut out of some solid opaque material, say, cardboard.

In front of the image are two revolving discs. The first contains a number of holes along its edge, cut in series of fives so that as it is revolved each hole transmits a strip of the image.

The light-flashes passed in this manner are interrupted by the second disc, which revolves in the opposite direction to the first and has a serrated edge, formed of spokes.

This disc is driven at a speed sufficient for these spokes to cut up the light impulses passed by the first disc into flashes of an audible frequency. These flashes are then focused on to a light-sensitive, or selenium cell.

Selenium and its Drawbacks.

In effect the light impulses have been changed into audible impulses, and if a telephone were connected across the output side of the cell a high pitched note would be heard.

This serrated disc has therefore overcome one of the principal difficulties, which by the way was mentioned in these columns a short while ago; that is the very definite "lag" of the selenium cell. Although this cell is very sensitive to light it is unable to follow very fast changes. When the impulses are of an audible frequency, however, it is able to follow these and definite current pulses result on the output side.

These pulses are amplified by four stages of low frequency amplifiers and one power stage which employs 200 volts on the plate.

These amplified impulses are passed into the ordinary microphone circuit of a wire or wireless transmitter and radiated in the usual manner.

The Reception Apparatus.

For detecting the audible impulses the usual valve detector is employed, the resultant impulses being again enormously amplified. (It should be made clear that distortion does not affect transmission or reception as the zero points of each wave must occur at recognised periods, no matter how distorted the waves may be.)

These amplified impulses are then passed to another revolving disc. On this in place of the holes which are cut in the transmitting disc there are a number of quick-acting lamps arranged in positions corresponding to the holes in the transmitting disc.

All of these lamps are wired up to a commutator to which the impulses are fed.

The lamps then light up in sympathy with the light flashes passed by the transmitting disc and reproduce the original image. It is necessary, of course, to synchronise the transmitting and receiving discs.

The Test.

I myself saw a cross, the letter "H," and the fingers of my own hand reproduced by this apparatus across the width of the laboratory. The images were quite sharp and clear, although perhaps a little unsteady. This, however, was mostly due to mechanical defects in the apparatus and not to any fault of the system.

Moving images may be transmitted by this means, and distance is no object, merely depending on the power of the wireless transmitter and the sensitivity of the receiver employed.

It is possible that machine-made apparatus on the lines indicated above could be made for some £50, which would be capable of transmitting letters and words clearly many miles through the ether, and all that appears to be necessary in order to reproduce and transmit moving pictures is more expensive and elaborate apparatus.

The inventor is confident that no technical difficulties stand in the way of the transmission of moving images by wireless.

Undoubtedly wonderful possibilities are opened up by this invention, its very simplicity and reliability placing it well to the front of many of the various complicated methods which have been evolved to do the same work. Now that the main principle has been communicated and proved it is not too much to expect that in the course of time we shall be able to see on the screen the winner of the Derby actually racing home watched by hundreds of thousands of his worshippers at Epsom.

Letter 1

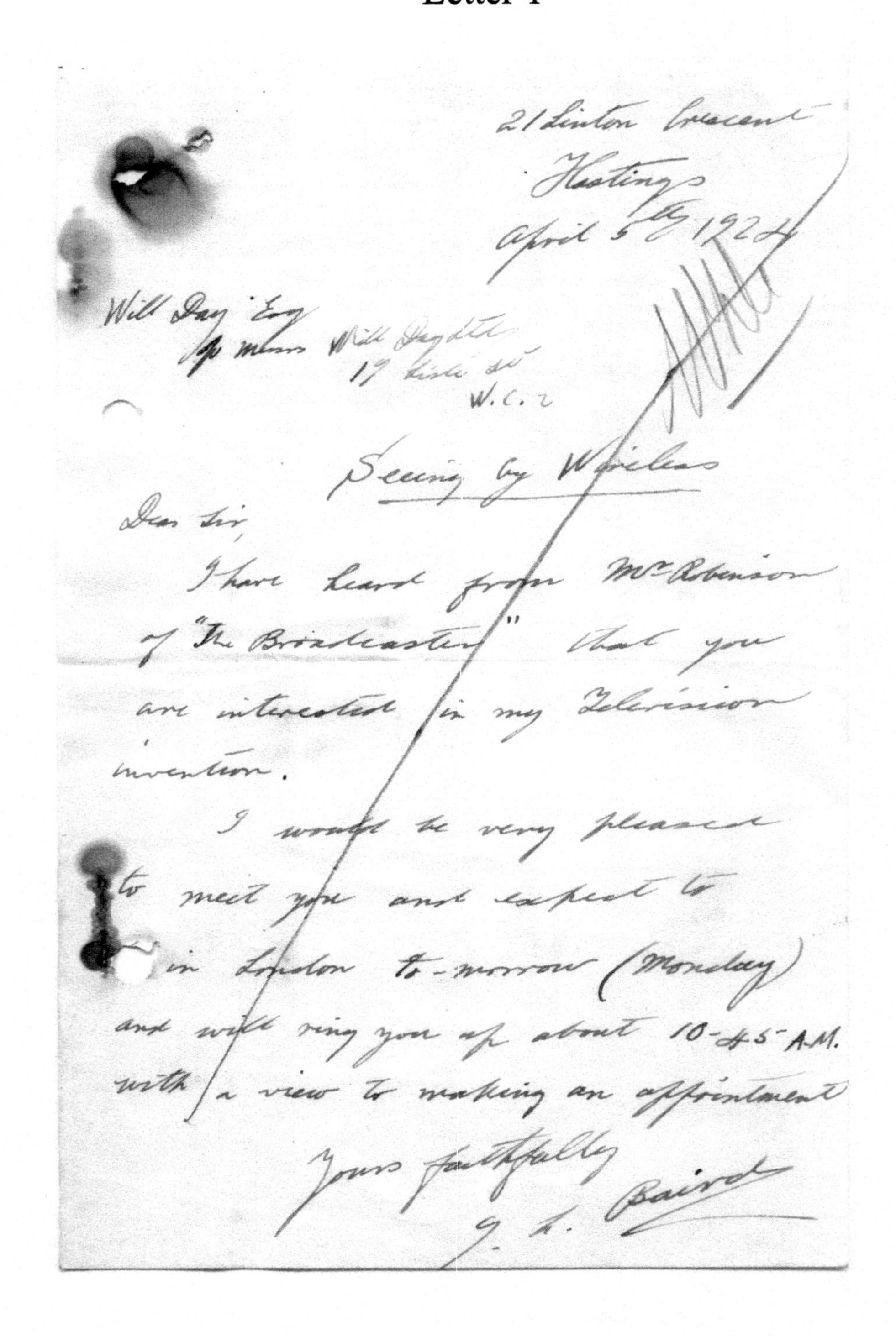

21 Linton Crescent
Hastings
April 5th 1924

Will Day Esq
of Messrs Will Day Ltd
19 Lisle St
W.C.2

Seeing by Wireless

Dear Sir,

I have heard from Mr Robinson of "The Broadcaster" that you are interested in my Television invention.

I would be very pleased to meet you and expect to [be] in London to-morrow (Monday) and will ring you up about 10-45 A.M. with a view to making an appointment

Yours faithfully
J. L. Baird

Letter 2

21 Linton Crescent
Hastings.
May 1st 1914

Mr Day

Dear Mr Day,

I am sending under separate cover a full size drawing of the second disc. and hope it will be clear

I will bring up the complete drawings for the model on Monday they are not finished yet

Letter 2 continued

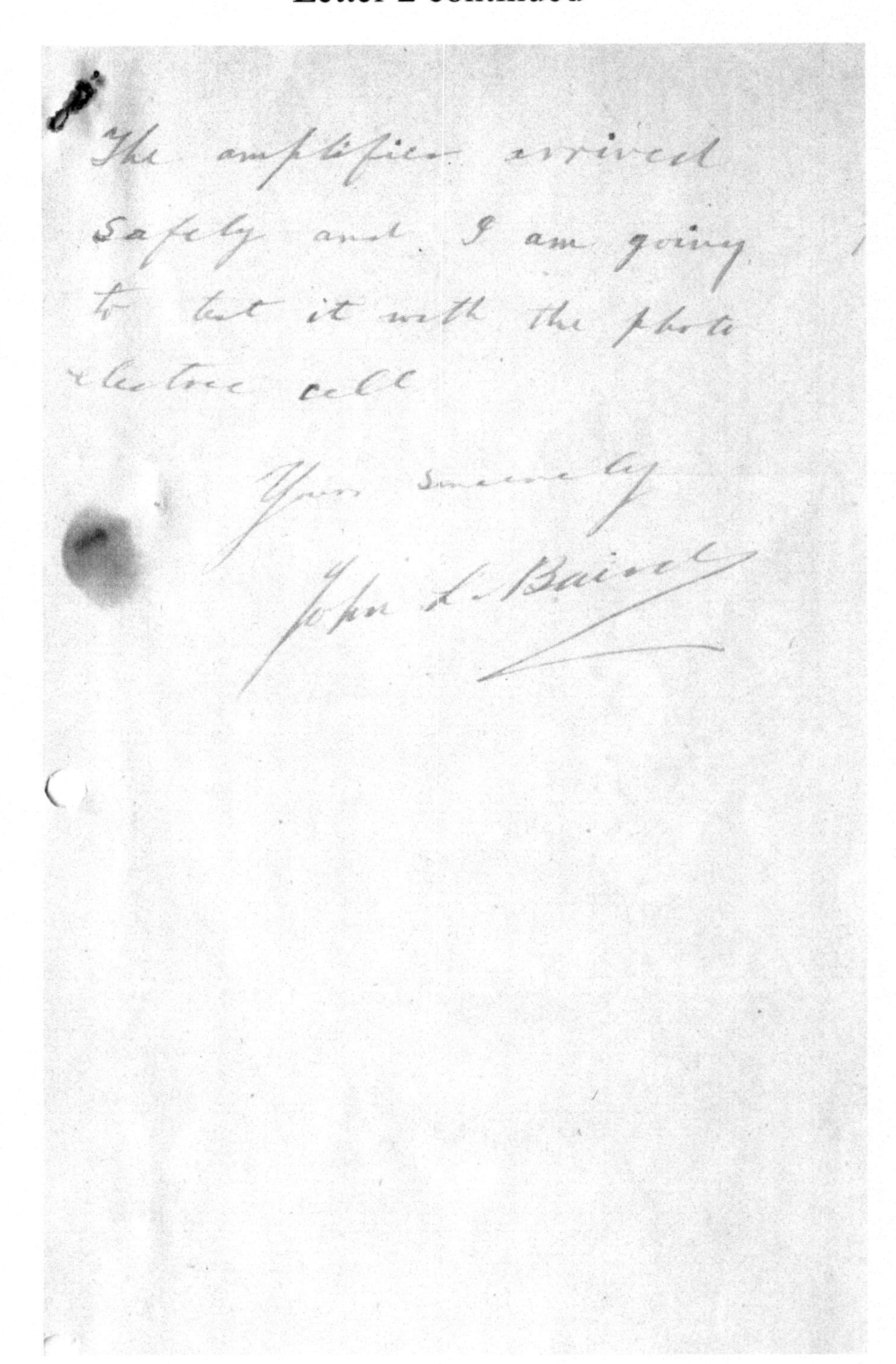

The amplifier arrived safely and I am going to test it with the photo electric cell

Yours sincerely

John L. Baird

Letter 3

May 2nd .1924 .

WD/VF

J .L .Baird , Esq .,
21 , Linton Crescent ,
Hastings .

Dear Mr Baird ,

Your letter duly to hand , and the drawing of the disc , which I have sent to my sheet metal worker this day , and trust he will make this satisfactorily to your requirements . It will be ready some time next week , and I am also glad to note that the amplifying panel arrived in safety .

I saw my friend Mr King yesterday , who gave the lecture on the eye before the Royal Photographic Society . He assures me that there are seven colour films at the back of the eye which receive the colour sensitive impressions and select through the fluids according to the colour impression received: but more than this - he is finding me an eminent professor's book on colour photography and the human eye , which may elucidate many mysteries , and I hope give you a clear , concise , idea as to what actually takes place in the human eye to receive the colour impressions .

Assuring you at all times of my best help in the great achievement we both wish to see accomplished , and hoping to see you one day next week ,

Yours faithfully ,

Letter 4

July 25th 1924.

WD/VF

John L Baird, Esq.,
21, Linton Crescent,
Hastings.

Dear Mr Baird,

I am sorry to see by "The Daily Chronicle" of today's date that you have again had a burn out with your high frequency current, and trust it has not been serious. You must learn that you are playing with a very dangerous element in using so much voltage and that it is likely to cause you a serious injury, if not a fatal one.

Surely you take proper precautions when handling this current, even if it were only using a pair of rubber gloves, which would eliminate a great deal of the trouble abd avoid any recurrence of the happenings recorded. I hope, however, to hear from you that you are progressing favourably and that no serious mishap has occurred.

I note in the report they say the instrument is badly damaged. I should be glad to know to what extent the instrument is affected.

Yours faithfully,

Letter 5

21 Linton Crescent
Hastings
28th July 1924

Dear Mr Day

Thanks for your letter I am not much injured – some burns on the hands otherwise all well, no damage has been done to the machine or any of the apparatus.

I will arrange to box in the H.T. and provide safety bars which will prevent a recurrence of the shocks

Letter 5 continued

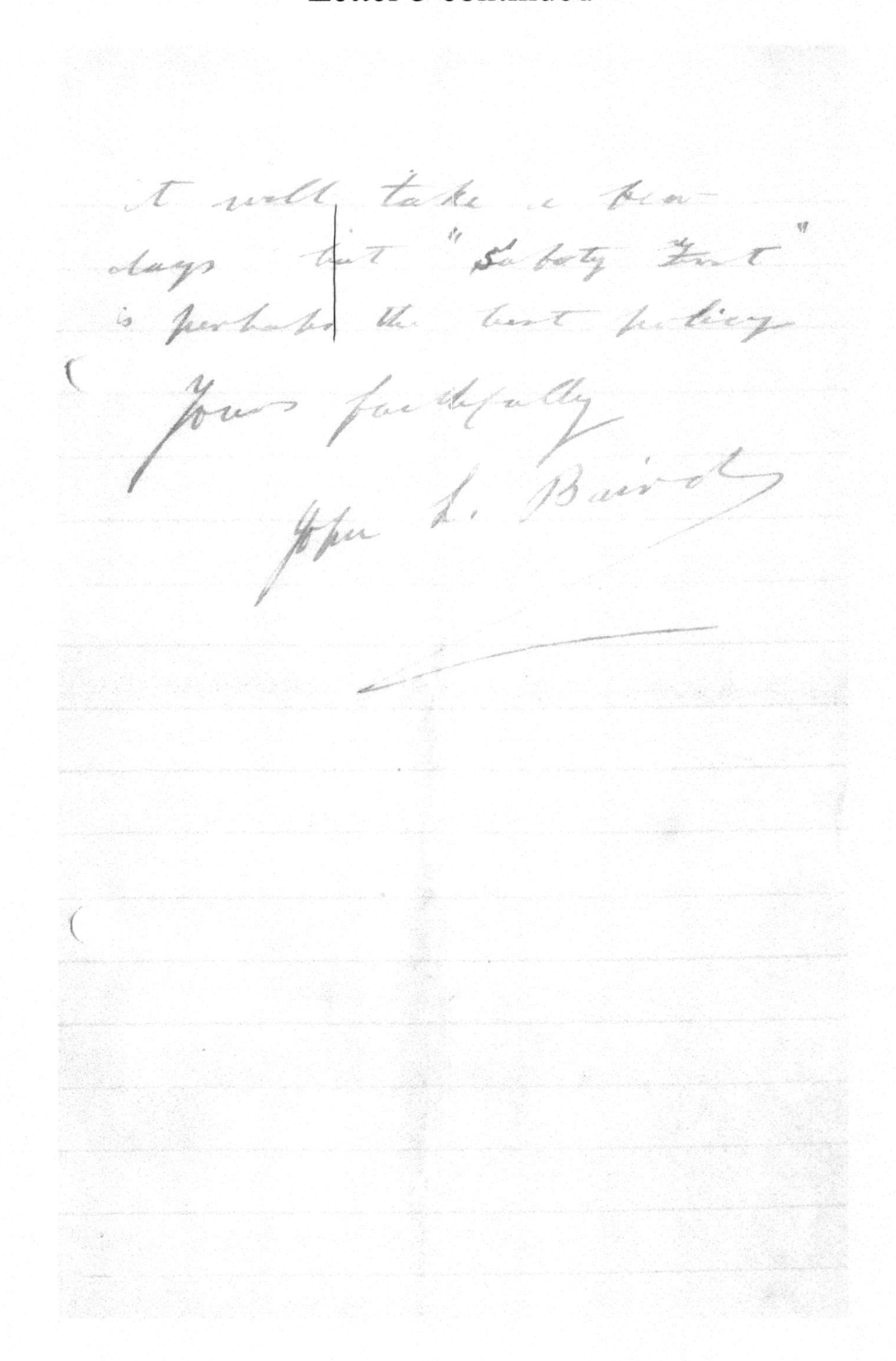

It will take a few days but "Safety First" is perhaps the best policy

Yours faithfully

pp L. Baird

Letter 6

September 6th 1924.

WD/VF

John L Baird, Esq.,
21, Linton Crescent,
Hastings.

Dear Mr.Baird,

I forgot to mention to you yesterday that you will have to be quite sure that the Television effect is produced absolutely by wireless without a circuit in operation, as it struck me after the display on Friday night last that the circuit was a closed one with wires.

You will, I am sure, see to this, so that there is no query regarding this matter when we come down on Monday afternoon next, as I do not want any disappointment and I feel sure a great deal depends on this display with regard to the future.

Trusting you got your motors in safety,

Yours faithfully,

SELFRIDGE'S

Present the First Public Demonstration of Television in the Electrical Section (First Floor).

Television is to light what telephony is to sound—it means the *INSTANTANEOUS* transmission of a picture, so that the observer at the "receiving" end can see, to all intents and purposes, what is a cinematographic view of what is happening at the "sending" end.

For many years experiments have been conducted with this end in view; the apparatus that is here being demonstrated, is the first to be successful, and is as different to the apparatus that transmits pictures (that are from time to time printed in the newspapers) as the telephone is to the telegraph.

The apparatus here demonstrated is, of course, absolutely "in the rough"—the question of finance is always an important one for the inventor. But it does, undoubtedly, transmit an instantaneous picture. The picture is flickering and defective, and at present only simple pictures can be sent successfully ; but Edison's first phonograph announced that "Mary had a little lamb" in a way that only hearers who were "in the secret" could understand—and yet, from that first result has developed the gramophone of to-day. Unquestionably the present experimental apparatus can be similarly perfected and refined.

It has never before been shown to the Public. Mr. J. L. Baird, the sole inventor and owner of the patent rights, will be present daily while the apparatus is working—in the Electrical Section at 11.30 a.m., 2.30 p.m., and 3.15 p.m. He will be glad to explain to those interested in details.

We should perhaps explain that we are in no way financially interested in this remarkable invention ; the demonstrations are taking place here only because we know that our friends will be interested in something that should rank with the greatest iuventions of the century.

SELFRIDGE & CO., LTD.

A SERIES of interesting demonstrations has been given at Messrs. Selfridge and Co., Ltd., London, W.1., by Mr. J. L. Baird, of an experimental apparatus of his own design for wireless "television" (*i.e.* the simultaneous reproduction at a distance of an image of a fixed or moving object). The inventor does not claim any great perfection for his results, but we have seen the production in the receiver of a recognisable, if rather blurred, image of simple forms, such as letters painted in white on a black card, held up before the transmitter. Mr. Baird has overcome many practical difficulties, but we are afraid that there are many more to be surmounted before ideal television is accomplished. In the transmitting apparatus, the object, strongly illuminated, is placed opposite a revolving disc provided with a series of lenses, each a little nearer to the centre than the last, which project a

series of moving images upon a selenium or other photo-electric cell, each a little displaced laterally from the last. This is the equivalent of passing the cell over the whole surface of the object in a succession of close parallel lines. The light thus reaching the photo-electric cell is rhythmically interrupted by a rapidly revolving slotted disc, and the result is that owing to the variations of resistance of the cell, undulations at an audio-frequency are produced in the current through it, whenever a bright part of the object is being dealt with. These are amplified and supplied to a simple wireless transmitter which is caused to emit corresponding signals.

IN the receiving section of Mr. Baird's television apparatus, the signals sent out from the transmitter are detected and amplified by very powerful valves until they are strong enough to light up a neon tube when a signal is received, *i.e.* when a bright part of the object is being dealt with by the transmitting apparatus. A disc with lenses or holes corresponding to the lenses of the transmitting disc is rotated synchronously with the transmitting disc, causing spots of light produced by the neon tube to appear upon a screen in positions corresponding to the part of the object being dealt with. With a sufficiently rapid rotation of the discs, a recognisable image of the object is produced. A duplicate of the receiving apparatus is provided at the sending end with its disc mounted on the same shaft as the transmitting disc, to enable the necessary adjustments to be made. Synchronism between the sending and receiving discs is obtained by a little alternator with a frequency of about 300 geared to the revolving system, which causes signals to be sent out by another wireless transmitter at this frequency. These are received and amplified at the receiving station to an extent enabling a similar little alternator connected to the receiving discs to be synchronised with them.

Electric Television.

INVENTION appears to be multiplying in regard to this interesting subject, and I hear that more than one inventor in Paris is employing, for receiving, the cathode ray arrangement that I believe I was the first to publish in a letter to NATURE of June 18, 1908. The ideas embodied in this arrangement had occurred to me several years prior to that date, indeed not long after the production of the Braun cathode ray oscillograph invented in 1897. I actually tried some not very successful experiments in the matter of getting an electrical effect from the combined action of light and cathode rays incident upon a selenium-coated surface, in which I was assisted by the late Prof. G. M. Minchin, himself a great authority on electric cells sensitive to light, and also by Mr. J. C. M. Stanton. The transmitting apparatus consisted of a home-made Braun oscillograph in which a metal plate coated with selenium was substituted for the usual fluorescent screen, the image to be transmitted being thrown by a lens upon the selenium surface, and the end of the cathode ray beam being caused electromagnetically to traverse the projected image. Experiments were also tried in receiving with a Braun tube which I purchased in Germany, but in its then 'hard' form it proved very intractable.

My ideas in regard to this cathode ray arrangement for the production of television were further detailed and illustrated in an address I gave to the Röntgen Society on November 7, 1911, and still further elaborated and brought up-to-date, with wireless methods applied, in a paper I read before the Radio Society of Great Britain on March 26, 1924.

My idea, which was to use cathode rays as employed in the Braun oscillograph, instead of moving material parts, both in the transmitting and in the receiving instruments, is, as I understand, only at present being applied for receiving, mechanical apparatus being still used for transmitting. I desire, however, to point out that when the cathode ray is also applied to transmitting it will be possible to dispense entirely with all moving material parts, as the alternating or intermittent electric currents employed for moving the two cathode ray beams synchronously at the transmitting and receiving stations respectively can be supplied by oscillating thermionic valves supplied by batteries.

In this way it should prove possible to have electric television of a satisfactory fine-grain description without the employment of any mechanical motion of material parts whatever, as cathode rays are practically without weight and inertia, and can be deflected with perfect accuracy and synchronism at almost incredible speeds, while the accuracy of oscillating valves properly tuned is also wonderful.

A. A. CAMPBELL SWINTON.

October 9.

NATURE

SATURDAY, JANUARY 15, 1927.

Television.

THE lecture delivered by Mr. J. L. Baird on Jan. 6 at the Physical and Optical Society's annual exhibition was very largely attended, and had to be repeated the same evening in order to prevent widespread disappointment. The attendance was a tribute to the intrinsic interest of the subject and to the expectations aroused by what little has become known of the lecturer's apparatus and achievements.

Mr. Baird did not add to the general knowledge by what he said and did on that occasion. The situation appears to be that he has achieved a certain success in transmitting instantaneous images divided into some thirty vertical strips, each strip showing more or less correct gradation in a vertical direction. The picture composed by the strips is therefore very coarse-grained in one direction while sufficiently graded in a direction at right angles to it. The most important point about the transmission is that it uses diffusely reflected light, and not the light transmitted, say, by a lantern slide. To the latter transmission Mr. Baird will not consent to apply the term 'television' at all, and in this we consider he is justified. The eye itself normally sees objects by diffusely reflected light, and television, to be worthy of the name and to accord with popular expectation, must do the same.

Now this proviso is a very serious obstacle to overcome. For the 'candle-power' of a brightly illuminated face is approximately unity, whereas in kinematograph projection it amounts to thousands. True television, as defined by Mr. Baird, is therefore at least a thousand times more difficult than the transmission of shadows of moving objects such as has already been achieved by several inventors, among whom Ruhmer was the first in 1907.

Mr. Baird appears to have succeeded in transmitting images of objects artificially illuminated to an extent equivalent to bright sunlight, and to have overcome the difficulties of synchronisation in transmitting and receiving within the same building. He also claims to have accomplished similar transmissions by radio over about ten miles, and to have transmitted living faces showing considerable detail. While giving him all credit for his very direct attack on a difficult problem, and for what he has demonstrably achieved, we must guard against an underestimate of the remaining difficulties. These will affect such things as synchronism, illumination, and detail.

It is comparatively easy to keep two motors running at the same speed, even at a distance. But when a difference of phase of only one degree is capable of spoiling definition, the maintenance of the correct phase becomes a formidable task. Granted that Mr. Baird transmits the equivalent of 30,000 signals per second as against the 300,000 required for satisfactory television, it is easily seen that the problem of synchronism may become sufficiently acute to bar further progress. The figure of 300,000, on which most workers on this problem are agreed, is based upon the fact that a face alone, to be recognisable, requires some 3000 graded elements. Most press portraits comprise at least 10,000 such elements; and in order to produce the illusion of continuity or motion, there must be at least sixteen transmissions of the whole picture per second.

From the information available—and it must be remembered that Mr. Baird has not disclosed the essential details of his method—it appears highly probable that little further progress need be expected along the lines chosen by him. The illumination seems to have been driven to its farthest limit, and the recent claims to have transmitted 'outlines' by infra-red rays mark no advance towards television with diffusely reflected light. The subdivision of the image by a disc of spirally staggered lenses seems incapable of extension, and the suggestion thrown out in the course of the lecture, that several photo-electric cells (and several radio-frequencies) might be used simultaneously, only serves to emphasise difficulties already known to exist.

It is to be regretted that, possibly on account of patent considerations, Mr. Baird has hitherto been unable to submit to a proper authentication of his claims by a learned society. For aught his recent audience could say, the invention might be a mere plaything, with no more resemblance to television than a toy engine has to the real thing. That is not the way to convince a sympathetic audience of experts who are expected to judge of the merits and prospects of an invention.

There are at least three pioneers in the field who seem to be on the verge of a complete solution of the television problem. Belin in France, and Alexanderson and Jenkins in America, are all approaching the problem by way of the transmission of photographs, which they accomplish in great perfection. Dr. Alexanderson, chief engineer of the Radio Corporation of America, claims to be able to transmit a complete photograph in one second, and points out that if sixteen successive photographs could be transmitted in one second, the problem of television (or, at least, of telekinematography) would be solved. He adds, however, that the difficulties, especially of synchronism, increase as the square of the speed.

It is well to remember that the earlier solutions proposed after the discovery of the light-sensitive property of selenium did not require the synchronisation of moving parts at the sending and receiving stations. But a multiplicity of wires, one for each picture element, is out of the question, and so is a multiplicity of radio-frequencies. Unless, therefore, some device such as Dr. Fournier d'Albe's acoustic resonator system can be employed for the simultaneous reception of a medley of signals, the question of synchronism, effective both in speed and phase, will arise; and when we realise that the synchronism required is that of two 'pencils' which traverse the picture completely in a sixteenth of a second, and in doing so describe several hundred lines, the mechanical difficulties may well appal us. The speed with which the stimulus can be applied at the transmitting end is great enough nowadays, for the action of a photo-electric cell shows no appreciable lag, and the use of a Braun cathode ray tube at the receiving end, first suggested, we believe, by Mr. Campbell Swinton, will solve the speed problem there.

There remains the problem of sensitiveness to illumination at the receiving end. Mr. Baird does not tell us what he uses. His results suggest a potassium photo-electric cell for ultra-violet, and either selenium or bolometer for the infra-red. If he has discovered any reagent of greatly superior power, that discovery alone would entitle him to our gratitude, and would constitute a valid claim even though all his other devices had been anticipated by others. The policy of withholding publication of an essential item does not commend itself to modern inventors. It savours too much of medieval practice, and usually defeats its own object of securing to the inventor the fruits of his invention.

It would be a source of satisfaction to us if one of our countrymen were the first to provide a practical solution of a problem of this magnitude; a solution such as the civilised world has been expecting for some years. If the solution has been reached without the scientific, engineering, and financial resources at the disposal of rival inventors, it will appeal very powerfully to our sympathy and imagination. But for the present, and on the evidence supplied, the scientific world will probably prefer to reserve its judgment.

Television.

THE article headed "Television" which appeared in NATURE of Jan. 15, contains the following statement: "a difference of phase of only one degree is capable of spoiling definition." Were this statement true, my television system, depending as it does on synchronism, would certainly, as the writer states, be faced with a very serious barrier. It is, however, a misstatement of fact. Phase difference between receiver and transmitter has no effect whatever upon definition, the whole effect being a displacement of the image as a whole.

Later in the article a statement is made: "The recent claims to have transmitted 'outlines' by infra-red rays mark no advance toward television with diffusely reflected light." This is an erroneous statement. I have on no occasion made claims to have transmitted 'outlines' by infra-red rays. What I have actually demonstrated is the transmission of real images of living faces in complete darkness, using diffusely reflected infra-red rays.

An open invitation was extended to members of the Royal Institution to witness these results, and on Dec. 30, 1926, some forty members of the Institution were given demonstrations at our laboratories. Among those who have witnessed demonstrations I may mention Dr. Russell, Mr. R. W. Paul, and Mr. Creed, who are, I think, sufficiently well known in the scientific and engineering world. In these demonstrations one party remained in a totally dark room; the second party, in a different, were then shown the faces of any of the first party who cared to sit in front of the transmitting apparatus in the dark room.

My discourse at the exhibition of the Physical Society is criticised and the statement made that it was "not the way to convince a sympathetic audience of experts." I was requested by Dr. Rankine, the secretary of the Society, to deliver a lecture suitable for a public audience interested in scientific matters generally, but not experts on television. The lecture was therefore of a semi-popular type and in no way intended for an "audience of experts."

While the writer of the article in NATURE appears to be dissatisfied, judging by the reception which the lecture was given and the appreciative letters which I have received, I am assured that opinion was not shared by the bulk of the audience.

I am further criticised for withholding technical details. The writer of the article is surely aware that my inventions are the property of a limited company. The disclosures by me of technical details likely to assist competing interests would therefore be a grave breach of trust to the shareholders. I may further add that we have demonstrated the invention to Government experts, and have received a letter from the Government requesting us to withhold publication of technical details.

The writer states further that: "There are at least three pioneers in the field who appear to be on the verge of a complete solution of the television problem," and mentioned Belin, Jenkins, and Alexanderson. The results which, according to press reports, they have demonstrated should be mentioned. Belin and Jenkins have succeeded in transmitting crude shadowgraphs; and Alexanderson, within the last few weeks, apparently claims to have achieved the same feat. This is a long way from television, and does not justify the statement that they appear to be on the verge of a complete solution of the television problem.

JOHN L. BAIRD.

Television Limited, Motograph House,
Upper St. Martin's Lane, London, W.C.2,
Jan. 19.

Television.

By Prof. E. TAYLOR JONES, University of Glasgow.

ON May 24 and 26 I proceeded, at the invitation of Mr. John L. Baird, to the Central Station Hotel, Glasgow, to witness demonstrations of television between London and this city. I was received by Mr. Baird's colleague, Capt. Hutchinson, who explained that the transmission was to take place over the telephone line, Mr. Baird, in his laboratory in London, being in charge of the transmitting apparatus.

The earlier apparatus devised and used by Mr. Baird has been described by him in the *Journal of Scientific Instruments* for Feb. 1927. A model of the original transmitting apparatus is in the possession of the University of Glasgow, of which Mr. Baird is a former student. The following additional information as to the method has been supplied by him:

"The method used in the London-to-Glasgow demonstration consisted in passing an image of the object being transmitted over a light-sensitive cell in a series of strips. The modulated current from the cell was transmitted over the ordinary trunk telephone line, and at the receiving station in Glasgow was used, after amplification, to control the light of a glow discharge lamp, a modified form of neon tube, giving a light of intense brilliance, being employed. By means of a revolving slotted shutter a point of light from this lamp was caused to travel over the field of vision in exact synchronism with the traversal of the image over the cell at the transmitting station, complete traversal taking place in about one-eighth of a second."

The receiving apparatus was set up in a semi-darkened room, the lamp and shutter being enclosed in a case provided with an aperture. The observer looking into the aperture saw at first a vertical band of light in which the luminosity appeared to travel rapidly sideways, disappearing at one side and then reappearing at the other. When any object having 'contrast' was placed in the light at the sending end, the band broke up into light and dark portions forming a number of 'images' of the object. The impression of side-way movement of the light was then almost entirely lost, and the whole of the image appeared to be formed simultaneously. The image was perfectly steady in position, was remarkably free from distortion, and showed no sign of the 'streakiness' which was, I believe, in evidence in the earlier experiments.

The size of the image was small, not more than about two inches across when the 'object' was a person's face, and it could be seen by only a few people at a time. The image was sufficiently bright to be seen vividly even when the electric light in the room was switched on, and I understand that there is no difficulty in enlarging the image to full size. I was told also that arrangements will soon be made for transmitting larger 'objects,' and for increasing the number of appearances of the image per second.

The amount of light and shade shown in the image was amply sufficient to secure recognisability of the person being 'televised,' and movements of the face or features were clearly seen. At the second demonstration some of those present had the experience of seeing the image of Mr. Baird transmitted from London while conversing with him (over a separate line) by telephone.

My impression after witnessing these demonstrations is that the chief difficulties connected with television have been overcome by Mr. Baird, and that the improvements still to be effected are mainly matters of detail. We shall doubtless all join in wishing Mr. Baird every success in his future experiments.

Nov.
BROADCAST
WRNY
STATION
25 Cents
Science and Invention
BUILD YOUR OWN
TELEVISION
RECEIVER
See Page 618
EXPERIMENTER PUBLISHING COMPANY,
FIFTH AVENUE, NEW YORK

Built and Described by the Staff

How To Build The S & I TELEVISION RECEIVER

A slight adjustment of the rheostats and the picture comes in clearly. This photo shows a complete television receiver connected to an ordinary radio set. The picture is seen in the cone.

A Television Receiver of Simple Design, Built Around an Ordinary 16-inch Electric Fan Motor

THE front cover illustration shows the simple television receiver designed and built by the editorial staff. The accompanying photographs and drawings show the appearance and the construction details of the television receiver, the apparatus pictured having, of course, to be connected to the output of a suitable radio receiving set. The ideal set for receiving television images from WRNY or other stations, is, for the broadcast wavelength of 326 meters, one comprising two or three stages of tuned radio frequency, a detector and at least three stages of resistance-coupled amplification. When a resistance-coupled amplifier is used, it will be found best to use aboue 250 volts at least on the last stage from either storage or dry "B" batteries. A good "B" eliminator may be used, but a special filter is usually necessary, to prevent "motor-boating" with a resistance-coupled amplifier.

PROPER MOTOR FIRST ESSENTIAL

THE first requisite for building this television receiver is a good 16-inch fan motor. If the television disc to be used (it should have 48 holes for reception from WRNY and 3XK; also 1XAY and WLEX of Boston; and 24 holes for reception from WGY, 2XAD, and 2XAF, G. E. Co., Schenectady), is quite light, a 12-inch fan motor may do the work. If you have direct current in your laboratory or other location where the apparatus is to be operated, then you will have no trouble in controlling the speed of the motor down to the 450 r.p.m. required for WRNY reception or the 900 r.p.m. required for reception from the other stations broadcasting television.

If you have to select or use an alternating current fan motor, then you will have to find out whether the motor can be slowed down to a steady speed of 450 r.p.m. If the A.C. motor happens to be of the type that has throw-out contact brushes, which open the starting winding after the motor has attained fairly high speed, you will probably find this sort of motor unfit for television purposes. If the motor is of the universal A.C.-D.C. type, with commutator and brushes, the armature being connected in series with the field, then you will find that this motor can be regulated as to speed very nicely by means of the series resistances shown in the accompany diagram. We strongly recommend a universal type motor if you are going to purchase one, as these have been found to regulate well with regard to the speed.

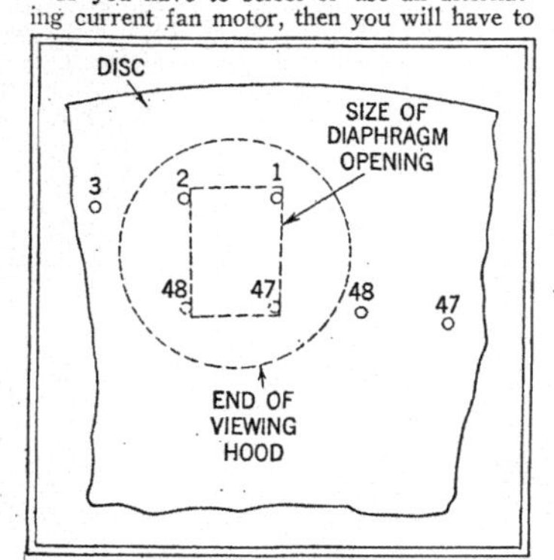

The method of laying out the diaphragm opening is shown clearly by the above drawing.

MOUNTING THE DISC

THE disc used in the television receiver here illustrated was a 48-hole 16-inch diameter bakelite disc of standard manufacture. This disc may be mounted and secured on a regular bushing provided with lock nuts supplied by the people who make the disc. In the present case, however, the perforated disc was mounted on the brass spider and hub which had originally carried the fan blades. The blades were removed from the legs of the spider and these were then flattened out in a vise and checked up on a lathe for alignment. A light cut may be taken across the face of the spider legs in the lathe, if one is handy. By drilling holes through the bakelite disc, it is readily secured to the spider by machine screws and nuts, or the holes in the spider legs may be tapped if the builder so desires. Care must be taken to see that the disc rotates as perfectly as possible in both planes of rotation, that is, flatwise and edgewise; in other words, it must not wobble and care must be taken to see that the spiral is rotated in a true manner. These two requisites are easily checked up by means of a machinist's surface gauge, or else by making up a gauge from a nail driven in a block of wood and holding this near the disc as it is slowly rotated by hand.

NEON TUBE MOUNTING

THE frame for supporting the neon tube behind the revolving television disc is simply constructed from light brass bar, measuring about 1/16-inch by ⅝-inch. Strap iron may be used if the builder happens to have this stock on hand. No dimensions are given for the height of the frame as many builders will want to use a different size disc than the one we used, and so the height of the frame and the dimensions of the metal composing it will depend upon the diameter of the disc, of course.

Examination of the drawings herewith will show that the neon lamp may be rotated, so that the front plate inside the tube may be placed exactly parallel with the perforated television disc. This is easily accomplished by the simple expedient of using a standard vacuum tube socket having a hole in the center, or what is known as the one-hole mount. By passing a machine screw through the center of the socket and putting a nut on top of the bakelite shelf, the socket and neon tube can be rotated as required. Two sub-base brackets or supports, available at any radio supply store, are used in building the top of the superstructure which carries the neon tube. Two well insulated wires lead from the vacuum tube socket down to the base of the machine. The connections to the socket for the average neon tube is to the plate terminal and to the diagonally opposite filament terminal. This can be determined by experiment after the machine is built, or else beforehand by testing the neon tube on your receiving set. The plate that faces the television disc is the one that has to be illuminated. In some neon tubes there is a large and small plate; the large square plate is the one that is to face the television disc.

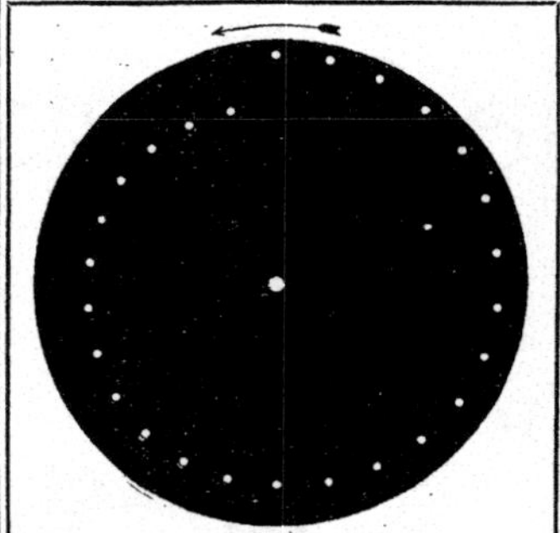

This indicates the arrangement of the holes and the direction in which the disc should rotate to receive television from station WRNY.

HINTS ON RECEPTION

WITH regard to the style of motor to use this is best of the series type; that is, with the armature and field winding connected in series. Small induction motors can be used, but do not regulate well in speed much below one-half their normal speed of 1750 r.p.m. If the picture image is observed and drifts toward the right, the motor is going too slow; if the picture drifts to the left, it is going too fast. The editor has found it advisable to regulate the motor speed to a point considerably above the desired value, and then to apply a piece of cardboard or a blotter against the surface of the disc to slow down the speed to the desired point. D.C. motors will regulate very well with the electrical rheostat arrangements shown in the circuit accompanying this article, however.

VIEWING HOOD AND LENS

THE viewing hood or visor shown on the machine herewith was built by cutting down a standard megaphone which can be purchased in any sporting goods store. The heavy metal ring at the mouth of the megaphone enabled the designers to secure it by means of three spring brass clips, soldered to the brass front plate shown in the drawings. It can be snapped off whenever desired. One of the accompanying drawings shows how the size of the diaphragm plate is determined, the rule here being that only one disc hole or perforation must be exposed at a time. A thin piece of leaf copper was used in the present case, from which to cut the diaphragm opening, and this was sweated to the brass front plate of the instrument. A fairly strong lens, about 2 inches in diameter, with a focal length of approximately 3½ inches, was procured for the purpose of helping to enlarge the image. This lens was secured inside the megaphone viewing hood by placing three machine screws through the megaphone shell and putting nuts on these, inside the shell. This is probably one of the best ways to build the viewing visor for any size television receiver, as the visor can always be snapped off the machine when it is to be moved to some other location.

STROBOSCOPE INDICATES CORRECT SPEED

ONE of the greatest problems the beginner in television reception will encounter is that of checking the correct speed. Of course the average machinist or electrician will not mind checking the speed frequently with an ordinary speed-counter, or possibly he may be so fortunate as to own a tachometer for the purpose. However, the average tachometer cannot be used with a small motor, as it takes too much power from the motor, and therefore slows the disc down and you do not know where you are at.

The method of using the *stroboscope* principle, with the black line disc noted on the front cover and in the present photographs,

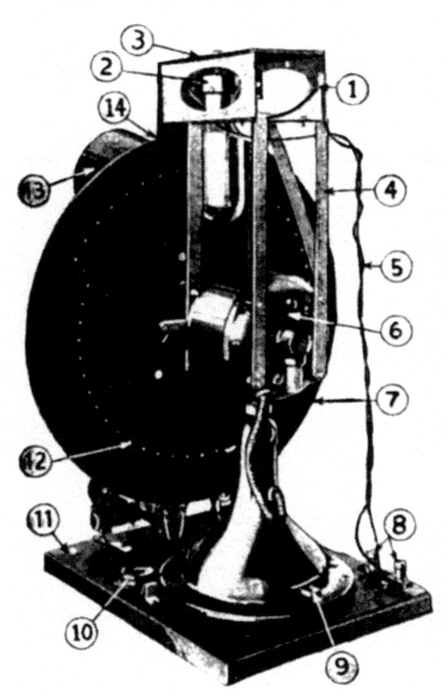

In the diagram at the left, 1 indicates a separation of the wires leading to the socket 2, affixed to top plate 3, which in turn is mounted on the uprights 4, screwed fast to the motor by the screws which hold the case in place. The wires 5, lead down to binding posts 8, which connect with the ordinary receiving set. 9 is the standard switch on the fan motor which receives its current through plug 10. 11 is a control button, 12 the holes in the television disc, and 13, the cone.

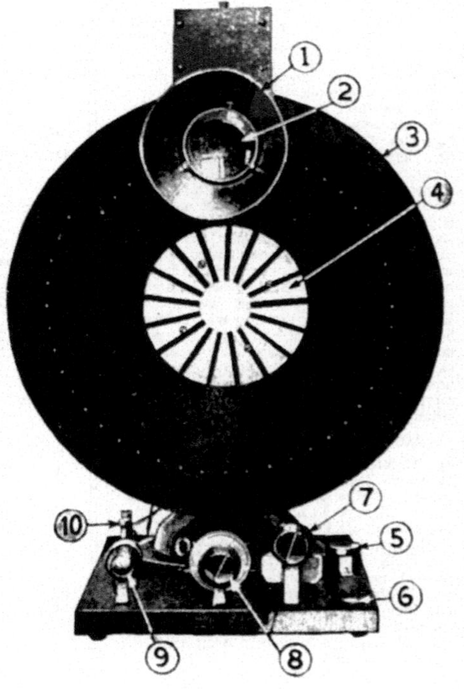

Right: 1 indicates the cone; 2, the lens; 3, the disc; and 4, the stroboscopic pattern; 5, attachment plug; 6, control button; 7, vernier rheostat; 8, main motor control; 9, neon lamp control; and 10, leads to the receiving set.

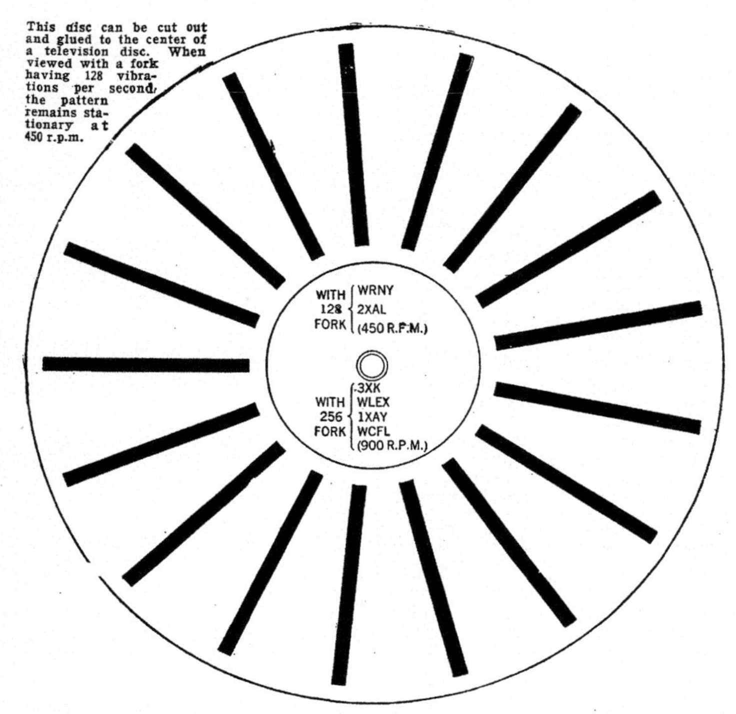

This disc can be cut out and glued to the center of a television disc. When viewed with a fork having 128 vibrations per second, the pattern remains stationary at 450 r.p.m.

together with a tuning fork of the proper pitch, was suggested by the Editor, Mr. H. Gernsback, and details were worked out by members of the staff.

For the benefit of those who are desirous of using the stroboscope principle for checking other speeds than those here given, the following table and formulae will be found useful.

STROBOSCOPE TABLE

R.P.M. of Shaft	*R. P. Sec.*	*Tuning fork frequency*	*No. of marks on chart*
60	1	128	128
120	2	128	64
180	3	128	42.6
240	4	128	32
450	7.5	128	17
480	8	128	16
900	15	256	17
1080	18	128 (72)	7.1 (4)
1260	21	128	6

These formulae will help to solve your problems: here N = Rev. per second of disc; F = freq. of fork per sec.; and M = number marks on disc. Then $N = F \div M$; $M = F \div N$; and $F = M \times N$.

The following pitch forks are available: 426.6, 256, 128, 288, 320, 341.3, 384, 480, 512.

For the benefit of the constructor we have provided herewith a good size reproduction of the stroboscope discs which can be cut out or else copied on to a piece of Bristol-board or drawing paper, and either glued or attached to the front of the television receiver. A tuning fork of the proper pitch may be obtained from music stores or from college laboratory supply houses, names of which will be furnished upon request from the editor.

For checking the speed of the motor at 450 r.p.m., a tuning fork giving 256 vibrations per second is necessary. This is used with a disc containing 17 black marks for the 450 r.p.m. specified. For other speeds, either a different fork has to be used, or else the number of lines on the stroboscope disc will have to be changed. All this data is contained on the drawings of the discs reproduced herewith.

All one has to do in using the stroboscope check for the proper speed, is to regulate the rheostats in series with the motor, and then repeatedly take a sight on the revolving black line disc through the legs of the vibrating tuning fork. The tuning fork is struck on the edge of the table or across the knee, and while vibrating, it is held a few inches from the eyes and twisted, so that the revolving disc is observed in a diagonal line passing under the corner of the upper fork leg and over the corner of the lower fork leg. This line of sight is shown in one of the accompanying diagrams.

While in most cases it will probably be found that the number of marks on the disc or else the vibrations of the tuning fork to be used will come out to an even figure, or at least that a suitable combination can be worked out for the speed desired, the calculation may show that an uneven number of marks will be required with any standard fork. Here, instead of using a number of radial black marks on the rotating disc, a spiral may be used and with this sort of design, any uneven number of convolutions such as 7½, 7-1/3, etc., may be employed.

HOOK-UP OF APPARATUS

ONE of the accompanying diagrams shows how the power clarostat (about 150 ohms maximum resistance) and the small 10 to 15 ohm variable resistance is connected in series with the motor. Across the small variable resistance a push-button is connected, and by pushing this button periodically, it becomes possible to keep the motor speed quite constant. In setting the speed of the motor in the first place, the rheostats are adjusted until the speed is a little below the 450 r.p.m. (if you happen to be "looking in" at WRNY's television signal), this factor being indicated when checking the speed with the stroboscope fork, by the fact that the black lines on the disc are seen to rotate slowly backward. If these lines rotate slowly forward or left-handed, then the speed of the motor and disc is above 450.

Rubber-covered wire or lamp cord may be used to connect the rheostats and the motor. The small clarostat at the extreme left of the motor baseboard is connected in series with the wires supplying the energy

(*Continued on page* 632)

WITH 128 FORK
WGY
2XAD
2XAF
(1260 R. P. M.)

This pattern can also be cut out and pasted to the center of a disc. When viewed with 128 fork, the design remains stationary at the speed indicated for the stations listed.

How to Build S. & I. Television Receiver

(*Continued from page* 620)

to the neon tube. The terminal posts to the neon tube circuit are mounted on a piece of bakelite, secured to the rear left corner of the baseboard. A rubber foot should be placed under each corner of the baseboard; this will allow the wiring to be simply placed against the wood and held in place with a few staples, if necessary. The clarostats are mounted on small right-angle brackets made from brass or iron. The push-button is placed in a tight-fitting hole, bored through one corner of the baseboard. The 110-volt supply for the motor circuit is brought into the apparatus, through an approved socket or receptacle, mounted on the righthand side of the baseboard, as shown in the picture.

OPERATING THE APPARATUS

WHEN the television signal is being received and the neon tube is connected to the output of the radio receiving set (and providing there is sufficient voltage used in the last stage—not less than 180) pulsations of pinkish light will be seen in the neon tube. If a sufficiently high voltage is used and the radio apparatus is properly adjusted with regard to the "C" bias, etc., then a pulsating pinkish light should be seen covering the whole neon tube plate which

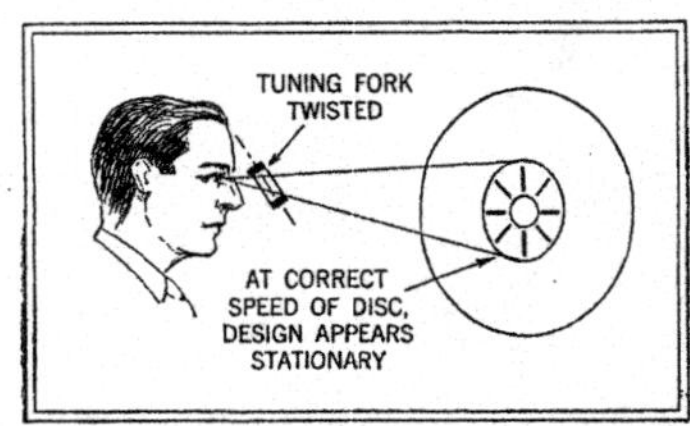

The tuning fork must be so twisted that either the upper or lower leg is closer to the eye. The aperture between the legs should be very small. The entire pattern can be viewed if the fork is held close to the eyes.

faces the rear of the television disc. If the pulsating glow is seen on the rear plate, then the wires leading to the neon tube must be reversed.

If you are "looking in" with the television receiver, and, having checked the time of the television broadcast with the newspaper program, you should first check the motor speed and make certain that it is revolving at the prescribed speed. As you look into the viewing visor, preferably in a darkened corner of the room, you will see successive lines of orange-colored light as the spiral of holes repeatedly scans the illuminated plate in the neon tube. If you see these bands of light, but they only form irregular splotches, then the chances are that your motor speed is either too high or too low, and a slight change in the rheostats should be made. It is well to recheck the speed of the revolving disc with the stroboscope fork after doing this, as you may change the speed too much.

Several things may happen if you are successful in building up a picture image with the machine; the image may be upside down or it may slowly drift across the viewing lens repeatedly. If the picture slowly drifts across the lens, then the motor speed should be momentarily accelerated by pushing the button connected across the smaller resistance. This presupposes that the motor is running slightly below the correct television speed for the station to which you are "looking in." You may have to change the small

(*Continued on page* 634)

HOW TO BUILD S & I TELEVISION RECEIVER

(*Continued from page 632*)

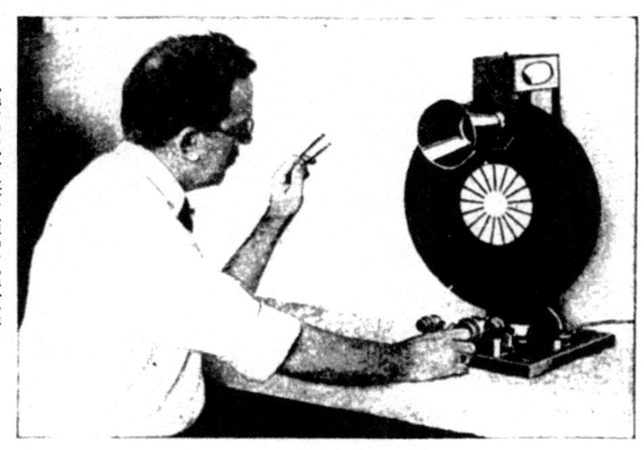

The photograph shows operator checking the speed of the television receiver by means of a tuning fork and a patterned disc. When viewed through the tines of a vibrating fork, the pattern remains stationary, exactly as you see it here.

variable resistance or even adjust the larger one slowly in order to make the picture stationary on the lens.

If the picture is upside down, then you are scanning the neon tube plate in reverse order; that is, from bottom to top, instead of top to bottom, and the disc must be taken off and turned around. In some cases it will be necessary to turn the disc around and also reverse the motor, or in still other instances, in order to rectify the picture image, the direction of the motor rotation will have to be reversed.

If the motor happens to be of the universal type, which means that it is usually a series-wound motor, then the direction of rotation can be changed by simply reversing the connections to the field or to the armature brushes. If the motor is an A.C. induction type, with a separate starting winding, then the direction of rotation is reversed by simply transposing the terminals from the starting winding. If the motor happens to be one of those types using copper shading plates, mounted on the tips of the iron stator poles, then the direction of rotation can be effected by remounting the shading plates on the opposite pole tips; or simpler still, the whole stator frame may be removed from the car-case or motor housing, and reversed in its position with respect to the same.

In some cases direction of rotation of the motor may be effected by sliding the shaft out of the rear bearing and then turning the motor around. This is rarely the case, but with some induction motors it is pos-

(*Continued on page 636*)

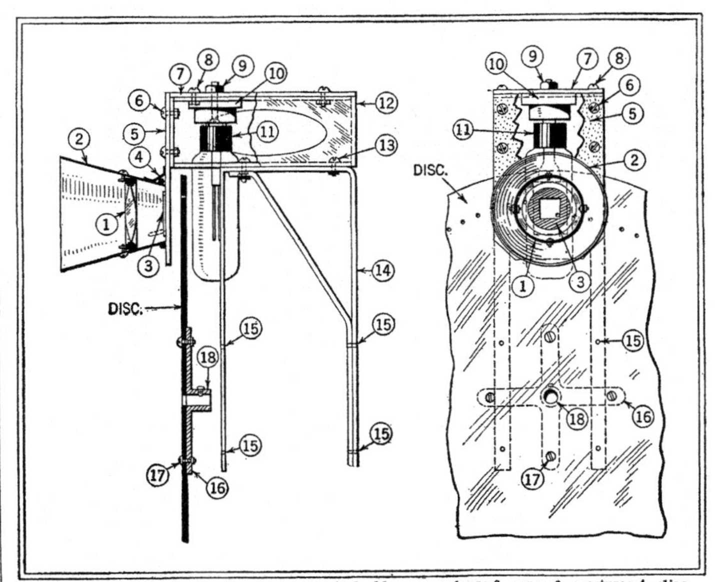

Further details of the television receiver. 1, double convex lens; 2, cone; 3, aperture; 4, clips for holding cone; 5, face plate; 6, screws for holding same; 7, top; 8, screws for bolting to plates 12; 9, single hole mounting of socket 10; 11, neon lamp; 13, bolts for holding plate 12 to upright 14; 15, holes for mounting uprights to motor; 16, mounting for disk held in place by screws 17; 18, shaft mounting.

HOW TO BUILD S & I TELEVISION RECEIVER

(*Continued from page* 634)

sible to do this, the rotor being secured to the shaft by a set screw.

The television set builder who is interested in the connections of the resistance-coupled amplifier, and other details connected with the radio receiving set, should read all about this matter, where complete diagrams are given with explanatory remarks, in the *Television* magazine, Volume I, No. 2. Various methods of connecting the neon television lamp are supplied by some of the manufacturers putting out these tubes. The common connection for the neon lamp, however, is in series with the plate and "B+" supply wire; in other words, it is connected in the same relative position as your loud speaker. Some of the neon lamps, however, are supposed to be checked carefully with a milliammeter, so that no more than a certain current in milliamperes is passed through them, in order to conserve their life. When using one of these more sensitive type neon lamps, it will be found necessary to connect a clarostat, or other fairly high variable resistance, in series with the "B" supply, before it reaches the neon tube. This series variable resistance in the neon tube circuit may have a range of 0 to 10,000 ohms. A fixed resistance of 10,000 ohms, with a variable ‿,000-ohm resistance, may be used. In the *Television* magazine, Volume I, No. 2, already referred to, details will be found for making your own rheostats for the speed control of the motor, as well as data for building an adjustable impedance for those using A.C. motors; the variable impedance being preferable to variable resistance control, where alternating current is used.

When all ready to listen in for a television signal, you will soon become accustomed to the peculiar whining note of the television signal proper; and if you follow the published program of WRNY, for example, you will receive the proper introduction by the announcer, and then you will make no mistake when you hear the television signals in your phones.

If you "listen in" to the station at first with a pair of headphones and plug them into the detector jack on your set, this is all right; but if you connect your headphones in the last stage wherein the neon tube is connected, be sure to connect a 1 micro-farad condenser in series with the phones, when connecting them in the place of the neon lamp, or across the neon lamp terminals. The television signal sounds in general like a buzz saw cutting through a plank, and the note continually changes as the person in front of the television transmitter moves about.

NEON TUBE NOTES

IN adjusting the neon tube circuit, it is the usual practice to adjust the "C" bias on the last amplifier stage, so that the tube just glows over the plate facing the rear of the revolving disc. In other cases the neon lamp is adjusted by raising the "C" bias potential on the last amplifier tube, so that

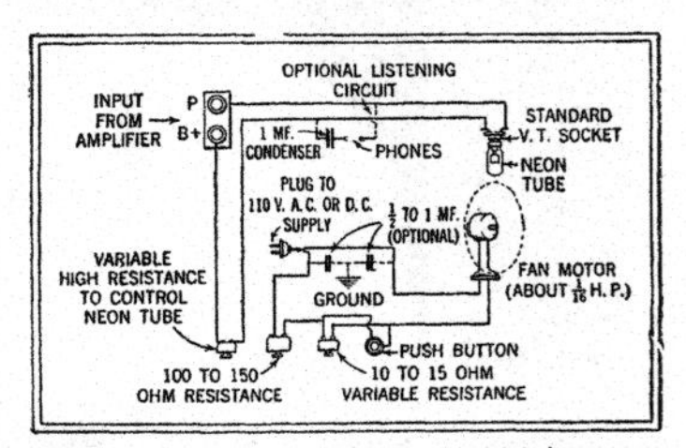

Above is a schematic diagram of the Science and Invention television receiver. A push button cutting a resistance out of the circuit speeds up the motor when necessary.

the neon lamp doesn't quite glow. In this case, when the television signal comes in, the lamp lights up as the television signal pulses are impressed on the circuit. In some cases it may be possible that you see the image in negative form instead of positive, i. e., you may see the image similar to a photographic negative. In this case the connections to one of the amplifier stages should be reversed; at other times it will be found that if the tuning dial of the set, or one of the dials, if it has more than one, may have to be moved from the right side of the peak of the carrier wave, so to speak, to the left side and vice versa. That is, if you had tuned the dial to say 45 degrees for maximum signal strength, and then detuned a little toward the left; you may have to detune toward the right of the peak, 45 degrees, in order to reverse the image.

Another reason for a reversal of the image from positive to negative is that a certain number of stages has to be used with a specified form of detector circuit; this is explained at length in an article covering an interview with Mr. C. F. Jenkins' radio engineer, which appears in *Television* magazine, Volume I, No. 2, page 8.

If you happen to see the television image on the lens right side up reversed, you will have to reverse the direction of rotation of the motor and also remove the disc from the shaft and turn it around with the other side out.

STEREOSCOPIC TELEVISION

Three-Dimension Images Now Obtainable

The illustration above shows the transmitter used for the production of stereoscopic images.

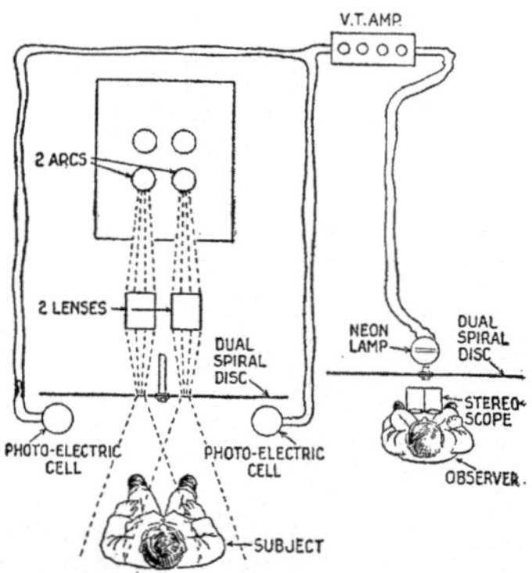

The above drawing shows in simple form the principle employed in the new television apparatus.

TELEVISION is progressing so rapidly that difficulty is experienced in keeping up with the new advances which are now announced almost daily. Radio movies and colored television were two of the big steps forward, and now comes the third, stereoscopic images. At the transmitting end of the system, two arc lights cast their beams through two lenses which concentrate the light and direct it through two separate spirals of holes. The holes in this disc are arranged alternately as illustrated and the subject is scanned from two slightly different angles, so that a three-dimension effect is obtained at the receiving end. A photo-electric cell arranged on either side of the subject, picks up the light impulses and changes them into electric pulsations. These are then amplified and broadcast in the usual manner. At the receiving end the usual neon lamp is used, in front of which rotates a spiral disc similar to that used at the transmitter. Two images are built up and when viewed through a stereoscope stand out in relief. The double spiral disc may be seen in the large photograph hanging on the wall. The two sets of holes are clearly seen. Unless the picture received is very small, it will be necessary for larger images to use a bigger neon lamp, that is, one having a larger sized cathode. In the future may we expect a combination of the three latest advances giving us three-dimension colored radio movies?

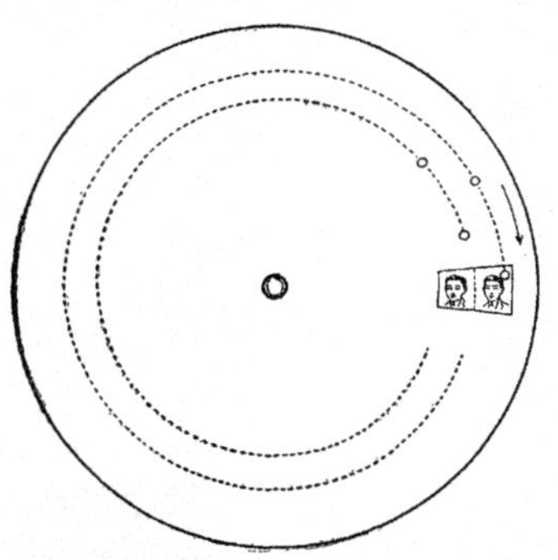

At the left is a drawing showing the construction of the scanning disc which is fitted with two spirals of holes.

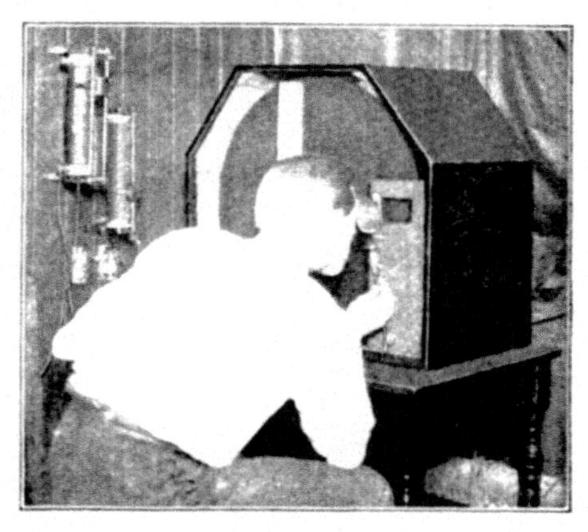

The photograph at the right shows a view of the receiver with the operator viewing the images through a stereoscope.

TELEVISION TIMETABLE

Station	*W. L. Meters*	*Disc Holes*	*Disc Speed R.P.M.*	*Pictures Per Sec.*	*Time*
WRNY—New York City	326	48	450	7.5	Every hour on hour when station is on air.
2XAL	30.91	Same as WRNY			
WGY—Schenectady	379.5	24	1260	21	Tues., Thurs., Fri.—12:30-1 P. M.
2XAD	21.96	Same	Same	Same	Tues.—10:30-11 P. M. (WGY and 2XAF)
2XAF	31.4				Sun.—9:15-9:30 (WGY and 2XAD) E. S. T.
3XK—Washington, D. C.	46.7	48	900	15	Mon., Wed., Fri.—8-9 P. M. E. S. T., Radio Movies
	186				
1XAY—Lexington, Mass.	51-62	48	900	15	No Regular Schedule
8XAV—Pittsburgh, Pa.	62.5	60	960	16	No Regular Schedule
4XA—Memphis	120-125	24	900	15	No Regular Schedule
9XAA—Chicago	62.5	48	900	15	Mon., Wed., Thur., Fri.—9-10 A. M. C. S. T.
6XC—Los Angeles	65.22-66.67	36	1080	18	Will Start Sept. 15th. Daily—10:30-11:30 P. M. P. S. T.
WLEX—Boston	62.5	48	900	15	
WCFL—Chicago	61.5	45	900	15	

Radio Movies Demonstrated

Motion Pictures Sent Via the Ether by New Television System

The transmitter used by the Westinghouse engineers for the radio moving pictures is shown in the above photograph. Note the film passing before the scanning disc. Each picture is scanned at the rate of sixty times each one-sixteenth of a second.

LEADERS in radio recently met at the famous Westinghouse plant to review the laboratory progress of what the layman might term imminent miracles of sight and sound transmission. The meeting was one of the most important in radio annals, since such work as television, facsimile radio, power tubes, photophone and broadcast motion pictures was not only reviewed but future progress in these lines definitely mapped. Of these developments, the photophone is the only one in which perfection has been attained, at present. The others are in various stages of development, ranging from the near perfection to embryonic laboratory experiments.

The most striking of all radio developments reviewed was the broadcasting of *motion pictures* which, transmitted on radio waves, were picked up on a receiver located in the television laboratory and reproduced before those assembled there.

It was the first demonstration of radio movies by this particular system, and possibly the most interesting of the many advances in the science of radio announced in the past year.

While radio movies are still in the laboratory stage, Mr. H. P. Davis, under whose auspices the demonstration was made, states that the event heralds the time when the radio listener will sit at home and have that most popular form of entertainment, motion pictures, projected by his own individual radio receiver.

The development of radio movies is a triumph of scientific engineering. Barely two months ago, the idea came to the mind of Dr. Frank Conrad, in charge of this branch of his company's activities, and the fact that he has brought the device to the laboratory stage in the degree of perfection witnessed a few weeks ago, is said to have set a record.

Radio movies are a step beyond previous developments in television and required the invention of a number of appliances in addition to a great deal of scientific calculation, synchronism of various high-speed mechanisms, and accurate control of light and radio waves.

GENERAL PRINCIPLES

ALTHOUGH the sending of moving pictures by radio, as may well be imagined, required many complicated and delicate pieces of apparatus, the principles of the art are not beyond ordinary comprehension.

Photography in its simplest form consists of the reproducing of spots of light and shadow in the same arrangement as they appear in the subject photographed. The screening of a motion picture, of course, requires that a roll of film be operated at a speed which sends sixteen pictures a second before a projecting beam of light. Because of the structure of the human eye, if a series of pictures follow each other at the rate of 16 or more per second, the human eye sees it as a single moving picture. (*Editor's Note: At a slower speed than 16 pictures per second, a fair picture is reproduced, accompanied by a certain amount of flicker. WRNY is using now about 7½ pictures per second, and favorable reports of successful reception have been received.*)

MOVIES PLUS RADIO

ALL this the broadcasting of radio movies requires, with the addition that the spots of light must be transformed into frequencies, some of which are in the audible range, transferred to a radio wave and broadcast as electrical energy. In receiving the pictures, the process is reversed, the electrical energy is picked up, and the frequencies returned to lights and shadows, which when viewed presents the radio movie.

In the first step of the process a pencil of light traverses each picture, or *frame*, as it is called, at the rate of 60 times each sixteenth of a second. This process produces a 60-line picture, as clear as the usual newspaper halftone illustration.

EACH PICTURE SCANNED

THE pencil of light is produced by a scanner, which is a disc with a series of minute *square holes* near its rim. The disc is so arranged that all light is excluded from the film except that which goes through the square holes. The disc turns very fast, and as it turns passes the beam of light across each *frame*, with the result that an individual beam of light touches every part of the *frame*.

The beam of light passing through the film falls upon an electric eye or photoelectric cell, which is not unlike an oversized incandescent lamp. Within the cell, however, is a metal whose electrical resistance varies with the light falling on it. Caeseum, a rare metal, is used in the Westinghouse cell. The amount of light falling on this cell determines the amount of current passing through it. The result is that each individual beam of light sends an electrical impulse which varies directly accord-

Above is a close-up view of the radio movie transmitter shown at the top of the page. The scanning disc has a series of minute square holes. As the disc turns, the beam of light passes across each frame.

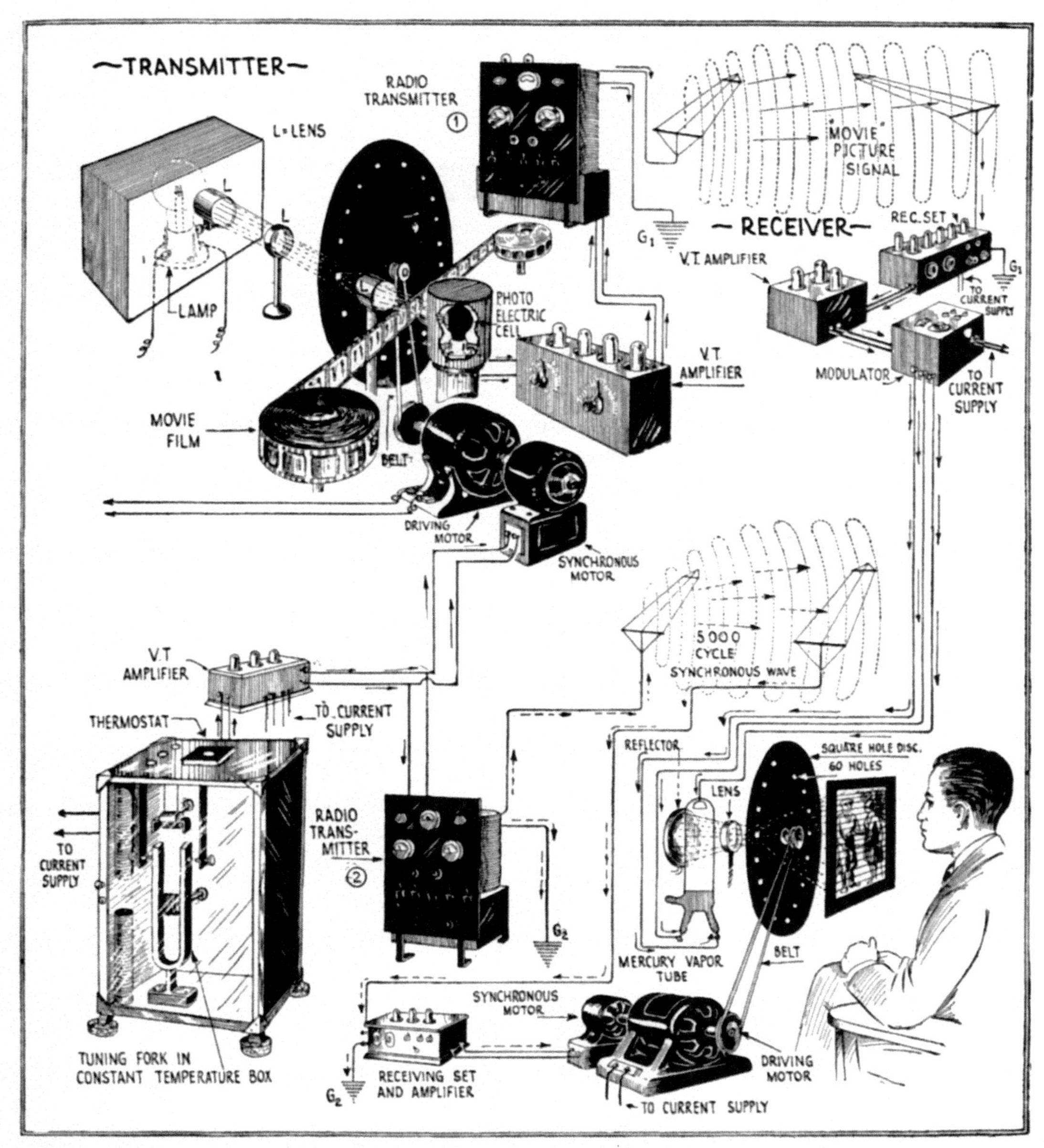

The above illustration shows in detail the apparatus employed in the projection of motion pictures by radio. In the demonstration described here, the signals were sent a distance of about four miles, that is, two miles from the laboratory to the transmitter by wire, and two miles back to the laboratory by ether waves. The lower part of the picture shows the main elements of the 5,000 cycle synchronizing circuit.

ing to the amount of light or shade in the film through which it passed.

LIGHT BEAMS BECOME RADIO WAVES

THE beams of light have now become electrical impulses and are sent on to the broadcasting station. Here the beams assume definite and varied frequencies, some of which are audible. Dr. Conrad states that these frequencies range from somewhere near 500 to approximately 60,000. Since the human ear is limited to frequencies of approximately 15,000, much of the radio movie wave is inaudible.

At the broadcasting station these frequencies are super-imposed on a radio wave and transmitted exactly as the ordinary music or voice. The radio signals can now be sent across a room, or across the continent. Their distance range is limited only by the broadcasting station's equipment (power).

In the demonstration here described, the signals traversed a distance of about four miles; two miles from the laboratory to the broadcasting station by wire and two miles back to the laboratory by radio.

To turn these radio waves back into light, an arrangement which permits the use of a *mercury arc lamp* is used. By this adaptation the weak radio currents control the action of the many times more powerful current operating the arc lamp. This action may be compared to the action of a radio tube, where the weak radio current on the grid of the tube controls the action

(*Continued on page 666*)

Radio Movies Demonstrated

(*Continued from page* 623)

of the independent and more powerful plate current.

Thus the mercury arc lamp goes bright or dim as fast as the current changes, and its light at any instant is in proportion to the light that the electric eye sees in the same instant. To return the dots of light to their original pattern, another revolving disc or scanner is also used which is similar to the transmitting scanner.

The use of a mercury arc lamp permits the radio pictures to be thrown upon a ground-glass or screen, the first time this has been done with television apparatus.

Both these scanning discs turn at exactly the same speed; the hole in the receiving disc must be exactly in the same relative position as the corresponding hole in the transmitting disc. In other words, they must be synchronized.

RADIO WAVE USED TO SYNCHRONIZE DISCS

From the transmitting equipment, which may be located in the broadcasting station, there is transmitted a constant frequency wave of 5,000 cycles. This wave is produced by a tuning fork, and transmitted over a special carrier wave from the broadcasting station. The constant frequency note is received on a special receiver and by means of special apparatus controls the speed of synchronous motors, which drive the scanning discs of both transmitting and receiving radio movie equipment. This unique method of controlling the equipment indicates, in a measure, the extent to which science must go in order to perfect—say television, or radio movies.

It is stated that the regular transmission of motion pictures from KDKA will begin shortly; also that the radio movie reception apparatus, when produced commercially, will be sold through the regular trade connections.

Part 3

BROADCASTING BEGINS

Experimental transmissions

PRACTICAL TELEVISION

Sir Oliver Lodge " Seeing-in."

[*Frontispiece*.

PRACTICAL TELEVISION

BY

E. T. LARNER

(ENGINEERING DEPARTMENT, GENERAL POST OFFICE, LONDON; ASSOCIATE OF THE INSTITUTION OF ELECTRICAL ENGINEERS; FORMERLY LECTURER IN ELECTRICAL ENGINEERING AT THE LONDON COUNTY COUNCIL HACKNEY INSTITUTE)

AUTHOR OF "RADIO AND HIGH-FREQUENCY CURRENTS," "ALTERNATING CURRENTS," ETC.

WITH A FOREWORD BY

JOHN L. BAIRD

LONDON

ERNEST BENN LIMITED

First Published in
1928
Printed
in
Great Britain

FOREWORD

THE present generation has seen the birth and growth of wireless telegraphy from a scientific novelty to a vast industry.

From the first experiments of Hertz in 1888 to the present day covers a period of forty years, and during those forty years it can be said that the whole outlook of the man in the street towards science has undergone a fundamental change. We have at the present time a great public interested, and intelligently interested, in scientific subjects, and a new literature has sprung up catering for this body of people. This literature deals almost exclusively with the many branches of wireless and depends for its appeal upon the listener-in. If no other benefit has been conferred upon humanity by the development of radio communication, the introduction of this interest in science to a public which had hitherto been apathetic would in itself be no small benefit.

The Shorter Catechism defines man's chief end as the glorification of God, the American Constitution more prosaically defines it as the pursuit of happiness, while in these days we might prefer to describe it as the pursuit of truth. Where better can we seek for truth than in scientific research? Sport, Business,

Art, Music, and all the other avenues into which man directs his energies, are tainted with commercialism, self-interest, passion, and emotion.

In introducing the public to science, wireless broadcasting has somewhat regrettably, but inevitably, given a preponderating, in fact an almost exclusive, interest to the study of phenomena connected with high-frequency electrical oscillations. Other branches of science unconnected with wireless have been almost completely ignored. To-day it would be difficult to find a household in which at least one member could not give a lucid distinction between a volt and an ampere, but the same state of affairs does not by any means prevail regarding optics, chemistry, and mechanics, and the young gentlemen who are so familiar with volts, amperes, microfarads, and henries are frequently totally ignorant of the most fundamental principles of those other branches of science, and would, for example, be unable to define the functions of a lens or a prism. Television, unlike Wireless, covers optics, chemistry, mechanics, in fact every branch of science, and introduces its devotee, not only to physics, but also to physiology, demanding, as it does, a knowledge of the physiology and psychology of vision; indeed, but for a purely physiological phenomenon, retentivity of vision, Television, as we know it to-day, would be an impossibility, and a study of the human eye is essential to a clear understanding of the principles underlying the electrical transmission of visual images.

FOREWORD

Photo-electricity is another branch of the subject which is of intense interest and importance. We are still in ignorance as to the true nature of light. In some respects it behaves as if it were a corpuscular emission; other phenomena would appear to prove conclusively that it is a form of wave motion. The key to the elucidation of this outstanding problem may well be found in the study of photo-electric phenomena.

In the present work the author deals very fully with the fundamental principles from which Television was developed, and deals with them in such a way as to interest the general reader without departing from strict scientific accuracy. It is to be hoped that the book will prove of the greatest assistance to those who are commencing the study of a subject which perhaps offers to the young scientific worker the most promising prospects of any avenue of research.

J. L. B.

1928.

CONTENTS

CONTENTS

LIST OF ILLUSTRATIONS

LIST OF ILLUSTRATIONS

LIST OF ILLUSTRATIONS

LIST OF ILLUSTRATIONS

TABLES

NOTE

THE author gladly acknowledges his obligations to the Cambridge Instrument Company, Ltd., and to the General Electric Company, Ltd. (of London), for much useful information freely offered, including the loan of electros and sketches, viz. Figs. 40, 41, 46, 47, and 48 from the Cambridge Instrument Company, Ltd., and Figs. 42, 43, 44, 45, and 84 from the General Electric Company, Ltd. Also to Messrs. Marconi's Wireless Telegraph Company, Ltd., for copies of pictures transmitted by radio, Figs. 2, 29, and 30; to the American Telephone and Telegraph Company and the *Bell System Technical Journal* (per Mr. H. E. Shreeve) for the use of illustrations given in Figs. 1, 13, 21, and 22; to the Editor of the *Wireless World* for permission to reproduce Figs. 53, 54, 55, 56, 57, 62, 67, and 68; to the Editor of *English and Amateur Mechanics* for permission to reproduce Figs. 14 and 15; to the Edison-Swan Electric Company, Ltd., for the loan of block, Fig. 20; to the British Thomson-Houston Company, Ltd., for the loan of blocks, Figs. 38 and 39; to the Standard Telephones and Cables, Ltd., for photograph, Fig. 65.

PRACTICAL TELEVISION

CHAPTER I

INTRODUCTORY

Sight is, beyond doubt, of all the senses with which man is endowed the medium through which he has gathered and continues to gather most of his ideas and knowledge of his surroundings and the universe. Hearing is almost equal in importance, but the general consensus of opinion would undoubtedly be in favour of sight, for by it the mind is stimulated to a greater degree than from any one of the other senses. Moreover, facts can be impressed on the mind by sight much more readily than by the spoken or written word. Actual seeing rivets the attention immediately, and there is a tacit comprehension of what is unfolded to view which is occluded by the barriers of language.

From the earliest times man has striven to improve his means of visualising objects and things; especially those things that are normally altogether unseen. There is little doubt that the use of the magnifying glass as an optical instrument was known to the ancient Greeks. Later, we know that a mechanical arrangement of lenses for seeing distant objects, known as the

2

telescope, was constructed. Its actual date of discovery is disputed, but as an astronomical instrument it may be said to date from the time of Galileo, who by its aid was able to see the spots on the sun, the mountains of the moon, and Jupiter's satellites.

From this stage to the next, when a device for utilising electricity in order that man might understand and see by signs what was happening at remote spots invisible to the eye, a considerable span of years elapsed, but succeeding developments brought forth the transmission of drawings and pictures which may be regarded as the "shadows of coming events cast before." To-day we have the actual seeing of persons and events rendered possible by means of apparatus that would have astounded the contemporaries of Galileo.

Tracing the development of the art as the various stages were reached, although very interesting, does not come strictly within the scope of this book, but for the benefit of those readers who desire to have for reference a short account of television research a summary of the early attempts will be given in the next chapter.

The Problem of Television.

The word television has now come into general use in the English language as a term descriptive of the instantaneous transmission of images of objects or scenes by telegraphy, either by wire or by wireless.

The transmission of photographs by telegraphy is

often confused with television, and it would be well at the outset to distinguish clearly between the two subjects. The telegraphic transmission of *photographs* is properly described as *phototelegraphy*, while the transmission of actual objects and scenes without the intervention of photography constitutes *television*. The transmission of sixteen photographs per second gives a moving picture at the receiving station, but this is effected by means of a film or kinematograph transmission and does not come under the heading of television, but under that of telekinematography.

In this book we propose to deal only with television, as although there is a certain connection between phototelegraphy and television, the methods at present used in phototelegraphy are not applicable to television.

The discovery of means whereby we can see distant objects and scenes outside the range of normal vision and without regard to intervening objects or other obstructions, such as, for instance, the curvature of the earth, has long been the ultimate aim of the scientific inventor. The application of electricity to seeing by wireless has occupied the minds of quite a number of experimenters who foresee in the wireless field of research the discovery of a new wonder that will not only interest and excite the imagination of the public, but will also have commercial potentialities. If it is possible to transmit sound and to talk over the ocean without the aid of any material conductor like copper wire, why should not sight be rendered possible in a similar manner? If electric waves can reproduce sound and speech in any part of America from a transmitting centre in Europe, why cannot they be made to reproduce rays of light, that is vision, on the other side of the Atlantic by the adoption of similar suitable means ?

First Stages of Solution.

The practical feasibility of such a problem having been duly recognised, it has been attacked in a variety of ways, the chief of which has been the pioneer work of the successful transmission of photographs both by wire and wireless — phototelegraphy as it is called.

Now there are no practical difficulties in the way of transmitting a photograph or transmitting views of actual objects by means of phototelegraphy over a considerable distance either with or without wires and reproducing a photograph or picture at the distance station which bears an exact likeness to the original, so that the achievement of successful phototelegraphy is certainly a step in the right direction towards radio vision.

A specimen picture transmitted over telephone lines by this means is shown in Fig. 1. A radio picture transmitted across the Atlantic without wires by the same means is shown in Fig. 2. It will be observed that more detail is possible with the former.

Several methods of phototelegraphy are already in good working order at the present day, notably those invented by M. Belin, Professor Korn, and others, and

Fig. 1.—Specimen of Photograph received over Telephone Lines by the American Telephone and Telegraph Company.

Fig. 2.—Specimen of a Drawing of a Baseball Player transmitted between New York and London by Radio.

[To face page 20.

short descriptions of the methods employed are given in Chapters II and V.

The great drawback with phototelegraphy, however, is the slowness of transmission—a small picture takes from five to twenty minutes to transmit, due to the fact that only small portions of the pictures can be taken one after the other until the whole picture is completed.

Preliminary Lines of Research.

To obtain television, the scientist and inventor must not stop at this stage, since the process of transmission is tantamount to picture transmission and not wireless sight of objects far away and out of range after the manner of seeing objects within the range of our eyes. The slowness of transmission, moreover, is accentuated where phototelegraphy is concerned in the additional time taken to develop and fix the light-sensitive films. By increasing the speed of this method, however, and making the transmission instantaneous, television in its true sense has been accomplished.

It is along the lines of this character, then, but with the object of finding out how to throw the image on to a screen giving life-size pictures similar to those seen at cinemas, that investigators have so far proceeded towards their goal. They have therefore directed their attention to the study of ways and means whereby existing appliances, both natural and artificial, could be copied for the purpose of bringing out the desired results. One of the very first fruits of their endeavours is given in Fig. 3.

Now in all inventions that come under the category of aids to physical faculties it has always been the practice of inventors to study first the methods nature provides—*e.g.* the lens of the eye is copied in the

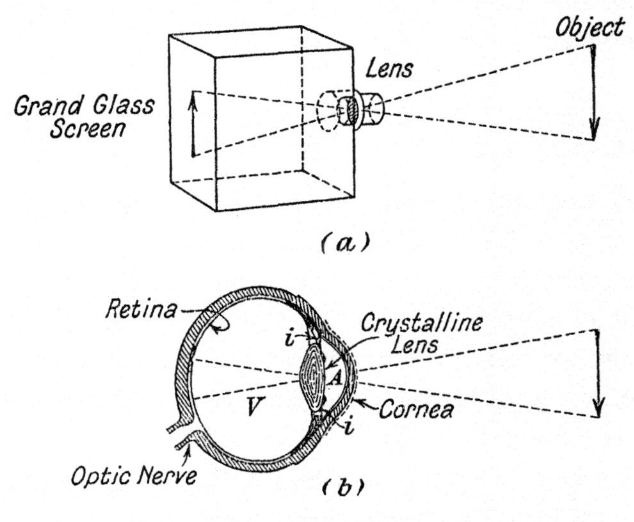

FIG. 4.—Comparison of Eye with Photo Camera.

telescope; the action of the drum of the ear is imitated in the construction of telephone receiver, etc. Nature has also been copied in the conquest of the air. Many years were spent by various experimenters in attempting to fly, but it was not until the careful study of the flying of birds had been undertaken that the problem of flight was solved. So, too, in wireless seeing by

FIG. 3.—One of the first Pictures received by Baird's Television Apparatus.

[*To face page 22.*

considering and copying the marvellous mechanism of the eye the possibility of television is achieved.

The Eye as a Model.

It will be well at the outset, therefore, to glance briefly at this remarkable piece of nature's mechanism, since nature's method of solving her television problem has formed the basis of many television methods.

The eye is essentially an optical instrument, and its principle of construction and working may be followed more intelligently perhaps by making comparison with an artificial eye like the photographic camera. Compare Fig. 4 (*a*) and (*b*).

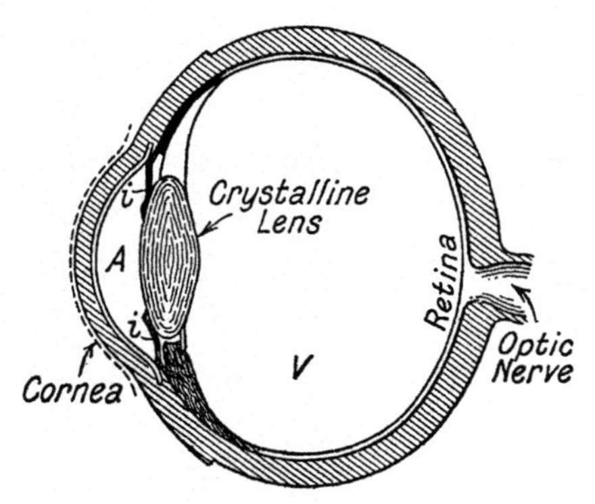

FIG. 5.—Vertical Section of the Eye. Front to back.

This comparison may be applied more closely by considering the construction of the eye in greater detail. In the vertical section of the eye front to back shown in Fig. 5 we have the *cornea* in front, behind which is the anterior chamber filled with a watery fluid called the "aqueous humour" and immediately behind which is the most important part perhaps, the *crystalline lens*. Behind the latter in the posterior chamber is the "vitreous humour"—a watery fluid similar to the aqueous humour. The walls of the eyeball consist of three coats, the inner coat is called the "*retina*," a delicate membrane which is really a fine network expansion of the optic nerve.

The structure of the eye may be likened to a photographic camera. The crystalline lens and retina of the eye correspond to the convex lens and ground-glass screen, respectively, of the photographic camera, by means of which the picture, or rather the millions of rays of light that proceed from the object to form the picture or image, are thrown on to the retina at the back. The latter takes the place of the ground-glass screen or sensitive plate in the photographic camera and is in communication with the brain by means of numerous sensitive nerve structures.

A clue to the solution of the problem is found in a close examination of the screen, which is called the retina. The surface of this is found to consist of a mosaic made up of an enormous number of hexagonal cells, and each of these cells is directly connected to the brain by a number of nerve filaments along which travel impulses which are dependent upon the intensity of the light falling upon the hexagonal cells. Exactly how these impulses are generated is not at present fully understood, but they are almost certainly due to the presence of a light-sensitive substance named "visual purple," which flows through the hexagonal cells. The images which we see are thus built up of an extremely fine mosaic of microscopic hexagons of varying degrees of light and shade. The number of these hexagonal cells is enormous. In a normal human eye there are several millions.

This arrangement has been imitated by devising apparatus whereby rays of light of varying brightness when reflected from an object are projected, through the intermediary of an electric current, on to a screen at the receiving station. The variations in the intensity of illumination of the object are controlled by means of a light-sensitive cell which sends out variations of electric current intensity to correspond with the light variations.

Persistence of Vision.

While on the subject of the human eye it may not be out of place to refer to the fact that the retina continues to feel the effects of the light rays after the object that caused the sensation of sight has been removed. This phenomenon is called *the persistence of impressions* or *visual persistence*—the impressions lasting approximately for one-tenth of a second. Examples of this are quite familiar to us all and advantage of the phenomenon is taken in television. Thus a glowing match-end when swung round in a circle appears as a bright ring of light, and not as a single bright spot changing its position every moment. Also the colours of a rapidly rotating disc that has several different colours blend into one colour, and an alternating current supply of electricity gives a steady light, the fluctuations of light that are actually taking place being too rapid for the eye to detect them. The reader will no doubt call to mind many other similar examples and illustrations that serve to show how easily the eye may be deceived.

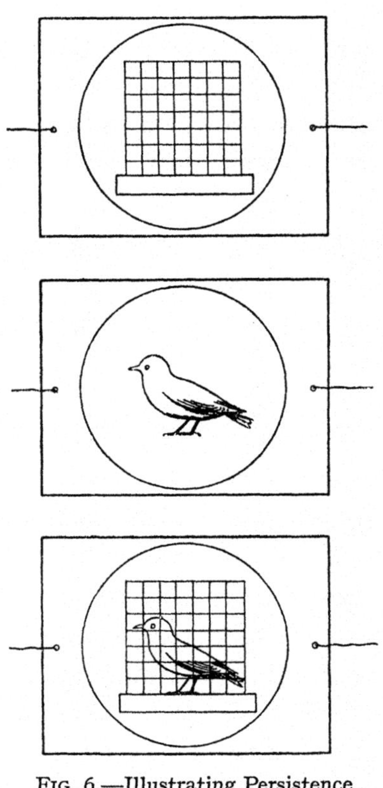

Fig. 6.—Illustrating Persistence of Impressions.

The persistence of impressions in the eye last quite a perceptible time, the images of the brightly illuminated objects if the latter are shown or presented for only one-thousandth of a second remain in the eye for a whole tenth of a second before they die away. If, then, a second impression is presented to the eye before the first has died away, the effect will be the same as that of seeing both at once. An experimental verification of this principle may be carried out by drawing on one side of a white card the outline of a bird-cage and on the other side a bird. If the card be held by means of two strings as shown in Fig. 6 and then twisted or twirled by blowing on the card so as to cause rotation, the effect will show a bird in the cage. Again, if a disc with a number of small holes perforated in it be placed in the slide of a lantern, the holes as separate spots of light will be seen on the screen; but if the disc is rotated rapidly upon a pin fixed at its centre, each little hole is no longer seen separately, but there is a continuous luminous line shown on the screen (Fig. 7). The eye, fortunately for the success of television, has a "time lag," and

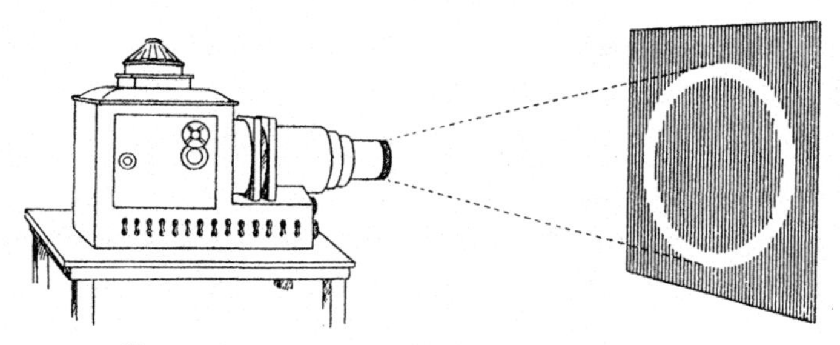

FIG. 7.—Another Illustration of Visual Persistence.

the images therefore need not actually be transmitted instantaneously; when they are transmitted at the rate of eight per second, the transmission appears to the eye to be instantaneous.

In television eight images per second are transmitted; these images, it should be clearly understood, are not photographs, but images of the actual living scene. The transmission of eight photographs per second would not achieve television, but would be the transmission of a cinematograph film, or telecinematography.

The Principle of the Telephone as a Guide.

While the marvellous mechanism of the human eye was being made the subject of study with a view to the construction of an artificial eye that would convey vision over distances outside the ranges hitherto obtainable, Graham Bell and others were at work on their methods of attaining the desired end. Graham Bell, however, as we know, perfected the telephone without achieving television, probably because the problem of hearing from afar was simpler than the allied problem of seeing from afar.

Immediately the details of the mechanism of the telephone became known, several workers in television research conceived the idea of adopting the principle of the telephone to guide them in their experiments. Hoping to solve the problem, scientists and inventors proceeded to imitate the electrical devices of the telephone transmitter and receiver.

They maintained that just as sound waves when they impinge on the sensitive transmitter (carbon granules) of the telephone, thereby altering the electrical resistance with each note and varying the electric current to be transmitted, so the different variations of light and shade of an object by means of a light-sensitive cell could be transformed into varying electric currents that could be transmitted to a distance; and just as the currents received by the telephone receiver cause the diaphragm to set up vibrations producing sounds in the ear that are faithful reproductions of the transmitted notes, so, in television, the transmitted electrical currents could at the receiving end vary the rays given off by a source of light and reproduce a photograph or picture of the object or scene transmitted.

By comparing the two diagrams in Fig. 8 which

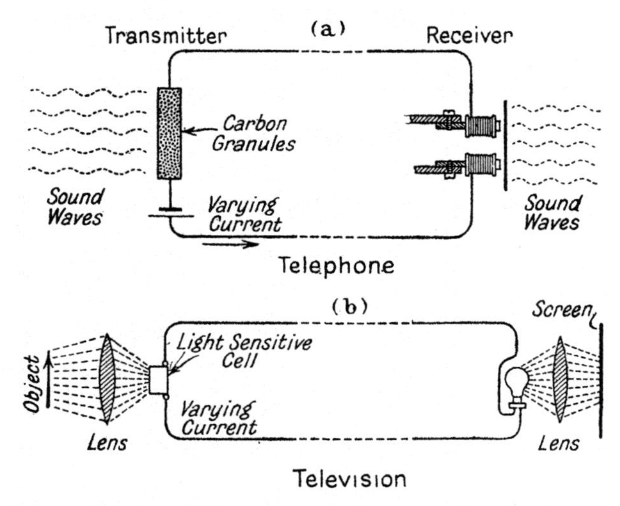

FIG. 8.—Similarity of Telephone and Television methods.

illustrate the principle on which the arrangement could be made, it will be seen how far their arguments were justified.

The chief difficulty that presented itself in their endeavours to accomplish this end successfully was the provision of an artificial eye that would transmit the image of the scene or object.

Finding an Artificial Eye.

A fortunate circumstance aided investigators while they were endeavouring to develop the idea of setting up an artificial eye. An element that had been known for a considerable time to chemists and electricians as one that had a very high electrical resistance as its chief property was discovered to be also very sensitive to rays of light. When subjected to bright light, its resistance dropped; when placed in the dark, its resistance rose. This element was Selenium, an element which up to that time had not been brought under notice very considerably. Here, then, were all the requisite essentials for a device which could be so operated that fluctuations of light intensity would produce variations of electric current. For obviously, if an electrical circuit in which the element Selenium was incorporated could be made to carry currents that varied in strength according to the variations in light and shade that go to form a picture, the problem would be solved. Other elements like potassium, rubidium, etc., have since been taken into the service of wireless seeing, especially in connection with photo-electricity and photo-electric cells. These light-sensitive constituents for the make-up of artificial eyes will be discussed in detail later.

Since the time when some sort of chemical eye was sought after, rapid strides have been made in the technique of television, so that to-day there are other aspects from which the solution of the problem may be viewed, and although no published results of attempts have yet appeared, if any discoveries have been made it is possible that one day light rays may be transformed into electricity and back again into light by one piece of apparatus. We have but to consider phenomena revealed to us by modern instruments and appliances like high-vacua pumps to recognise this possibility.

Take, for example, the exploration of the very wide band of electromagnetic waves consisting of eight octaves between the ultra-violet rays and the *X*-rays. The rays in question are so extremely absorbable by matter that special vacuum spectrometers have to be brought into use for their detection; their existence would still be unknown to-day but for the application of the principle employed in the high vacuum pump. It is quite possible that in the future fresh discoveries and appliances will enable us to extend the field of investigation and bring additional knowledge to bear on the subject of wave-frequencies and the relation of light rays to electricity so that science will be able to transform the one into the other.

While on this subject, we may glance at the wave-lengths, or rather frequencies, of light waves as compared with the wave-lengths or frequencies of the wave-bands that are utilised in electric transmission, that is, the propagation of electric currents either in conductors or otherwise. Electric waves are exactly like light waves in that they can be reflected, refracted, polarised, absorbed, and diffracted; they differ,

TABLE OF WAVE-LENGTHS

X-RAY WAVES ←	LIGHT WAVES			UNKNOWN RADIATION	RADIO WAVES →
INVISIBLE	INVISIBLE	VISIBLE	INVISIBLE	INVISIBLE	INVISIBLE
	ULTRA VIOLET		INFRA-RED		
OCTAVES →					
	0·1μ 0·2μ	0·4μ	0·8μ 10 μ	100 μ 1 mm.	10 mm. WAVE LENGTH ↓

1 Micron written 1μ = a thousandth part of a millimetre (mm.)

however, from light waves, in that their length may be inches, feet, yards, or even miles, whereas light waves measure a few millionths of an inch only.

We have reason to believe from the electro-magnetic theory of light that the velocity of light is the same as that of electricity, and this velocity has been estimated at 186,400 miles per second or 30,000,000,000 cm. per second in air or vacuum. Hence, knowing frequency the wave-length can be deduced at once from

wave-length = velocity/frequency

i.e., $\lambda = v/f$

where λ is the wave-length, v the velocity, and f the frequency.

In the case of an oscillatory electric spark lasting, say, one five-millionth part of a second as seen by the eye, there may be twenty successive oscillations each lasting only one-hundred-millionth part of a second, the frequency therefore is one hundred-million a second; dividing 30,000,000,000, by 100,000,000 we get 300 cm. (or 10 feet approx.) as the wave-length.

Now there is a very wide range of wave-lengths or frequencies between the latter kind of waves and those of all kinds of light, visible and invisible. This may be seen at a glance by inspection of the values given in the accompanying table.

How can we transform or convert these very small light waves into electric waves and *vice versa?* At present we do not know. Two principal obstacles stand in the way of achieving this object, first,

3

the extremely absorbable nature of light rays, and, secondly, the impossibility at present of measuring them. There is also another drawback, namely, the difficulty of propagating light waves over a considerable distance, even when obstacles do not intervene, due to loss of energy during the process of propagation.

CHAPTER II

HISTORICAL

INVENTORS and others, inspired by brilliant and (to them) original ideas, often throw away a surprising amount of time and labour, because the historical side of their subject is not or cannot be studied. A little time spent in looking briefly at attempts made in the past would save many an inventor from the pitfalls which so often beset him.

One great advantage of studying or glancing briefly at the historical aspect of the problem at issue is that it gives the inventive mind a good perspective view before setting out. It shows the technical difficulties and sometimes the commercial aspects whereby one may guard against the deficiencies of preceding attempts and the more quickly surmount the difficulties that have confronted other inventors. Moreover, as one of the objects of a book should be to inspire or excite action, the following summary of what has been already essayed in the direction of television is given in the present chapter.

The Pioneers.

The name of the early inventors is legion, and we can only give a few of the more prominent workers whose endeavours to transmit pictures and images by means of electrical apparatus reveal to us that experiments with a view to transmitting pictures and writing by electrical means took place so far back as 1847. The researches that have now been going on for more than a quarter of a century have gradually unfolded ideas that have culminated in the desire to view contemporaneously far-off events that are happening, just as past events are pictured on the screen at cinemas.

The first indications of a trend in the direction of discovering means for electrical seeing are evidence in

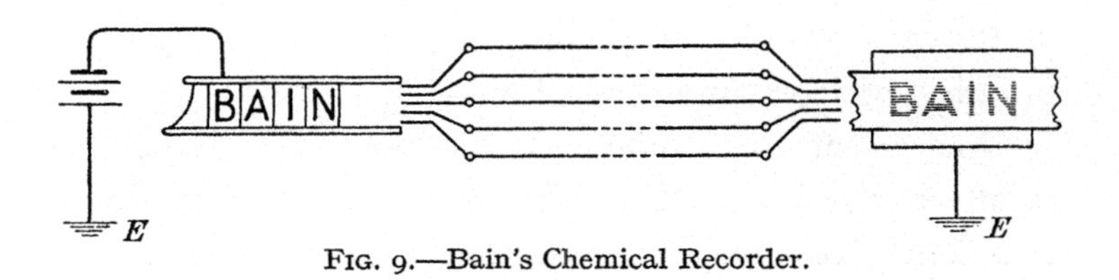

FIG. 9.—Bain's Chemical Recorder.

the embryo stage, when drawings, contours, and diagrams were transmitted over wires by electric current in 1862 by Abbe Caselli. This system was in practical operation for several years between Paris and Amiens. It was really a more elaborate form of Bain's chemical recorder (1842), and Bakewell's apparatus (1847), paper treated with cyanide of potassium being marked by an electric current through an iron point or stylus. Bain's apparatus is interesting in that he used a form of copying telegraph as shown in the illustration, Fig. 9. The letters to be transmitted were actual metal types set up in a composing stick and connected to an earthed battery. Five metal brushes, made up of several narrow springs connected to the same number of lines were passed over chemically prepared tape resting on an earthed conducting plate. As and when the currents were received over the line, marks were made on the paper and a copy of the type faces at the transmitting end was obtained. It was necessary for the two brushes to move in synchronism.

Bakewell in 1847 followed in the wake of Bain

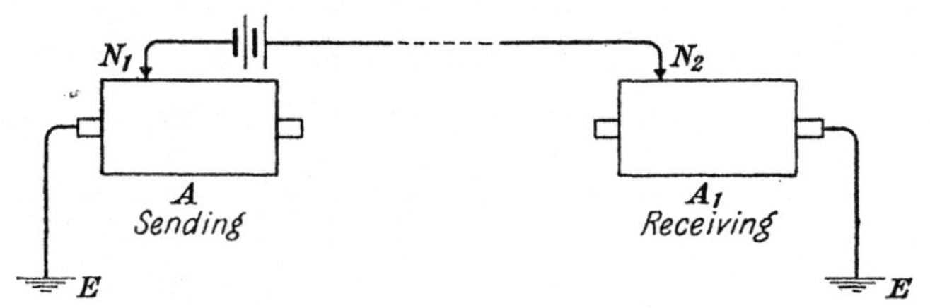

FIG. 10.—Bakewell's Apparatus.

with apparatus something similar in character. He set up at the two ends of a telegraph wire two metal cylinders (A and A_1, Fig. 10) which revolved in synchronism, while metal needles, N_1 and N_2, connected to a battery traced out spirally at one end a sketch drawn in shellac-ink, and at the other end a replica on chemically prepared paper was produced.

At the sending end the sketch was drawn on a sheet of tinfoil wrapped round the cylinder, and a sheet of chemically prepared paper was placed on the receiving cylinder. When the needle N_1 came in contact with a shellac line no current flowed, but when the needle was in contact with tinfoil (and, of course, the cylinder) a current passed through the circuit, causing a chemical mark on the paper on A_1.

Edison improved on these methods in 1873 by substituting a punched tape for the inconvenient composing stick with metal type (see Fig. 11) and previously in 1865 Messrs. Vavin and Fribourg produced a kind of monogram by their method as shown in the same sketch, the elements of which could be arranged to form any letters it was desired to send. Mimault, a year later than Edison, also followed the system of having a chemically prepared receiving tape. A printing lever brought the tape in contact with an insulated block containing 49 pins, a suitable selection of pins connected to a battery being capable of printing any desired letter.

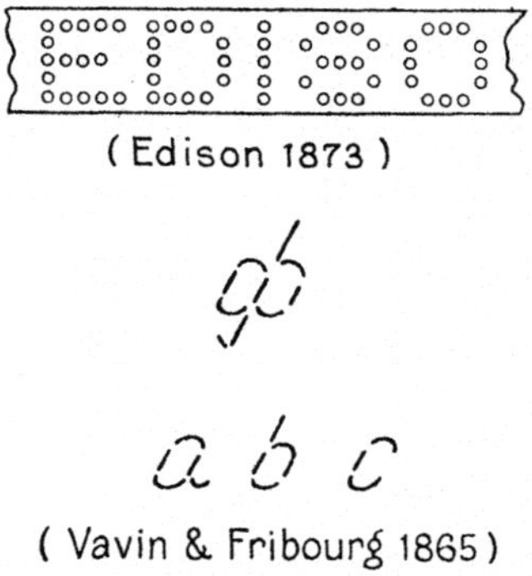

(Edison 1873)

(Vavin & Fribourg 1865)

FIG. 11.—Edison's and Vavin and Fribourg's Apparatus.

The diagram in Fig. 12 shows the connections for sending the letter "T."

We thus see that from the crude endeavours of Bain and others to communicate electrically with distant points by means of metal letters, hearing and seeing from remote parts of the world as achieved to-day have been developed. The growth may be regarded in the light of a family tree, Bain's work

FIG. 13.—Showing how a Picture may be drawn by means of Lines of Varying Width.

[*To face page* 39.

being the stem from which sprang branches, notably telegraphy with its long and short signals in code, and picture or photograph transmission with its "parallel lines" of varying width corresponding with the dark and light portions of the picture (see Fig. 13)

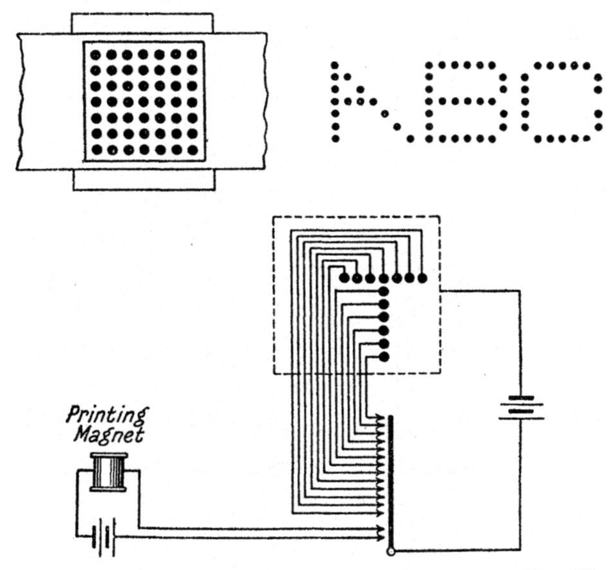

Fig. 12.—Mimault's Method and the Arrangement for Signalling the letter "T."

from which latter television eventually became an offshoot.

This idea of transmitting drawings and writing to distant points—the forerunner of television—continued to produce inventors as late as 1900–1901. The Telautograph, an instrument for reproducing actual handwriting at a distant station, and the Telepantagraph, an instrument for transmitting actual drawings over considerable distances by means of telegraphy, were placed under experiment in the Electrical Research Laboratories of the G.P.O. during that period.

It should not be overlooked, moreover, that even the various types of oscillographs that have come into use for recording purposes may be regarded as simple elementary forms of televisors.

The Advent of Selenium.

Reverting to earlier times, stimulus was given to television research in 1873, when it became known that Selenium, "the moon element," possessed the property of exhibiting great sensitivity to light rays, the announcement of this fact being first communicated to the Society of Telegraph Engineers (now the Institution of Electrical Engineers) by Mr. Willoughby Smith. The notification of this property led to the adoption and the use of Selenium in nearly all subsequent television research.

Many attempts were made during the following decade to enlist this newly-discovered property of Selenium in the service of television, but it was not until M. Senlecq in 1877 brought out his "telectroscope," mainly modelled on Caselli's apparatus (Fig. 14), but with Selenium as an ancillary operating device, that any real theory of television was disclosed. In this apparatus the principle of reproducing by electromagnetic means the effects of light and shade that were cast on an object was adopted. This was to be accomplished by projecting the image to be transmitted on to the ground glass of a camera obscura, tracing it out by means of a Selenium point; when transmission took place variations of current in the latter were to cause the effects of light and shade to appear at the receiving end. The invention was regarded as so important a step in advance of its

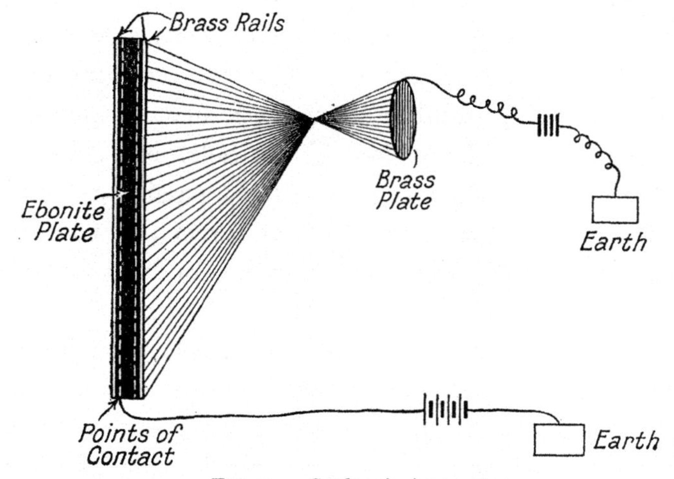

FIG. 14.—Senlecq's Apparatus.

predecessors that several other workers, amongst them Graham Bell, the inventor of the telephone receiver, brought out similar inventions and apparatus. It is perhaps noteworthy and remarkable that attempts were being made to transmit pictures electrically at the same time that attempts were being made to transmit sound electrically, both problems being the subject of investigation by the same genius, Graham Bell.

In 1880, Messrs. Ayrton and Perry and also Professor Kerr published an account of their proposed television system. Selenium cells were employed in this system, all of which were worked with a multiple wire (and multiple cell) arrangement, each wire contributing a small portion of the picture until the whole was pieced together in mosaic form.

Ayrton and Perry's apparatus consisted of many wires for transmission purposes, each wire having a Selenium cell connected to it. At the receiving end, a corresponding number of magnetic needles operated in unison with the action of the Selenium cells, so that each time a needle moved light passing through an aperture was either shut off or allowed to pass on, a kind of mosaic of the picture being built up by this means.

In Kerr's apparatus, the same principle was adopted at the sending end, but at the receiving end, instead of the magnetic needles, electro-magnets with silvered ends were illuminated by a polarised beam of light. The currents received through the electro-magnets rotated the plane of polarisation according to the amount of light sent out.

Employment of Thermo-Electricity.

It was in 1880 also that Middleton, of St. John's College, Cambridge, announced his invention of apparatus in which thermo-electric couples instead of Selenium cells were employed in a multiple wire transmitter. Corresponding thermo-electric couples were fixed at the receiving end, which received the radiant heat sent out at the transmitting end in the form of reflection from the bright-polished surfaces of the thermo-couples.

In America in the same year, not only did Graham

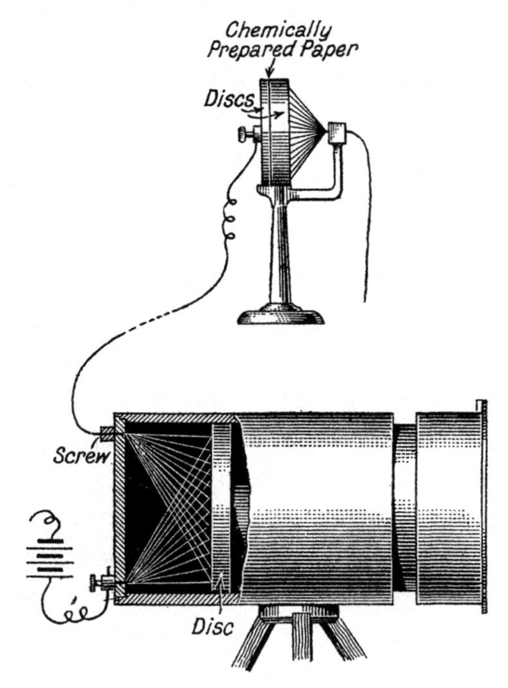

FIG. 15.—Carey's Apparatus.

Bell take out patents for television apparatus, but also Connelly and McTighe, of Pittsburg, and Dr. Hick, of Bethlehem, Penn.—the apparatus invented by Hick was called the "Diaphote." Carey, an American, in the previous year had also brought out apparatus of the multiple wire type (Fig. 15), with the

Selenium cell arrangement at the transmitting end and an incandescent receiver containing carbon or platinum elements (see *Sci. Am.*, Vol. 40, page 309, 1879).

Shelford Bidwell, an experimenter on the properties of Selenium, in 1881 showed how that element could be applied in improving the Photophone and in telephotography. He contributed an account of "Practical Telephotography" in *Nature* in 1907 (see Vol. 76, page 444).

Wireless Photography.

A gap occurs until more recent times, probably owing to the disappointing results experienced with Selenium in television pure and simple as apart from picture photography. It is noteworthy that in 1908 Knudsen made the first practical step towards the transmission of photographs by wireless.

De Bernouchi, of Turin, also had previously devised a system for this purpose. He employed a method whereby the intensity of a beam of light was varied by passing it through a photographic film on to a Selenium cell, the resistance of the latter varying at each instant accordingly (Fig. 16).

Cathode-ray Systems.

Impetus was given to the pursuit of the transmitting pictures by wire by Ruhmer (1891), Rignoux and Fournier in France (1906), Szczepanik in Austria, and Rosing in Russia (1907), who sought to achieve perfected results. Rosing made use of the Cathode-ray in his apparatus and Mr. Campbell Swinton, in this country at about the same time, designed quite independently a Cathode-ray system of television. These are mentioned as marking important steps towards the solution of television. Improvements in phototelegraphy have been going on up to the present time with remarkable results.

In addition to the attempts made by the foregoing,

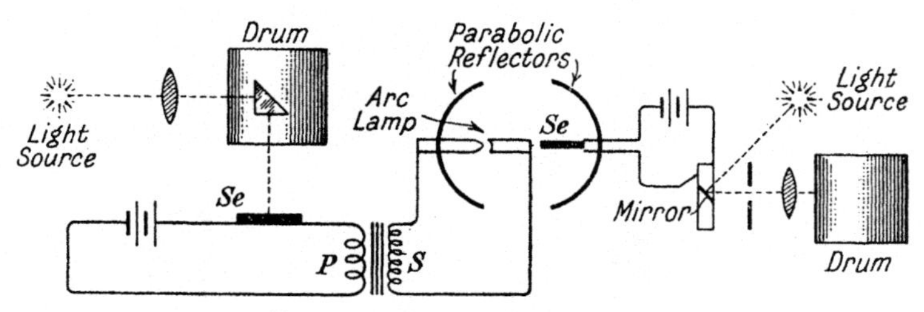

FIG. 16.—Bernouchi's Method.

the pioneer inventions of Korn (1907) and of Knudsen (1908) perhaps deserve special notice.

Korn's Experiments.

The first method of successfully transmitting photographs was that of Professor Korn in Germany, who made use of a circuit arrangement of which the essential features are given in diagrammatic form in Fig. 17.

The apparatus invented by Korn was afterwards set up with certain modifications and improvements by the Poulsen Company for the wireless transmission of pictures. In view of the distinct advantage which characterises this invention from all preceding attempts, a description of its working is given below, although space forbids any details beyond the merest outline.

In the figure, *D* is a revolving cylinder or drum on which is wrapped a metal print, and when the stylus, Z, traverses this, the receiving apparatus records by means of a string galvanometer denoted by *H*, a copy of the picture. A high frequency alternator or

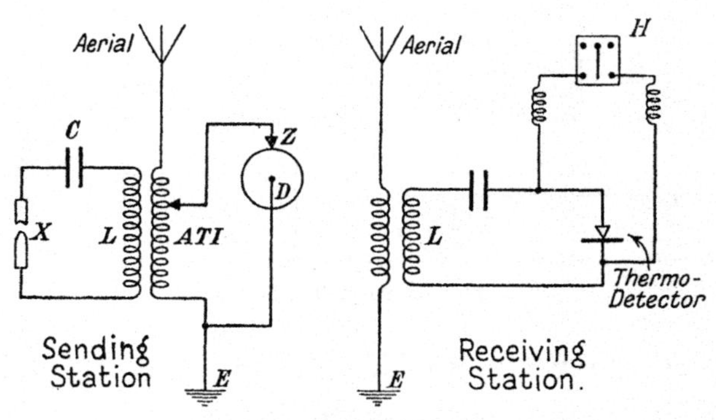

FIG. 17.—Diagram of Korn's Circuit.

Poulsen arc generator, *X*, generates continuous or undamped waves. Tuning is effected by means of the *ATI*, the inductance, *L*, and Condenser, *C*.

For transmission, Korn's apparatus consists of a glass cylinder or drum, which revolves in a box on a threaded shaft, thus giving it also a vertical motion as shown in Fig. 18.

A source of light, L_1 (for example, a Nernst lamp), has its rays cast by means of a lens system so that they intersect on the surface of the cylinder or drum,

C_1, whence they are passed on to a prism of 45°, P_1. In revolving, the cylinder on which is wrapped the photographic plate thus causes different small portions of the picture to come under the intercepting rays; the entire picture is therefore traversed by the light beam during the process. As often as the light variations occur, so are they reflected by the prism on to the Selenium cell, Se_1, which in turn suffers

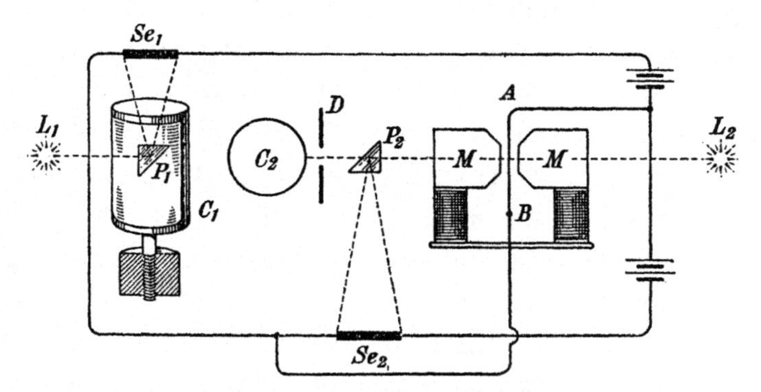

FIG. 18.—Arrangement of Korn's Apparatus.

variations in its resistance, thereby giving rise to the varying electric currents that are sent out to the receiving apparatus.

The Poulsen–Korn System.

In this apparatus as operated in the practice of phototelegraphy by the Poulsen Company, a string galvanometer of the Einthoven type (Fig. 19) was introduced as a means for directly reproducing the picture. The Einthoven galvanometer is the simplest form of oscillograph. Its essential part is a stretched conducting fibre or wire placed in a strong magnetic field. This fibre is made very fine; if of silver or tungsten a diameter of 0·02 mm. or less is the general rule, or if of silvered glass a diameter from 0·002 to 0·003 mm. is quite common. The figure shows the general arrangement of mounting the fibre in the magnetic field.

It is usual to have a Pointolite lamp (Fig. 20) for the source of illumination, the light being concentrated on the fibre by the main and substage condensers, but an over-run "gas-filled" lamp or arc lamp can be used as an alternative.

One of the principal features of the Poulsen–Korn apparatus is a flat silver ribbon, *A*, *B* (Fig. 18), five milli-inches wide and one milli-inch in thickness, which is stretched between the poles of an electro-magnet, *M*, *M*. Light rays from a source, L_2, are brought to a focus by means of a lens, and it should be mentioned that the pole pieces of the electro-magnet are tunnelled for the purpose of allowing the light rays to converge to a focus. Normally this silver ribbon cuts off all rays of light, but if the least current traverses *A*, *B*, the magnetic field that is acting causes it to be repelled. The degree to which the ribbon is shifted depends on the strength of current passing whether the movement be very small or otherwise, and the light is thus allowed to reach a second lens, by which it is focussed on to the surface of the receiving photographic film. Like the arrangement at the transmitting end, this film is wrapped round a revolving

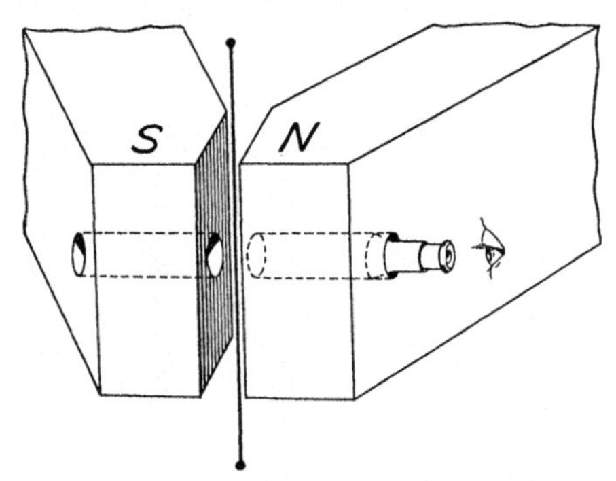

FIG. 19.—Diagrammatic Illustration of the Einthoven String Galvanometer.

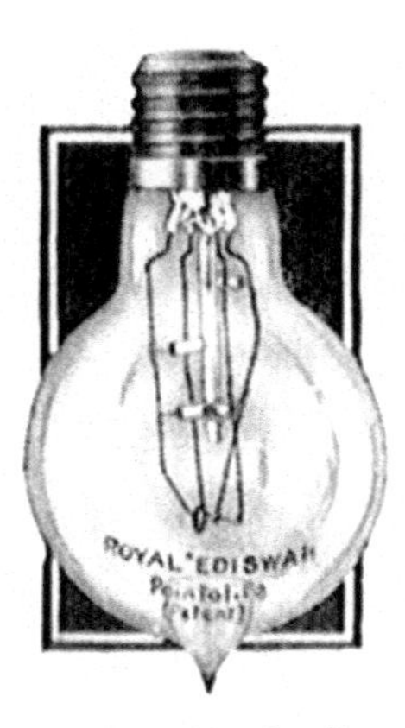

FIG. 20.—Pointolite Lamp.

[*To face page* 48.

drum, C_2, in a cabinet or camera from which all light is excluded. Since the intensity of light varies according to the density of the image, the amount of light falling on the film corresponds likewise. The film is developed in the usual way and a copy of the original obtained.

The Transmission of Pictures over Telephone Lines.

The Bell Telephone Laboratories Inc. of America have worked out a much more efficient system of phototelegraphy based on that of Korn's invention. As will be seen by the following description of working, it differs very little from Korn's except in so far as up-to-date improvements are added. For instance, whereas Korn used a Selenium cell, a photo-electric cell now takes its place. Other improvements have also been introduced, the chief of which is a much more accurate method of keeping the two cylinders in perfect synchronisation. This has been rendered possible by regulating the speed of the special type electric motors used and with the aid of phonic wheels controlled by tuning-forks operated electrically.

The Transmitter and Receiver.

Transmitting by this method depends first of all on having a negative of the photograph prepared from which a positive is made on a celluloid film 5 by 7 inches wrapped closely round a hollow glass cylinder. The latter is mounted on an arrangement so that it travels longitudinally at the same time that it is rotating round its axis. A source of light throws a spot of light on to the cylinder with its film, and this spot travels over the film in a spiral direction on account of the motion already referred to. Inside the cylinder is a photo-electric cell. Obviously the spot of light while traversing the film gives out light in varying intensity according to the lightly- or darkly-shaded portions of the photograph and thus varies the amount of current received from the cell which is connected locally by wires to the main lines. This action is tantamount to modulating the current transmitted in accordance with the degree of light or dark shading of the picture.

In the apparatus at the receiving end there is a device termed a "light-valve" having a small orifice which is normally covered entirely by a very thin metallic ribbon, the latter being subjected to the influence of a magnetic field. At the same time, a very strong beam of light from an electric lamp is directed through the orifice mentioned. Now when the incoming current traverses the ribbon, an electro-magnetic field is set up around the ribbon, and this, acting in conjunction with the magnetic field already generated, causes the ribbon to shift, thus allowing a certain modicum of light to pass through the aperture. Obviously the movement of this ribbon being governed by the varying incoming current will correspond with the amount of light and shade of the picture at the sending end. Hence a cylinder on which a blank film is wrapped, provided it revolves in exact synchronism with the cylinder at the sending end and is placed so that the light beam strikes it, will have

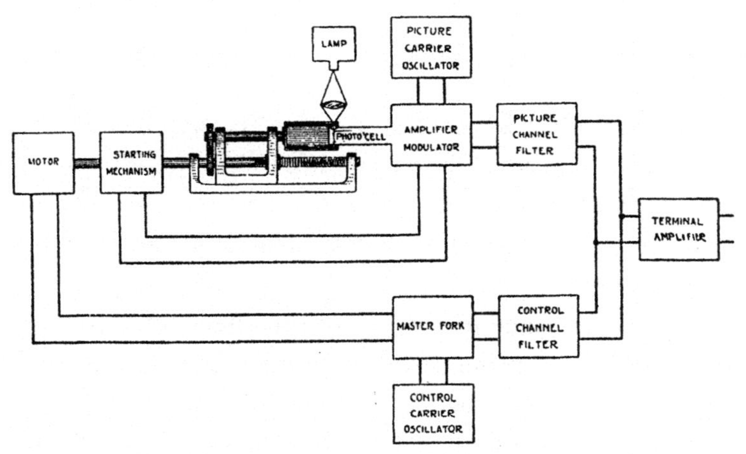

Fig. 21.—Schematic Diagram of Sending End of Apparatus.

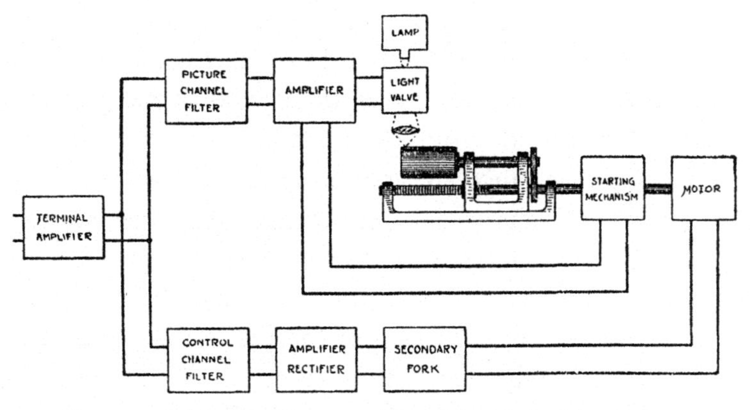

Fig. 22.—Schematic Diagram of Receiving End of Apparatus.

reproduced on it a negative picture or photograph which is an exact likeness of the original.

The essential parts of the electrical circuits used are shown in the schematic diagrams Figs 21 and 22, which are taken from the *Bell System Technical Journal.*

Pictures by Wire.

Pictures, drawings, photographs, letters, including *X*-rays, radiograms, finger or thumb prints, cheques, etc., may all be transmitted in this manner over short distances, the actual time of transmission lasting about seven minutes.

In long-distance line working the effects of capacity and inductance come into play, and it is necessary to adopt a circuit similar to that used for radiotelephony by using an oscillator valve and a modulating valve. At the transmitting end, therefore, a frequency of 1300 cycles per sec. generated by an oscillator valve has superimposed on it much lower frequencies (controlled as before by the lights and shades of the picture being transmitted) from a modulator valve.

In this case, since an alternating current is sent, the negative will show on close examination variations in the thickness of the lines of the picture traced out, corresponding to the changes of current during each cycle, but the reproduction of the picture is not interfered with in any way on this account. This method gives about 65 lines to the inch and militates somewhat against the use of a ribbon for plates where reduction, enlarging, or retouching is necessary. Another method is to have an aperture of fixed measurements and to allow the light to fall on the film in a diffused manner, when 100 lines to the inch may be selected and a half-tone picture produced.

It is absolutely necessary that perfect synchronisation of the cylinders both at the sending and at the receiving ends should take place. In order to ensure this, phonic wheels controlled by tuning-forks operated electrically are employed with the cylinder mechanism at each end.

The accuracy and efficiency of synchronisation now achieved are such that received photographs when sent over a line in this manner are indistinguishable to the naked eye from their originals.

The transmission of pictures and photographs by wire is carried out in America at the present time by the American Telephone and Telegraph Co. Amateurs in the United States of America who are in the possession of licences are allowed to use certain wave-lengths for the transmission of pictures.

Knudsen's Experiments in Radio Photography.

The first attempts made in wireless phototelegraphy over any considerable distance were those of Knudsen in 1908. As a matter of historical interest, a diagram of the apparatus used is given in Figs. 23 and 24.

The transmitter consisted of a camera with lens behind which was placed, between the lens and the plate, a line screen for splitting up the photograph into parallel lines. The photographic plate had to be specially prepared with a thick gelatine film. In this kind of plate the shaded portions of the picture dry less quickly than the transparent or lightly-shaded parts. Hence, when iron filings or dust is sprinkled

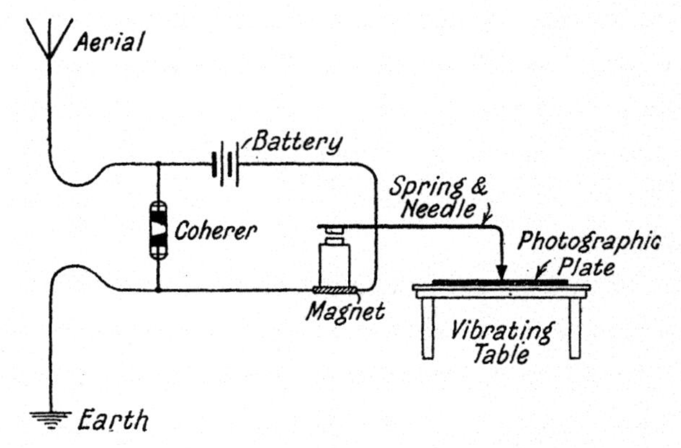

FIG. 23.—Knudsen's original Receiving Circuit for Radio Photography.

over it, the shaded portions retain more iron than the lighter portions. A steel point in conjunction with a

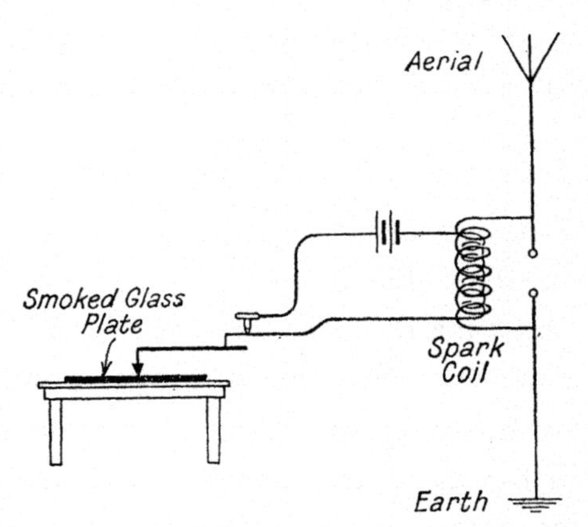

FIG. 24.—Knudsen's Transmitter Circuit.

spring is made to traverse this picture in iron dust, making contact and passing a current from a battery where and when necessary, the currents being transmitted by wireless to the receiving end, where a similar steel point and spring actuated by a small electromagnet travel over a smoked glass plate. A print may be taken from the picture scratched out on the smoked plate in a manner similar to that adopted with an ordinary photographic negative.

The results obtained by this apparatus were extremely crude and it was found to be quite unsuitable for wireless transmission.

It is interesting to recall that Mr. Marcus J. Martin some years ago (1916) devised a system of radio photography by means of an instrument which he designated the "Telephograph."

A metal print of whatever had to be transmitted was necessary, and in preparing this print a screen having a number of ruled lines was used, the effect being to break the picture into parallel lines. The time for a complete transmission of a picture 5 by 7 inches with 50 lines to the inch took twenty-five minutes.

In working the apparatus the metal line print was wrapped round the drum of the machine, during the revolution of which a stylus made contact (or otherwise) with the lines of the print according to the light or dark lines into which the picture was split up.

By means of an optical arrangement at the receiving end variations of light from a Nernst lamp were received on the film on the drum at that end according to the point of contact of the stylus of the transmitter, *i.e.* whether tracing over a conducting or an insulating strip on the metal print. By this means a positive picture was received from which a photograph could be reproduced. The obsolete syntonic system with Carborundum detector was used in the wireless portion of the apparatus. Further details of this interesting apparatus are given in Mr. Martin's book, " Wireless Transmission of Photographs."

Thorne Baker Apparatus for Wireless Phototelegraphy.

With this apparatus pictures can be broadcast by wireless which may be picked up by means of a simple form of apparatus adapted to a two-valve receiving set.

The transmission of a picture by this method is effected by first of all copying the photograph to be transmitted through a photo-mechanical screen. The latter is ruled with close parallel lines which cause the photograph to appear made up of thick and thin lines, the different shading in the image represented by dark and light parts of the photograph corresponding to thick and thin lines respectively. A photograph of this character is printed on sheet copper after the latter has been sensitised with fish-glue treated with bichromate of potash. The unexposed portions of this kind of plate can be got rid of by washing the plate in water, when the image will be clearly shown in fish-glue lines. It will be noted that these fish-glue lines are non-conductors of electricity. The picture is then wrapped round a revolving metal drum provided with a steel needle or stylus. As the drum revolves a current of electricity flows, and whenever the steel point touches a fish-glue line the current is broken, the wider the line the longer the break and *vice versa*, so that the degree of shading in the photograph will be represented by the width of the lines.

The transmitting circuit has the condenser in the valve circuit, which is short-circuited by a transmitter of this description and thus stops oscillation, radiations through the ether being controlled by the photo print on the drum. For receiving purposes, the apparatus at the distant end is provided with a revolving metal drum round which is wrapped the sheet of moistened paper chemically prepared on which the photograph is to be traced. This is done by a platinum stylus which traces a spiral path as the drum revolves.

It may be remarked that the stylus and cylinder are joined to the terminals to which either headphones or loudspeakers are ordinarily connected.

The actual printing is done by electrolytic action on the paper wrapped round the revolving drum. A dot, or rather series of dots, appears as very minute stains less than 1/200 of an inch in diameter in some cases, dependent on the density or shading of the lines in the photograph.

Transatlantic Radio Pictures.

In November 1924 experiments took place at Radio House in connection with Marconi's Wireless Telegraph Co. by means of apparatus invented by Mr. R. H. Ranger and developed by the Radio Corporation of America, whereby photographs were transmitted between Radio House, London, and New York. Briefly, the apparatus is constructed very much on the same lines as that of Korn's and others,

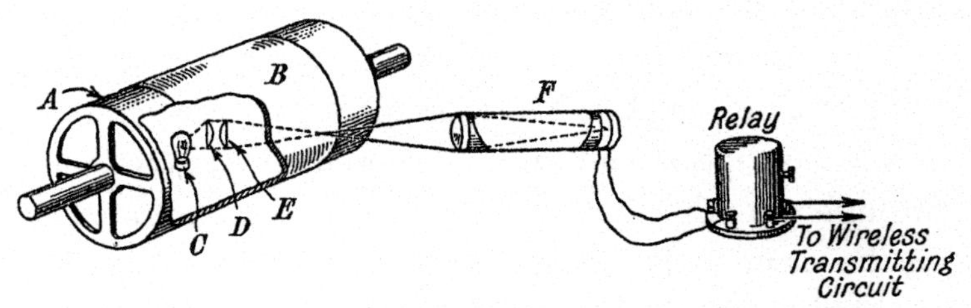

FIG. 25.—Diagrammatic Sketch of Ranger's Transmitting Cylinder.

the same principle of working and the employment of revolving cylinders, photo-electric cell, etc., being adopted. Fig. 25 shows the revolving glass cylinder, *A*, on which is wrapped the photographic film, *B*, the "Pointolite" lamp, *C*, inside the cylinder acting as a very powerful source of illumination. *D* and *E* are condensing and focussing lenses, respectively, the former for converging the rays to a point on the film whence it is directed to *E*, which focusses the beam on to *F*, the light-sensitive cell, placed some distance away from the cylinder.

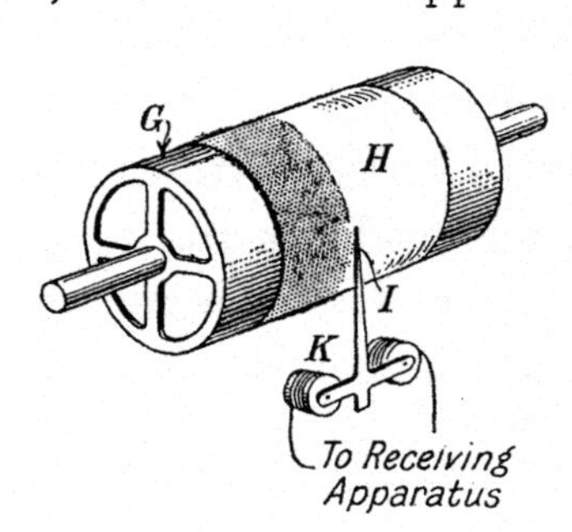

FIG. 26.—Diagrammatic Sketch of Ranger's Receiver.

The same principle of dark and light patches of the films controlling the variation in current is followed. The variations have their effect on the receiving apparatus, where in Fig. 26 *G* is the receiving cylinder,

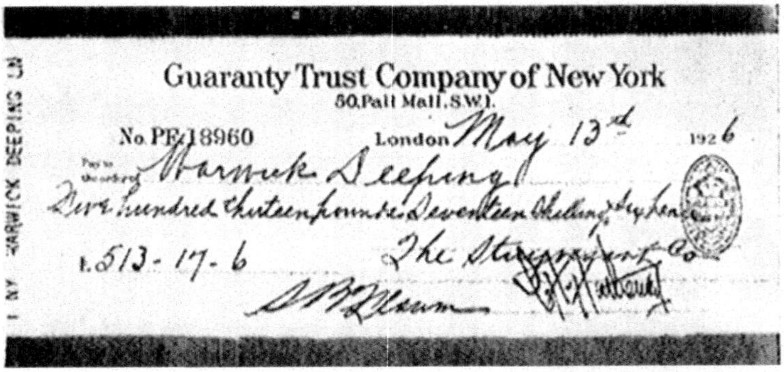

Guaranty Trust Company of New York
50, Pall Mall, S.W.1.

No. PE 18960 London May 13th 1926

Pay to the order of Warwick Deeping

Five hundred thirteen pounds Seventeen Shillings Sixpence

£513-17-6 The Stuyvesant Co

Fig. 27.—Specimen of Print sent by Radio.

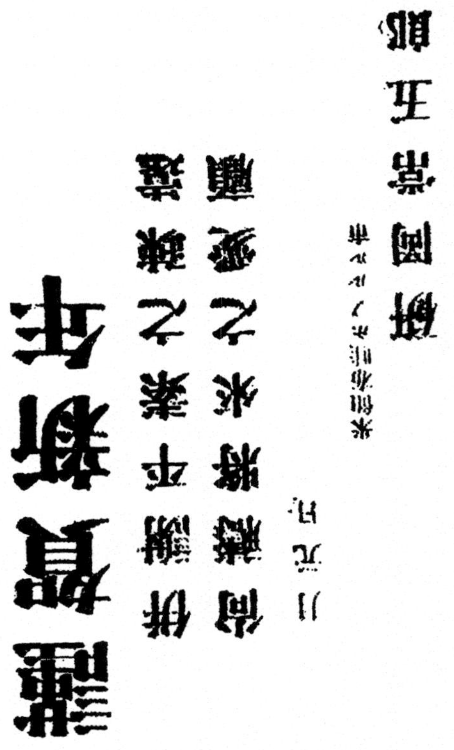

Fig. 28.—Specimen of Print sent by Radio.

[*To face page* 59.

on which is revolving the paper, *H*, and *I* is a self-inking stylus or pen, worked by an electro-magnet, *K*, whose action is controlled by the received currents. The pen makes small dots or blanks in a series of lines according to the dark and light patches of the pictures. The spacing between the rows of dots is nearly 1/200th of an inch for each revolution of the cylinder. The image is therefore reproduced in the form of dots or lines and spaces. When telegraphing a picture or photograph by wireless, it should be borne in mind that in all the methods employed the current sent out by a photo-electric cell is superposed on the alternating current carrier wave sent through the ether at the same time.

Since the date mentioned, important improvements have been effected in this system, which at the time of writing is being operated successfully by the Marconi's Wireless Telegraph Company, Ltd., in co-operation with the Radio Corporation of America. A recent specimen drawing received by radio in London from New York was given in Fig. 2. Additional reproductions of original prints sent by radio across the Atlantic are given in Figs. 27 and 28.

Recent Research.

Mr. A. A. Campbell Swinton, F.R.S., in a lecture before the Rontgen Society and also at a meeting of the Radio Society of Great Britain some time ago, suggested the use of Cathode-rays, and showed that the difficulty of securing extremely rapid and accurate motion of parts could be got over by making use of immaterial substances like these rays. He further described a new device for analysing the image under transmission and making use of a controlled Cathode beam which could be appropriately influenced by magnetic fields.

The apparatus for working out his idea on these lines, however, does not appear to have been constructed so far, and therefore any practical results that might be obtained by this means are not known.

During the last five years a crop of fresh inventors and aspirants has arisen. Most of these have brought forward new ideas, developments, and improvements, not to mention the scrapping of older devices, the Selenium cell, for example. Notwithstanding the sensitiveness of Selenium to light rays, it has been found that photo-electric cells respond in a far better manner to the enormous speed of signalling that is involved in all forms of transmission of vision. Hence in all recent developments the photo-electric cell is relied on for good results.

Television research has been carried out on the Continent by MM. Belin and Hollweck, Dauvillier, and numerous other workers, chiefly along the lines of the Cathode-ray.

Shadowgraphs have been successfully sent both by MM. Belin and Hollweck and by M. Dauvillier in France, and in Austria Denes von Mihaly claims to have achieved the transmission of shadows, using a complex apparatus of oscillating mirrors.

In America, C. F. Jenkins has successfully sent

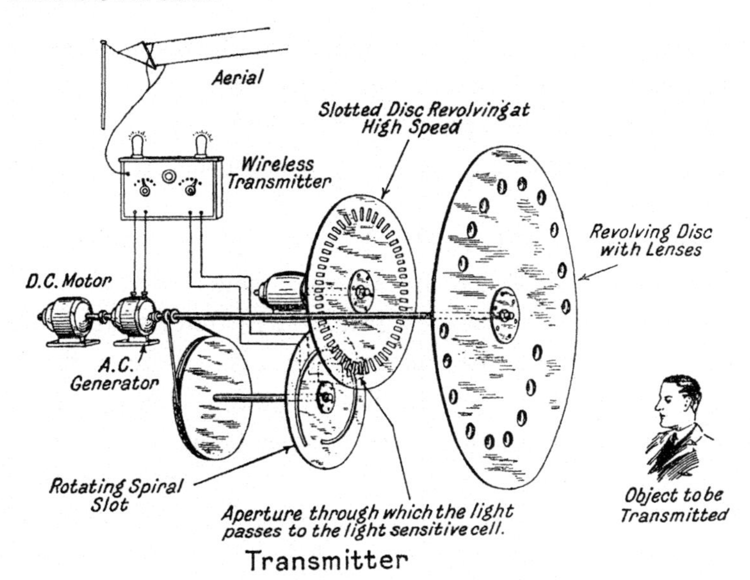

FIG. 29.—Mr. Baird's original Transmitting Apparatus.

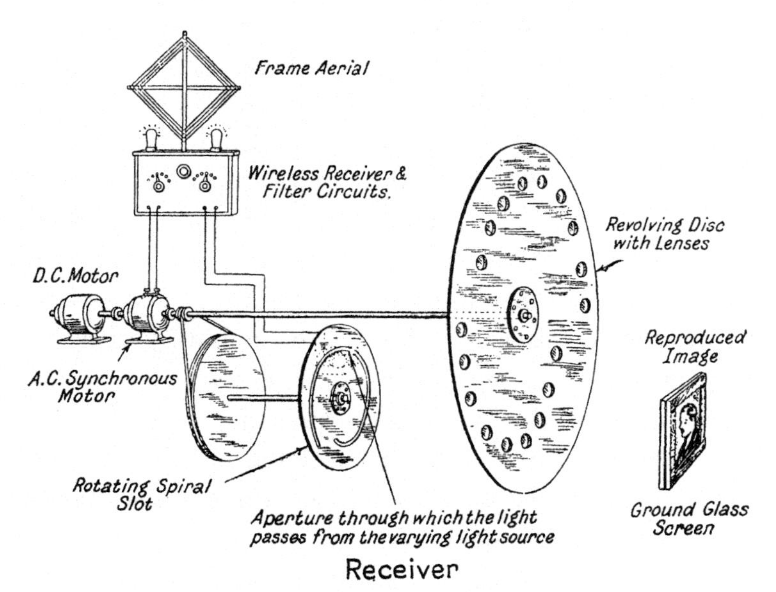

FIG. 30.—Mr. Baird's original Receiving Apparatus.

shadows, and Dr. Alexanderson, the Chief Consulting Engineer of the General Electric Company of America, has described his apparatus with which he hopes to achieve results, although he has so far given no demonstrations.

In 1923, Mr. J. L. Baird in Great Britain and also Mr. C. F. Jenkins in America demonstrated the transmission of " Shadowgraphs " by television, but in 1926 Mr. Baird followed up his demonstration of the transmission of outlines by giving a demonstration of true television, real images being shown by diffusely reflected light. Figs. 29 and 30 show the earliest forms of his transmitting and receiving apparatus. These results and the apparatus with which they were achieved are fully described in the present volume in Chap. VIII.

The American Telephone and Telegraph Company in May 1927 gave a successful demonstration of true television, the first given outside of England. Their work will, however, be dealt with in a separate chapter.

The general method employed by these workers is in each case the same. Elemental areas of the image being transmitted are cast in rapid succession upon the light-sensitive cell. The pulsating current is transmitted to the receiving station, where it controls the intensity of a light spot traversing a screen in synchronism with the traversal of the image over the cell. The means employed to obtain this end vary, however, very considerably.

CHAPTER III

SELENIUM AND THE SELENIUM CELL

Selenium.

THE history of the element Selenium is so closely associated with the history of television that a short account of the properties of this remarkable element may well form the preliminary to succeeding chapters of this book. No account of television, not to mention the early and ineffective systems of picture transmission that were tried, could be intelligently understood without an account of the properties of this electrical eye.

Selenium or the " moon element " was first discovered by the Swedish chemist, Berzelius, in 1817, in the red deposit formed in vitriol chambers. There are several allotropic modifications of this element, but for making a " cell " the grey or crystalline variety is selected, since this form conducts electricity. Its resistance to the passing of a current of electricity, however, under normal conditions is very high; when heated, its electrical resistance is increased. When exposed to rays of light, it instantly lowers its resistance from 15 to 30 per cent., according to the intensity of the illumination, but increases its resistance when the rays are shut off. The red rays appear to give the maximum effect. This property has been made use of in certain determinations of feeble photometric intensities like those of the light received from the stars. On account of its property of sensitivity to light rays, it has therefore been employed considerably in television research work, being made up in the form of a " cell " for the purpose.

Although a member of a chemical family (Sulphur, Tellurium and Selenium) having all the same characteristics, Selenium, however, appears to be the only member of the family that is light-sensitive. It forms compounds with other elements in the same manner, for instance, that sulphur does; it assumes the same allotropic modifications and is capable of the same kind of transformation under the action of heat, etc., but apparently it is the predominant partner so far as light-sensitiveness is concerned.

Ordinary Selenium is red or dark brown in colour, according to the amorphous condition in which it occurs, but on being subjected to heat it turns bluish-grey in colour. It vaporises at about 655° C., and its vapour burns with a blue flame producing a characteristic penetrating odour which has been compared with that of decayed horse-radish (Thorpe).

A Chemical Eye Operated Electrically.

As we have already seen, the early inventors endeavoured to construct artificial telegraph eyes by substituting Selenium for visual purple and building an artificial retina out of a mosaic of Selenium cells, each of these cells being connected by wires to a shutter. This shutter opened when light fell on the cell connected with it and allowed a spot of light to fall on a screen. In this way each cell controlled a spot of light, the image being reproduced as a mosaic formed of the spots.

While the baffling problem of " seeing from afar "—television—was in process of solution on the lines of nature's marvellous mechanism of the eye, another cognate problem, namely, " hearing from afar "—telephony—was actually solved by Hughes, Graham Bell, and others, by following the lines adopted by nature in the construction of the ear. The appearance of the Bell Telephone immediately stimulated workers in television research to devise mechanism that represented artificially the action of the eye. In this respect, as we have also seen, they were fortunate in having the extraordinary properties of Selenium brought to their notice, namely, its enormous resistance when an electric current is passed through it, and its remarkable sensitivity whereby its resistance is lowered when it is exposed to light rays.

This latter phenomenon first came under observation quite accidentally. The element was known to be a metal possessing enormous resistance, and this property made it useful in the construction of the high resistances used in telegraphy; such resistances were employed in the early days at Valentia, the terminal station of the Atlantic Cable, a little village in the West of Ireland. One afternoon the attendant, Mr. May, was surprised to see his instruments behaving in a

5

very erratic manner. It was a day of bright sunshine, and the sunlight fell occasionally upon his Selenium resistances. He found that every time the sunlight shone on the Selenium the needle of the galvanometer moved. The phenomenon was investigated and the light-sensitive properties of Selenium were disclosed.

Selenium thus provided a means of turning light into electricity, and the scientists of those days were quick to see that in Selenium they possessed a chemical eye which could be used for transmitting vision.

Make-up of the Selenium Cell.

All the apparatus invented in the early days depended for transmission of vision on the provision of some form of chemical cell that produced variations of electric current in the circuit in which it was joined when variations of illumination were thrown on to the cell. In the make-up of the cell Selenium was employed. The alkali metals, such as potassium, sodium, calcium, rubidium, have under certain conditions the property of sensitivity to light rays and of thereby producing an electric current, but these will be considered later.

As already mentioned in the previous chapter, the discovery that the resistance of Selenium altered considerably on exposure to light was communicated to the Society of Telegraph Engineers by Willoughby Smith. The announcement led to the construction of so-called Selenium " cells " by Shelford Bidwell, Graham Bell, Sabine, Minchin, and others.

Graham Bell was the first to make a practical application of this newly-found property of Selenium. He built up a Selenium cell by arranging two metal plates, one having fixed to it a number of studs, these studs being slightly smaller in diameter than a series of holes in the second plate. The two plates were fixed together so that the studs entered the holes but did not touch the second plate, the two plates being insulated from each other. Molten Selenium was then poured in to fill the interstices and annealed. Annealing is necessary because in its amorphous state Selenium is not sufficiently sensitive to light, the grey metallic variety only being conspicuously light-sensitive. When therefore the cell is ready to be annealed, it is placed in an oven and heated to a temperature of 180° C. for about five minutes, when the transformation from the amorphous to the greyish variety should be complete. It is then slowly allowed to cool.

The process of annealing renders the cell more sensitive, but to obtain the best results it should be protected from moisture by placing it in a vacuum receptacle, or waxed to a plate of mica or glass. The early forms of cells were often painted over with a transparent varnish to protect them from the damp. Moisture decreases the resistance of the cell; it functions more efficiently when thoroughly dried.

Shelford Bidwell also constructed a form of Selenium cell. This form of cell may be made by taking a small sheet of ground-glass and spreading over it a very thin layer of purified amorphous Selenium by means of a hot glass rod. If now four strands of bare wire, either nickel or platinum, are wound round the plate so that the whole of the surface is covered and two alternate strands are then removed, the other two strands are left separated for the whole of their length by a space equal to the diameter of a wire. The cell can then be annealed in the ordinary way.

Another method of making a Selenium cell is to spread a very thin layer of platinum on a glass plate and then to scratch a zigzag line across the platinum by means of a fine steel point. Selenium in the molten condition is then spread over the platinum layer and converted into the crystalline metallic form by means of heat, the zigzag line dividing the platinum layer into two separate parts or plates. Electrodes can be connected one to each part, the fine line of Selenium acting as a varying resistance when incident light is thrown intermittently on the cell. It may be remarked that the light on this form of cell should fall as nearly perpendicular to the surface as possible to increase its sensitivity. Any parts of the cell left unilluminated naturally increase the resistance and decrease the change in fluctuations of current through the cell.

In its simplest form a Selenium cell originally was made up of brass and mica plates on which Selenium was deposited. Thin, rectangular plates of brass and mica were clamped together alternately in a frame by bolts. Selenium was rubbed over the plates and the whole heated on a sand bath whereby the Selenium

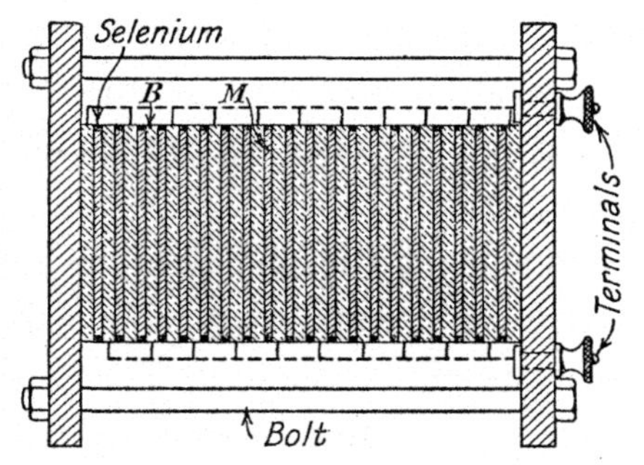

FIG. 31.—Early form of Selenium Cell built up of alternate sheets of Brass and Mica with surface layer of Selenium between the Brass Plates.

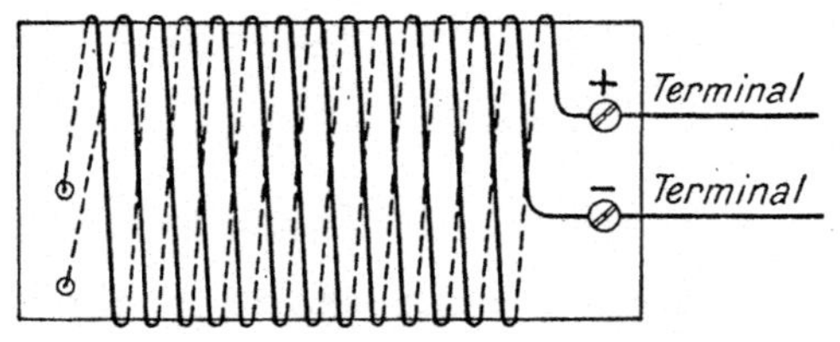

FIG. 32.—Early form of Selenium Cell.

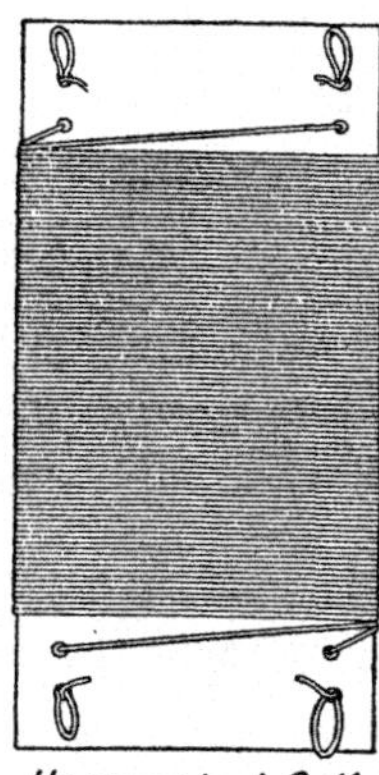

FIG. 33.—Modern form of Selenium Cell (unmounted).

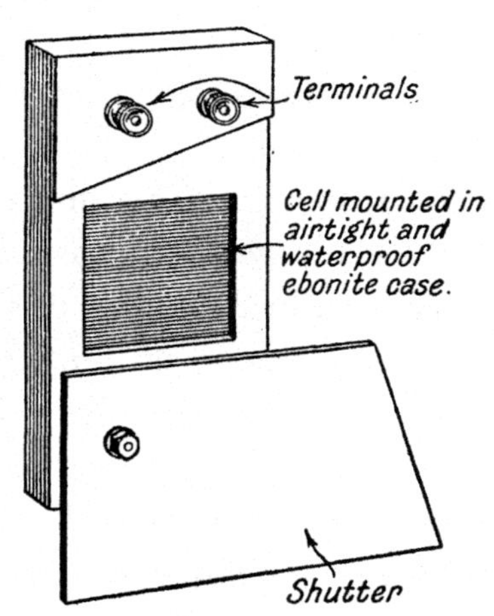

FIG. 34.—Modern form of Selenium Cell (mounted).

melted and filled the narrow spaces between the brass plates. The essential features of the construction are shown in Fig. 31 (see also Fig. 32 for another early form of construction).

The more modern form of Selenium cell consists of a steatite, porcelain, slate, or marble slab on which a thin coat of crystalline Selenium is deposited. On this steatite slab are wound bi-spirally two or more gold, platinum, or copper wires. The mineral steatite is favoured as a dielectric in modern practice, as, apart from its high insulation properties, it is practically non-hygroscopic, which is an important essential in all Selenium cells. The modern type of Selenium cell is shown in Figs. 33 and 34.

Performance and Behaviour. Inertia and Lag.

Recently published results show that the maximum sensitivity of the most efficient type of Selenium cell is that of being capable of detecting an illumination of 10^{-5} metre-candle, a limit comparable with that of the human eye.

The response of a cell to illumination is not the same for all wave-lengths of the visible spectrum; it is greatest in the region of the red rays, the response for the remainder of the spectrum varying as the square root of the stimulus-value.

The disadvantage attending the use of the Selenium cell in television work is the time factor or slowness in recovering its resistance after exposure to illumination, the period taken for recovery increasing with the intensity of the illumination to which it has been exposed. Spontaneous change of sensitivity with lapse of time can largely be overcome by mounting the cell *in vacuo*.

In responding to rapid changes in the amount of

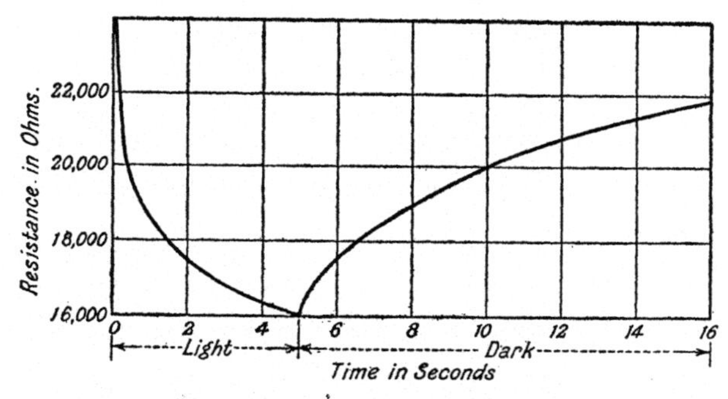

FIG. 35.—General Curve showing Inertia or Lag of a Selenium Cell.

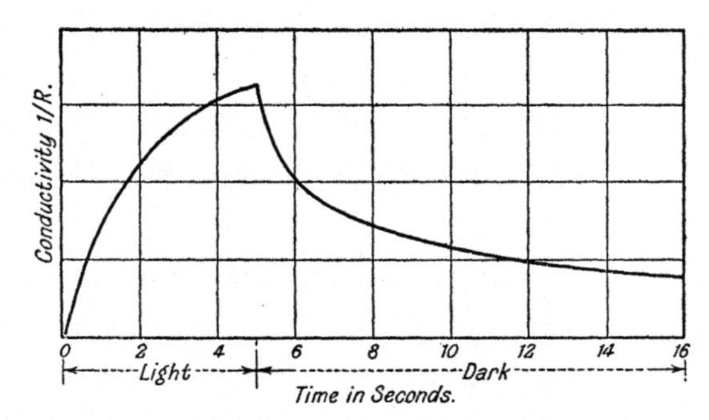

FIG. 36.—General Curve for behaviour of a Selenium Cell.

illumination falling on it, a Selenium cell exhibits a certain amount of inertia or lag, which is detrimental to the instantaneous effects demanded in television work. The higher the resistance of the cell the smaller the inertance. The effects are readily seen by reference to the two charts shown in Figs. 35 and 36.

The higher the resistance of a Selenium cell the greater the ratio of sensitiveness. The sensitiveness of a cell is the ratio of its resistance in the dark and its resistance when illuminated or

$$\text{Ratio} = \frac{\text{dark resistance}}{\text{light resistance}}.$$

The ratio of dark resistance to light resistance of a cell of average type may vary between the limits of 2–1 and 5.

The performance of such a cell may be shown in a general manner by the accompanying chart (Fig. 36). Taking the time exposure to light (in seconds) as abscissa and *RI* as ordinate, we see that the value of the curve rises suddenly at the first, then only gradually rising afterwards until at 4·5 seconds it reaches a maximum. On shutting off the light at this maximum it does not drop at once to a minimum, but takes several seconds to do so.

While the Selenium cell has been used extensively as a light-sensitive cell, it has in later phototelegraphic and television experiments been displaced by the more satisfactory photo-electric cell. The advantage possessed by the latter cell over the Selenium cell is that its action is certain and instantaneous, there being no evidence of lag or fatigue at any time during its period of action.

While Selenium has been discarded by modern inventors on account of the tardy manner in which it recovers its normal resistance value, Dr. Fournier d'Albe holds that the lag in its action is purely a relative term. He contends that there is really no lag in its action, although there is a time interval between the stimulus given to Selenium and the final effect obtained. The process is a chemical one exactly the same as that of an instantaneous photographic plate, the stimulus is given to the Selenium and the chemical action commences at once, and he illustrates the truth of his assertion by referring to his Optophone experiments whereby 600 signals per second were transmitted.

This rate of signalling, however, is not sufficient for television, hence a disadvantage arises in the use of a Selenium cell for the instantaneous responses required; no light-sensitive cell is suitable for television work unless it responds instantaneously to the rapid variation of light and shade; the very fact that a curve can be obtained showing the lagging effect when subjected alternately to "light" and "dark" proves its unsuitability. An ideal sensitive cell should show no curve of lag.

CHAPTER IV

PHOTO-ELECTRICITY AND THE PHOTO-ELECTRIC CELL

General Description.

ALTHOUGH Selenium and Selenium cells have been and still are very much to the fore in experimental television research, they have to a certain extent been superseded by the photo-electric cell. This cell is capable of detecting the light of a candle two miles distant and of stars that would have otherwise been undiscovered but for its action. The flashing of light on its surface, moreover, need only last a millionth of a second. Photo-electric cells of cadmium and the alkali metals—sodium, potassium, rubidium, and calcium—are now used for the photometry of visible and ultra-violet light. Credit for the discovery of the photo-electric cell is really due, in the first place, to Hertz, who in 1888, when carrying out his famous Hertzian waves researches, found that the sparks set up by his apparatus passed more readily when rays of ultra-violet light lay in their path. This phenomenon led, through the work of Hallwachs, Elster, Geitel, and others, to the construction of glass tubes or bulbs from which air was evacuated, for the purpose of primarily demonstrating the action of light in producing electricity. Subsequent developments

Fig. 38.—The Langmuir Mercury Vapour Pump.

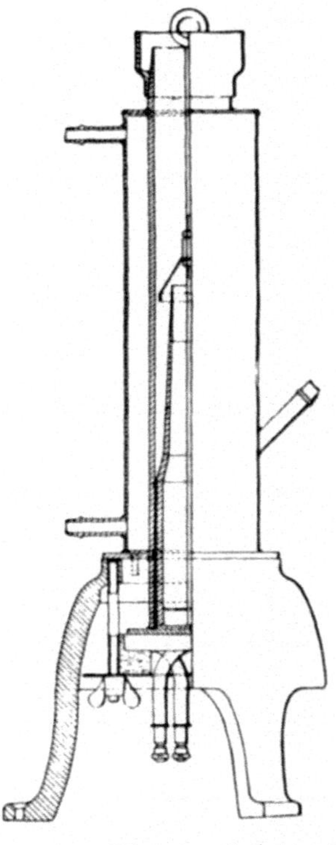

Fig. 39.—Diagram showing Internal Construction of the Langmuir Pump.

[*To face page* 75.

in the design of such glass tubes and treatment with alkali metals brought about the evolution of the photo-electric cell. A cell of this kind depends for its action on the emission of a stream of electrons at the surface of the metal with which it is coated, usually an alkali metal such as rubidium, sodium, or potassium.

A suitably designed glass tube of the form shown in Fig. 37 is exhausted of all air by means of a special air pump. It may be remarked here that the pump

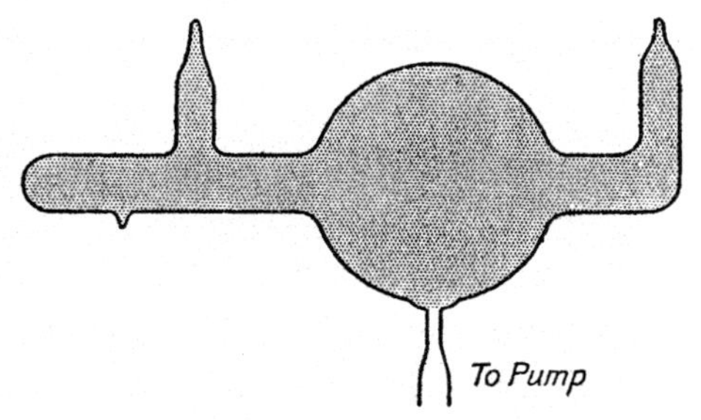

FIG. 37.—Glass Bulb of typical Photo-electric Cell.

that has displaced all others for high vacuum work is the Mercury Vapour Pump first devised by Gaede in 1915, the ordinary mechanical pump being useless for obtaining a high degree of exhaustion. Many improved forms of mercury vapour pump have since been developed; the present-day method of attaining high vacua is by means of the type known as the Langmuir Mercury Vapour Pump, an illustration of which is given in Figs. 38 and 39.

Certain portions of the tube are silvered and thinly layered with a deposit of the alkaline earth used (rubidium, for example). The alkali metal is made much more effective if heated in hydrogen gas at a temperature of 350° C. The hydrides so formed are clear, colourless crystals, and on being bombarded by Cathode-rays become brightly coloured.

It is necessary for the photo-electric cell itself to be kept in a wooden box, light-tight except at one aperture for the admission of light rays that fall on the metallic coating. Freedom from parasitic effects is obtainable when the metal within the cell is prepared and maintained within a vacuum of the highest order.

The current given off by such a cell is extremely small, one-hundredth of a micro-ampere only being obtained when a 100 c.p. lamp is placed at a distance of half a foot from the cell. The lower band of invisible light waves, that is, the ultra-violet rays, produce greater sensitivity in the cell than the ordinary visible rays.

The Electrodes, Anode and Cathode.

Illustrations of a photo-electric cell (The Cambridge Instrument Co.'s type) are given in Figs. 40 and 41.

A fairly high voltage, about 250 volts, is necessary to work the cell of which the points of connection or terminals are marked *A* and *C*. The point of entry of current at *A* is called the *Anode*, and *C*, where the current leaves, is called the *Cathode*.

The axial electric field is not one of uniform value throughout the length of the tube. At the anode *A*

FIG. 40.—The Cambridge Inst. Co.'s Photo-electric Cell.

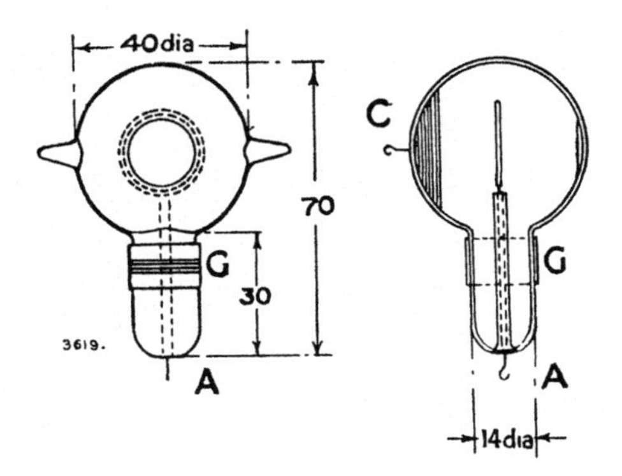

FIG. 41.—The Cambridge Inst. Co.'s Photo-electric Cell.
(Internal Construction.)

[*To face page* 76.

there is a sharp fall of potential dependent on the current passing and the nature of the residual gas. The electric field fluctuates in value in passing from anode to cathode.

It is at the cathode *C* that the electrons originate by positive ray bombardments. The electric field is of high value very near the cathode, but it drops suddenly at the cathode itself.

Other features and phenomena connected with a vacuum or discharge tube are discussed in the next chapter, where the discharge tube is considered from the standpoint of a cathode-ray tube rather than a photo-electric cell, since interesting researches in television have been made with cathode-rays from time to time.

The Two Classes of Photo-electric Cell.

The photo-electric cell, like the wireless valve and the electric lamps employing tungsten filaments, is the outcome of high vacua research. There are two classes of this type of cell, namely, (*a*) the Vacuum type and (*b*) the gas-filled type. These we shall now proceed to describe. In the first-named class, the vacuum cell, the current passing through the cell is that which is liberated by the direct action of the light on the sensitive cathode. In the second class—the gas-filled cell—this current is magnified by the passage of the primary electrons through the gas with which the cell is filled and in which they produce secondary electrons. At the same time, the presence of the gas makes it possible and useful to " sensitise " the cathode still further during preparation, by making it the cathode in a discharge in hydrogen, so that the primary current is greater than it is from an unsensitised cathode. Gas-filled cells are therefore much more sensitive than vacuum cells; the same light may give several hundred times as much current. They are therefore greatly preferable to vacuum cells for all purposes in which sensitivity is of prime importance. Vacuum cells are preferable only when the incident light is to be measured accurately, so that the same light must always give exactly the same current and the current must be nearly proportional to the light.

Potassium in Vacuum Cell of the General Electric Company, Ltd. (of London).

As an example of a vacuum photo-electric cell take that manufactured by the General Electric Company, Ltd., the internal construction of which is shown in Figs. 42 and 43. The shaded portions are silvered, and on them is deposited by condensation from vapour a thin layer of either rubidium, potassium, or sodium. This surface has not been " sensitised " by the Elster–Geitel method of a discharge in hydrogen, the cells are completely evacuated and are not filled with gas. This method of preparation, though it sacrifices sensitivity, gives perfect reliability.

The cell must be enclosed in a light-tight box with a single opening against which is pressed the window

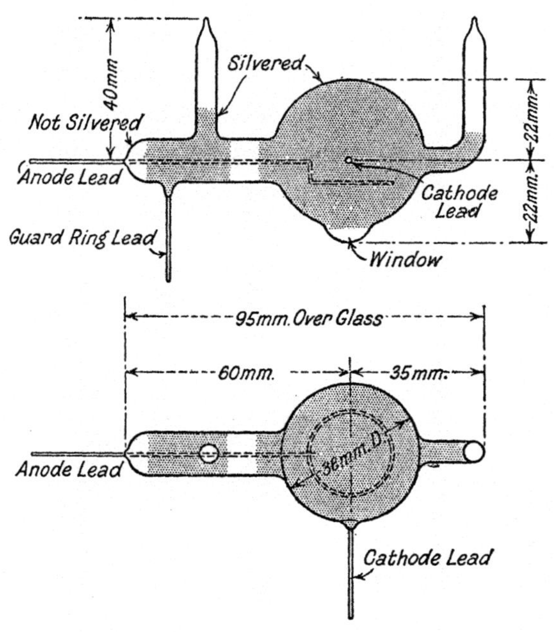

FIG. 42.—Internal construction of the General Electric Co.'s Vacuum-type Photo-electric Cell.

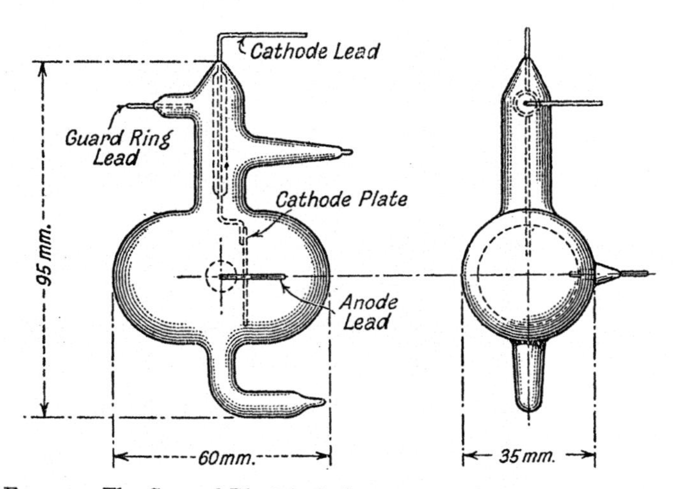

FIG. 43.—The General Electric Co.'s new type of Photo-electric Cell.

of the cell. The cathode is a plate supported in the middle of the bulb as shown in the figure; the silvered surface of the bulb acts as anode.

Usually the cathode is connected to the driving potential, the anode to the apparatus.

The G.E.C.'s Potassium in Argon Cell.

As an example of the gas-filled type, take the

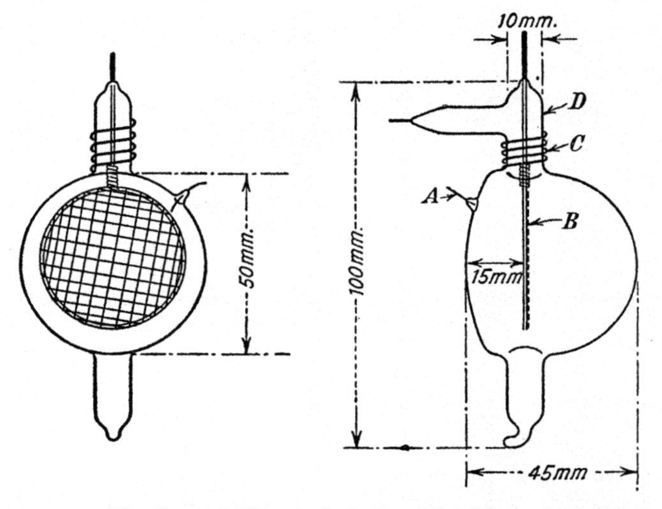

FIG. 44.—The General Electric Co.'s Gas-filled Photo-electric Cell.

General Electric Company's cell shown in Fig. 44. The surface of sensitised potassium is deposited on the silvered cup, *A*, which acts as cathode; the anode is the gauze, *B*. Since the currents are usually much greater than those in vacuum cells, the internal guard ring is omitted; but an external guard ring of wire, *C*, wrapped round the tube, *D*, and connected to earth is sometimes desirable. The cell is filled with argon to a pressure of about 0·15 mm.

The Cambridge Instrument Company's Potassium in Helium Cell.

The internal construction of a typical gas-filled type of photo-electric cell is well exemplified in the photo-electric cell shown in Fig. 41. The cell is of the pattern originally developed at the Clarendon Laboratory, Oxford, potassium being deposited in a sensitive colloidal form on the silvered walls of the glass bulb. The electrons emitted when light falls on the cell are caught on a ring-shaped anode, *A*, and the current, which is proportional to the incident light, is taken by insulated leads to the apparatus. The current can be amplified by using an accelerating potential which causes ionisation by collision in the rare gas helium with which the bulb is filled at an appropriate pressure. Leakage is prevented by guard rings consisting of a strip of tinfoil wrapped around the outer surface of the cell and a platinum wire ring around the inner surface. The two rings are connected together and are maintained at the potential applied. The cell is fitted with a window (20 mm. in diameter, 20 mm. radius of curvature), which reduces to a minimum errors due to non-parallelism of the incident beam of light.

Sensitivity of the Photo-electric Cell.

A 60-watt gas-filled lamp with its filament 15 cm. from the vacuum type cell will give a current of the order of 10^{-8} ampere. The absolute sensitivity (*i.e.* the ratio of photo-electric current to luminous flux entering the window) varies by a factor not greater than 2 from cell to cell with the same active metal. To this light the rubidium cells are, on the average, rather less sensitive, the sodium cells rather more sensitive, than the potassium cells.

The sensitivity increases with the frequency of the light. Thus, for lamps of the same candle-power, the response to a vacuum lamp is somewhat less than that to a gas-filled lamp. The ratio of illumination to photo-electric current varies to a factor of less than 2 between a vacuum and gas-filled lamp, it varies less in this range for the rubidium cell than for the sodium cell, the potassium cell being intermediate. The variation of sensitivity with frequency of the light is very closely the same for all cells filled with the same metal. The sensitivity is independent of the temperature within atmospheric limits.

Variation of Current with Voltage.

Fig. 45 shows the variation of the current through a G.E.C. vacuum type cell with a given voltage between the electrodes. It will be seen that the current is not saturated, even with several hundred volts. The absence of saturation is not due to residual gas and ionisation by collision, but to the form of the active surface which consists of fine drops from the

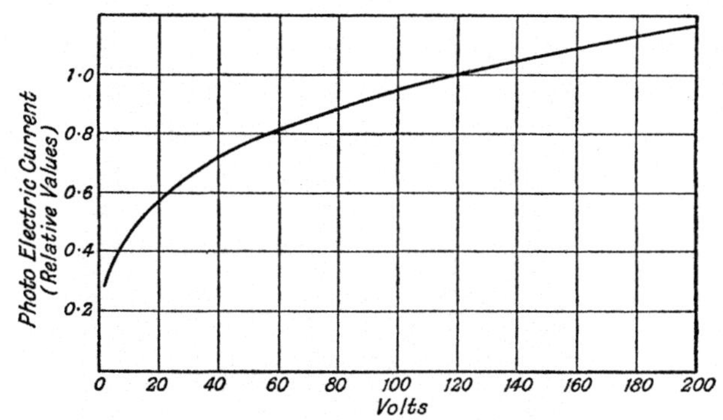

FIG. 45.—Curve of Current-voltage Variation (General Electric Co.'s Vacuum Cell).

interstices between which the electrons have to be dragged. The form of this curve varies somewhat

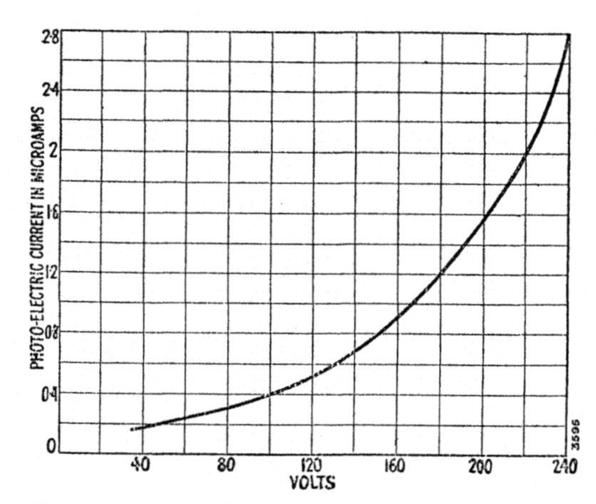

FIG. 46.—Current-voltage Curve of Cambridge Instrument Co.'s Cell.

from cell to cell, but never differs greatly from that shown.

The Cambridge Instrument Company's Photo-electric cell described on p. 81 when exposed to illumination from a tungsten gas-filled lamp (100 candle-power) at a distance of 7 cm. (5 mm. aperture) shows a current-voltage curve like that given in Fig. 46. It will be observed that the sensitivity increases four-fold between 100 and 200 volts, the slope of the

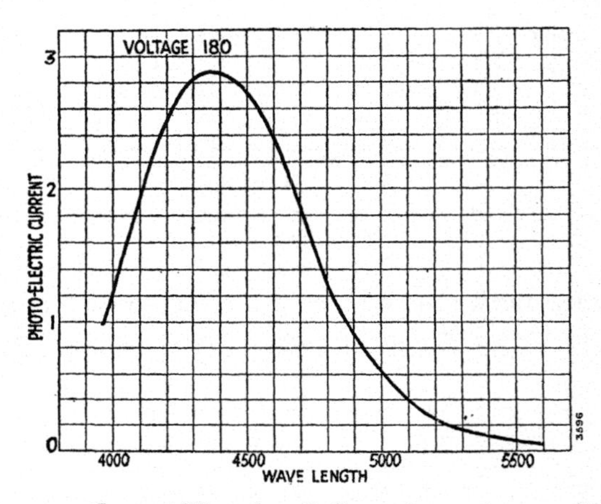

FIG. 47.—Current Wave-length Curve for an Average Cell.

curve at 200 volts being about 2·3 per cent. per 2 volts. It is essential for the voltage to be kept constant.

The ratio of photo-electric current to luminous flux entering the window of the cell varies by a factor not exceeding two from cell to cell; cells can be selected in which this relative factor is practically negligible. The sensitivity varies with the frequency of the incident light, and reaches a maximum value (due to the "selective effect") when the incident light has a certain fixed wave-length. The maximum value varies slightly from cell to cell, but remains constant for a particular cell. Fig. 47 shows a current wave-length curve for an average cell; the ordinates are arbitrary units, corrected for spectroscope deviation and reduced to uniform intensity across the spectrum. Temperature changes within the normal working range (0° to 50° C.) do not appreciably affect the sensitivity.

A resistance of about 50,000 ohms, as shown in Fig. 48, should be inserted between the cathode, *C*, and the high tension supply, in order to safeguard the cell should a spontaneous luminous discharge occur owing to an unusually large potential being accidentally applied. In general, it is safe to use a potential up to about 250 volts, but the actual potential at which glow discharge occurs depends to some extent on the particular cell. An indication that the potential for glow discharge is being approached can be gained from the sensitivity voltage curve, which can be constructed from the figures supplied with each cell. When the sensitivity doubles with an increase of, say, 10 volts, it indicates that the critical value is being approached, and it is not advisable to raise the potential much further. The potential can be conveniently applied by means of ordinary wireless high tension batteries.

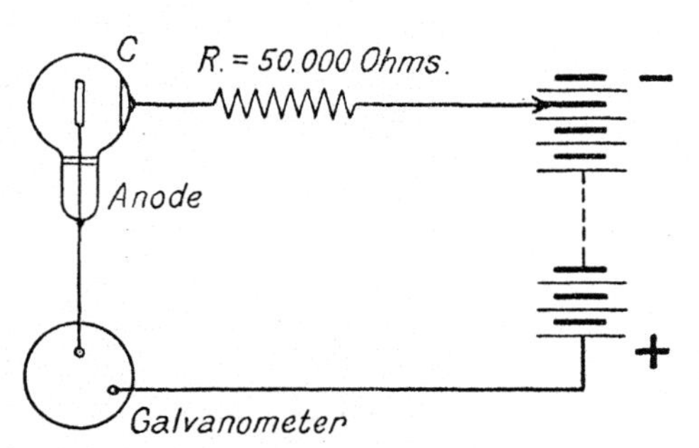

FIG. 48.—Connections for measuring Photo-electric Current.

Amplifying the Photo-electric Current.

Since photo-electric currents are so minute, it is necessary to employ a three-electrode (triode) valve amplifier in all practical applications, as the currents obtained are of the order of one-hundred-millionth of an ampere.

The three-electrode valve altered the whole aspect of the problem of television by giving a means of amplifying the most minute currents to almost any extent, and an attempt was made to use the valve in conjunction with the photo-electric cell for that purpose, but although the three-electrode valve provides an immensely powerful amplifier, the amplification obtainable is limited. A stage is reached when irregularities in the emission from the first valve become audible, and if the signal is not heard before this point is reached further amplification is useless. Enormous amplification is obtainable before this stage is reached, but the response from the cell is very far below the limit. With the ordinary potassium cell, a further amplification of approximately a thousand times would be required.

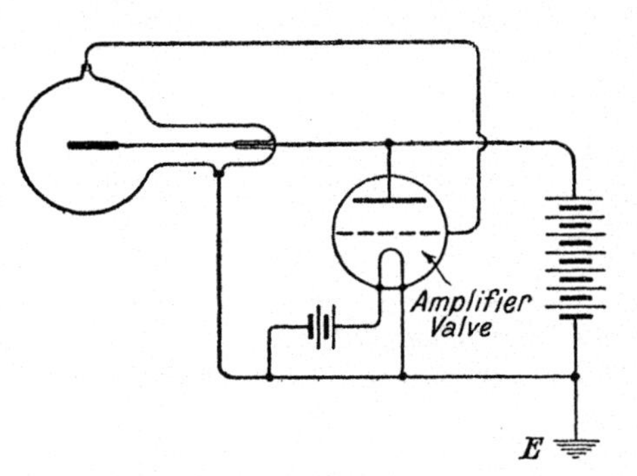

FIG. 49.—Method of connecting Amplifier to a Photo-electric Cell.

Amplifying the Photo-electric Current.

Since photo-electric currents are so minute, it is necessary to employ a three-electrode (triode) valve amplifier in all practical applications, as the currents obtained are of the order of one-hundred-millionth of an ampere.

The three-electrode valve altered the whole aspect of the problem of television by giving a means of amplifying the most minute currents to almost any extent, and an attempt was made to use the valve in conjunction with the photo-electric cell for that purpose, but although the three-electrode valve provides an immensely powerful amplifier, the amplification obtainable is limited. A stage is reached when irregularities in the emission from the first valve become audible, and if the signal is not heard before this point is reached further amplification is useless. Enormous amplification is obtainable before this stage is reached, but the response from the cell is very far below the limit. With the ordinary potassium cell, a further amplification of approximately a thousand times would be required.

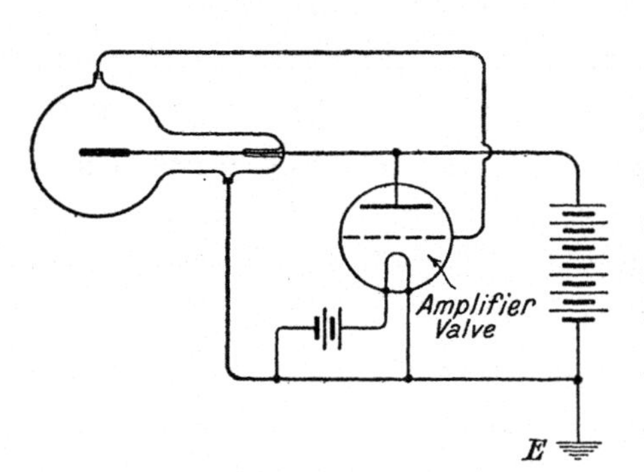

FIG. 49.—Method of connecting Amplifier to a Photo-electric Cell.

ordinary photographic plate, cæsium cells, sensitive in the infra-red region, or lithium and sodium cells, sensitive in the ultra-violet region of the spectrum and quartz cells are also made.

Disadvantages of the Photo-electric Cell.

Although photo-electric, a cell is rapid and instantaneous in its action, it does not always respond satisfactorily to the very small and limited amount of light available where television is concerned.

Shadows may be sent by its aid, for with shadows the light from any powerful source can be directed straight on to the photo-electric cell, but in television, where the objects or scenes concerned reflect only a very small and limited amount of light, the results are poor.

At very large amplifications the intrusion of parasitic noises due to battery irregularities and other causes sets a practical limit to the amplification obtainable. By great care, this limit can be extended, but even then a further limit arises in which the noise due to irregular emission of the valve filament makes its appearance.

Zworykin's Cell.

This is a three-electrode (or four-electrode) type of photo-electric cell in which the stream of electrons emitted is treated in the same manner as that of a thermionic tube and in consequence of its special construction increased amplification of the current is thereby obtained. By combining a three-electrode valve in the same vacuum bulb capacity effects are somewhat reduced. It is claimed that the cell is an improvement over the ordinary combination of photo-electric cell plus three-electrode valve amplifier.

The construction of the cell is illustrated diagrammatically in Fig. 50, where the plate, grid, and fila-

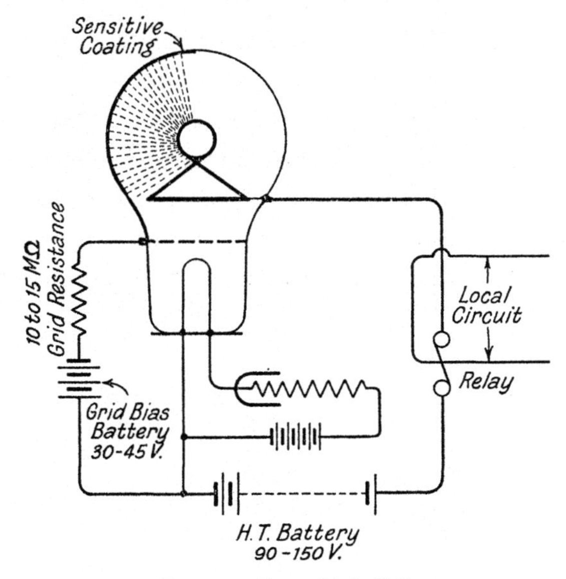

FIG. 50.—Zworykin's Cell.

ment of the valve portion of the cell are shown by the usual symbols. The valve is of the dull emitter type and not gas-filled. The upper section of the glass tube takes the form of a spherical bulb and on the inner side of this bulb a sensitive coating of potassium hydroxide is spread. This coating constitutes the anode of the photo-electric cell and as shown in the drawing is connected to the grid of the amplifier valve below. An electron collector in the form of a wire ring, is connected to the plate of the valve. Both coating and collector are carefully shielded from the valve portion of the cell so that light from the filament may not fall on the coating. The method of connections is given in the figure.

The cell is also made with four electrodes. When so constructed the filament is surrounded by an open mesh grid, which latter is surrounded by a grid of much finer mesh. These three parts are enclosed by the plate, the two grids and the plate being mounted coaxially. In this form of cell the fine-mesh grid is in electrical contact with the coating of potassium while the plate is connected to the ring collector.

CHAPTER V

CONTINENTAL AND AMERICAN RESEARCHES

IMMEDIATELY subsequent to the discovery of the properties of Selenium and the recognition that the human eye might be followed as a model on which to design some practical means for seeing over long distances, several inventors on the Continent and in America constructed apparatus on the lines already indicated. A brief account of the early and later attempts, showing how they contributed their quota towards the progress of research and the ultimate evolution of television will now be given.

Two lines of approaching the problem presented themselves:

Either (1) *by imitating the construction of the eye very closely and having a large number of Selenium cells thereby forming a mosaic of the scene.*

Or (2) *by having one cell only and causing the illuminated elemental areas of the scene to fall in rapid succession on this one cell.*

Ernest Ruhmer's Attempts.

Ernest Ruhmer, whose brilliant pioneer work in connection with wireless telephony is so well known, constructed and attempted to realise television by employing the first-named principle. Stencils of letters or simple objects were placed in front of a wall composed of a number of separate Selenium cells and forming a kind of screen, the idea being to build up the picture in mosaic form. Each cell was exposed to a certain amount of light according to the dimness or brightness of the light that happened to fall on it. Electric currents were sent out by these cells to a distant receiving screen of corresponding pattern and form and these received currents controlled the intensity of a light that illuminated the receiving screen, on which a luminous image appeared as a facsimile of the original. Although Ruhmer continued his investigations from 1901 till 1912, all his attempts at true television failed, chiefly through the inertia of the Selenium cells used and the prohibitive cost of such apparatus due to the number of cells required. He did, however, succeed in the transmission of crude shadows of simple objects such as letters of the alphabet, using 25 Selenium cells to form a crude mosaic of 25 spots of light.

Rignoux and Fournier.

These two French scientists constructed a similar machine to that of Ruhmer; it was intended only to demonstrate a principle, and had no pretensions towards presenting an instrument for television. The transmitter consisted of a wall covered with Selenium cells, 64 fairly large cells being used. From each of these cells two wires ran to the receiving screen, which was constructed with 64 shutters, each shutter controlled from its respective cell, and thus when a strong current from a brilliantly lighted cell at the transmitter arrived at the receiving station its corresponding shutter was opened and light fell on to the corresponding part of the receiving screen. By covering the transmitting wall with large stencils, shadowgraphs of letters of the alphabet and geometrical figures were transmitted and could be recognised. The enormous number of cells, wires, and shutters required made the practical and commercial development of such a scheme quite unthinkable.

The Problem from a New Angle.

Many other workers were attracted by this system of building up a mosaic, but the thousands of cells, shutters, and wires necessary prevented the adoption of any schemes under (1) and an endeavour was made to solve the problem on the principle mentioned under (2). Instead of using a separate cell for each point of the picture, it was proposed to use only one cell, every point of the picture to fall in succession on this single cell, and the varying current from the cell to be transmitted to the receiving station, there to control a point of light traversing a screen exactly in step with the traversal of the image across the cell. The point of light was to be bright at the high lights, dim at the half-tones, and completely out at the black parts of the image, the process to be carried out with such rapidity that, owing to persistence of vision, the eye would see, not a succession of spots, but the image as a whole instantaneously.

A great number of devices in order to achieve the end in view on these principles was invented. It will be possible to describe only a few of the most representative.

Szczepanik's Apparatus.

Fig. 51 shows the apparatus of Jan Van Szczepanik,

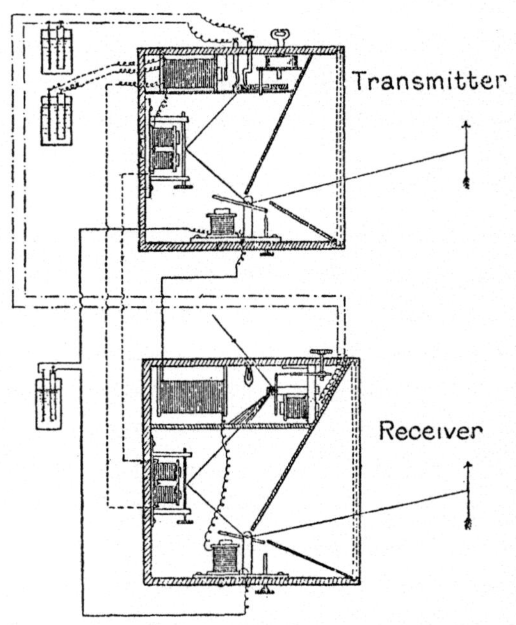

FIG. 51.—Szczepanik's Television Apparatus.

in which the image traverses the cell. At the transmitting station two mirrors are employed, vibrating at right angles to each other, the image being projected by the lens, first on to one mirror and from this mirror on to the second one, which in turn reflected it on to a Selenium cell. The result of the combined motions of the mirrors was to cause the image to travel over the cell in a zigzag path, and the current from the cell was transmitted to the receiving station where it controlled the intensity of a spot of light, this point of light being reflected in a zigzag path across a screen by means of two mirrors vibrating at right angles in the same way as the mirrors of the transmitter.

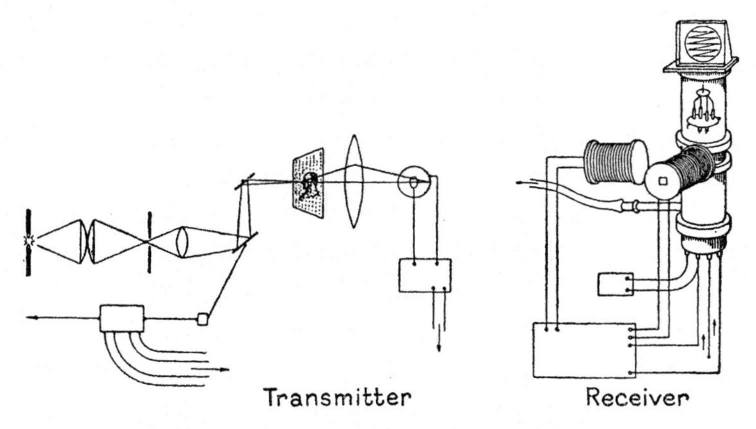

FIG. 52.—Apparatus of MM. Belin and Holweck.

MM. Belin and Holweck and the Cathode-ray.

Fig. 52 shows the apparatus of MM. Belin and Holweck, who employ at their transmitting station two mirrors vibrating at right angles to each other in a manner somewhat similar to that suggested by Szczepanik, these mirrors causing the image to traverse a potassium photo-electric cell. The current from this cell controls the intensity of a Cathode-ray at the receiver, the ray being caused to traverse a fluorescent screen by magnets which are energised from an alternating current transmitted from a motor that moves the mirrors at the transmitter.

Synchronism is obtained by the use of two synchronous motors, one synchronous motor having a low periodicity drives the slowly vibrating mirror, and the other with a high frequency drives the rapidly vibrating mirror. At the receiving station, the currents from these motors pass through two coils at right angles to each other and in close proximity to the Cathode-ray. The combined action of these coils causes the Cathode-ray to traverse the fluorescent screen in a zigzag path, corresponding to a path of the image over the cell at the transmitter.

Using this apparatus in conjunction with a potassium photo-electric cell MM. Belin and Holweck succeeded early in 1927 in transmitting simple shadowgraphs. They were, however, unable to demonstrate television owing to difficulties in obtaining sufficient response from the cell with reflected light.

M. Dauvillier's Apparatus.

M. Dauvillier has conducted extensive experiments along similar lines, using potassium cells and the Cathode-ray for reception purposes and two vibrating mirrors as the exploring device. He describes his difficulties as being chiefly those connected with the potassium electric cell, and, like MM. Belin and Holweck, while he was able to transmit shadows, he found it impossible to transmit by reflected light. Writing in the *Proceedings of the French Academy of*

Science in August 1927 he states: "No object normally illuminated from the exterior diffuses sufficient light to make an impression on the apparatus, and it would have to be a thousand times more sensitive to make it utilisable."

Mihaly's "Telehor."

In Austria, Denes Von Mihaly uses a device which he calls a "Telehor," both to explore his image at the transmitting end and to traverse the receiving screen at the receiving station. The principle employed may be explained by means of Fig. 53. Essentially the apparatus consists of a little mirror, *P*, actuated by an oscillograph, the oscillograph being also given a further motion at right angles, so that the mirror oscillates with movements of vibration of different frequencies about two perpendicular axes—horizontal and vertical. By means of a powerful lens system, a very small image of the object to be transmitted is thrown upon this minute mirror, which is set at 45° to the optic axes. The beams of light after forming the image are reflected through a right angle upon a diaphragm and the motion of the mirror causes the

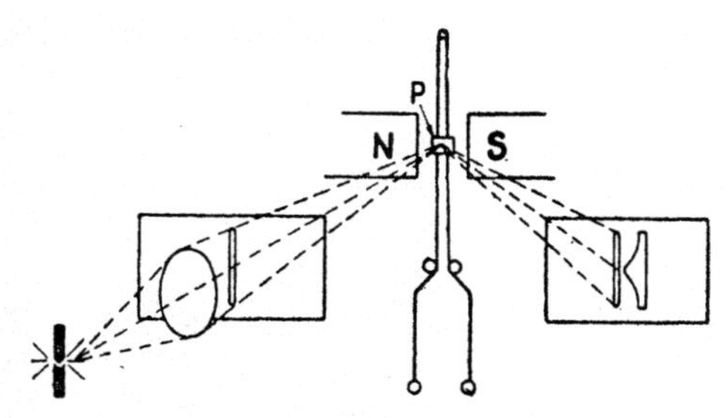

FIG. 53.—Explaining the Principle of Mihaly's Telehor.

7

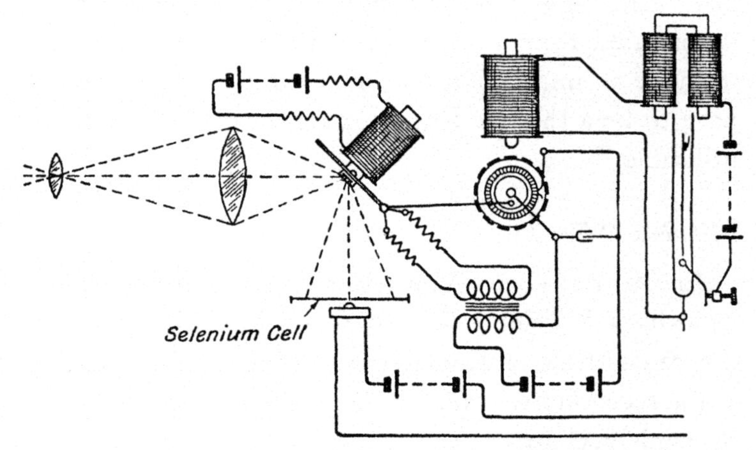

FIG. 54.—The " Telehor " Transmitter Circuit.

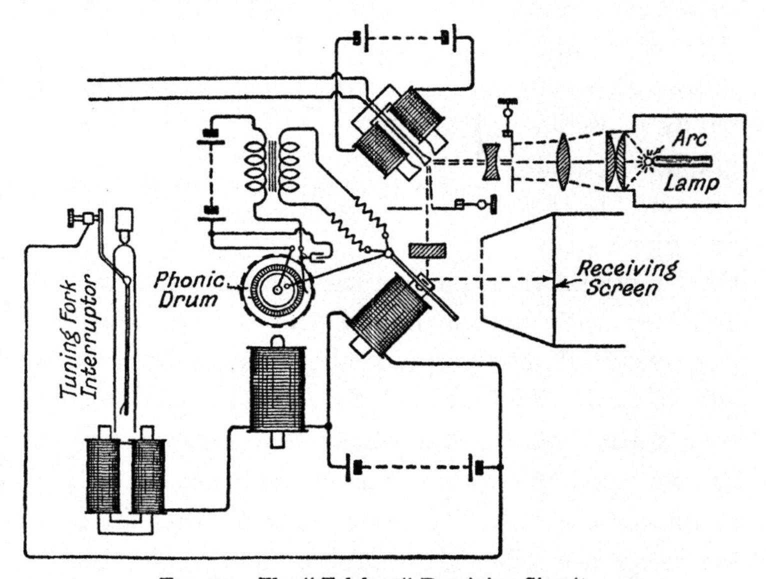

FIG. 55.—The " Telehor " Receiving Circuit.

image to traverse a slot in this diaphragm, behind which is a light-sensitive cell. The diagram of the transmitter circuit is given in Fig. 54.

At the receiving station the current from the cell controls the motion of a mirror mounted upon an oscillograph (Fig. 55). The vibration of the mirror causes more or less light to pass through an aperture in a diaphragm. An image of the aperture is caused to traverse a screen by an optical system similar to and vibrating in synchronism with the transmitting device. Synchronism is obtained by the use of two special devices: the tuning-fork interrupter, and the "phonic drum" of La Cour (Figs. 56 and 57).

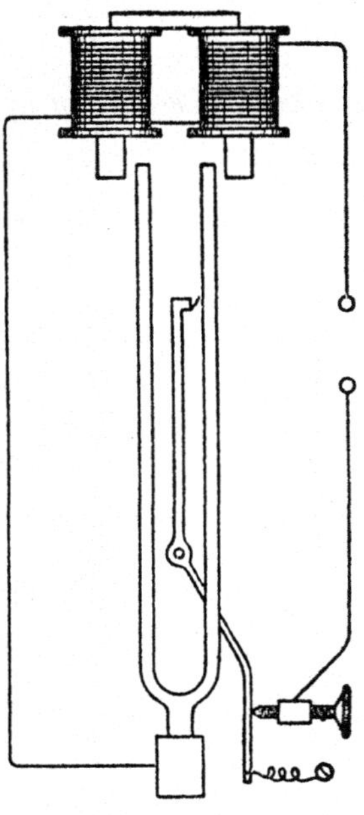

Fig. 56.—The Tuning-fork Interrupter.

Mihaly claims to have transmitted simple geometrical silhouettes, but his optical device, from the description given, appears to be unsatisfactory in principle. No clear image, but rather a blur, would be formed by the combina-

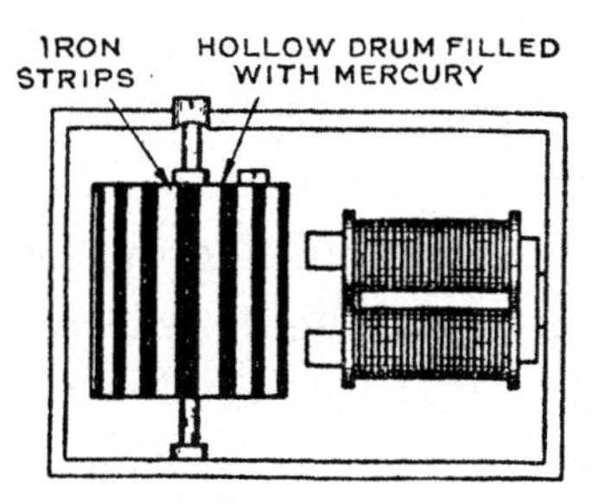

Fig. 57.—La Cour's Phonic Drum

tion described, as the image projected upon the mirror would be out of focus on the diaphragm. This, in conjunction with the inertia of the Selenium, may possibly account for the unsatisfactory results obtained.

Jenkins' Television of Shadowgraphs.

Fig. 58 illustrates the principle on which the apparatus used by Messrs. Jenkins and Moore works.

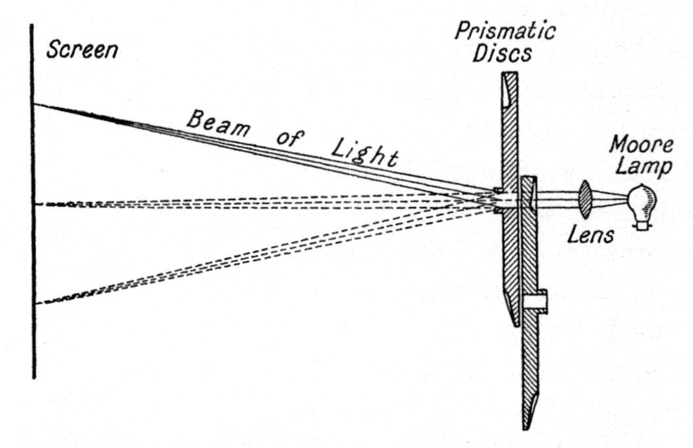

FIG. 58.—Explaining Principle on which the Jenkins' Apparatus works.

In the United States, Mr. Jenkins, whose name is well known in connection with phototelegraphy, has, in conjunction with Mr. Moore, succeeded in transmitting shadows.

In transmitting the shadow of an object, we can use unlimited light, whereas, in transmitting images of the actual object itself, only an infinitesimal light is available. To provide a light-sensitive device capable of responding instantaneously to this infinitesimal light was the outstanding problem to be solved in order that television could be regarded as achieved. This, the apparatus of Messrs. Jenkins and Moore has failed to accomplish. The important feature of the apparatus, however, and one which deserves particular notice is the prismatic disc used for exploring purposes, a remarkably ingenious device invented by Mr. Jenkins. This consists of a circular glass plate, the edge of which is ground into a prismatic section, the cross-section varying continuously round the circumference. As the nature of this device deserves special mention, it will be more fully discussed in the next paragraph.

Jenkins' Prismatic Disc.

In the early attempts at television mentioned in this book, a rotating glass cylinder in which was placed the source of light was usually made use of, the cylinder being rotated and given a longitudinal motion at one and the same time. The photographic film was wrapped round the revolving cylinder so that the rays from the source were always directed radially. As a flat screen placed vertically is used in the Jenkins' system, obviously the employment of a revolving glass cylinder is out of the question and some form of glass prism for bending rays of light is necessary. This latter therefore takes the form of a specially made combination of disc and prism known as the Jenkins' Prismatic Disc. Fig. 59.

The disc itself is made of a mirror glass, the prismatic lens being ground into the face of the disc, the latter thus having its own support on the shaft on which it is mounted.

The prism has its base inward from one end to a point midway round the periphery of the disc, thence it has its base outwards round to the end of the half-section, the slope being gradual from start to finish, first one way and then the other. Since this prism varies in thickness, it is really better in functioning than many single lenses; a beam of light is by its action swept across or oscillated from one side to the other of the picture or screen. The device will be better understood by reference to the diagram. A beam of light passing through such a disc is bent backwards and forwards as the disc revolves, and by using two of these discs at right angles, one to give a lateral movement and the other to give a perpendicular, the image is made to traverse a potassium photo-electric cell at the transmitting station. The current from this cell is transmitted to the receiving station, where it controls the light from the lamp invented by Mr. Moore. This lamp changes its intensity instantaneously in proportion to the current, and its varying light is caused to traverse a screen by a device similar to that at the transmitter.

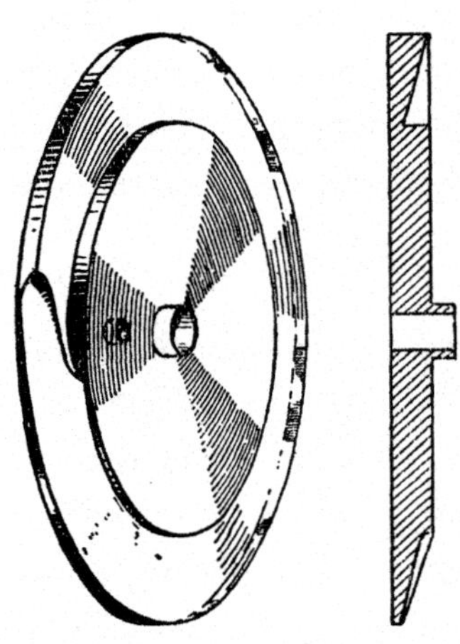

FIG. 59.—Jenkins' Prismatic Disc.

The Moore Lamp.

The Light Source is a special type of electric glow discharge lamp invented by Mr. Moore (Fig. 60), this lamp giving a light spot of great brilliancy. Using this apparatus in conjunction with a Potassium Photo-electric Cell, Messrs. Jenkins and Moore were able to transmit shadowgraphs successfully. It is placed outside the light-tight box or cabinet in which the other receiving parts, such as discs, etc., are contained. The lamp is a modified form of Neon tube the discharge being concentrated in the small hole in the central electrode.

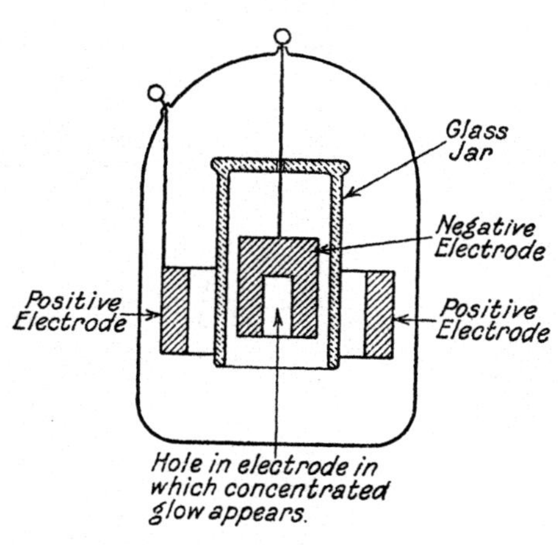

FIG. 60.—The Moore Lamp.

When the received currents are passing they vary the brilliancy of the discharge in accordance with the degree of intensity sent out at the transmitting end. The rays are focussed on to a ground-glass screen after passing through the prismatic discs and so produce the picture.

Method of Operation.

Two prismatic discs are used in transmission, each revolving at different speeds, and one disc is set at right angles to the other, the effect of the light rays being to draw very close lines across the picture. This action is represented in Fig. 61 as a side view or elevation.

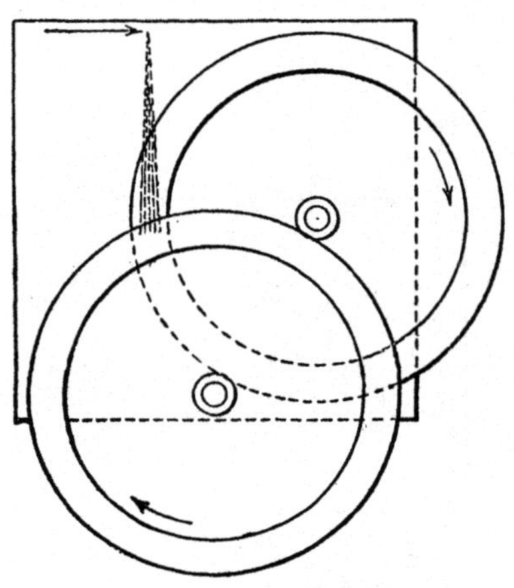

FIG. 61.—Illustrating how the Beam of Light traces out Lines across the Screen.

When a magic lantern is used for the transmission of pictures, it is obvious the picture to be transmitted must be a transparent one, *i.e.* one depicted on a slide, to transmit an actual scene natural or artificial light reflected from the scene must be used.

For transmitting motion pictures, however, the magic lantern is replaced by a motion picture projector and the light rays transmitted by the motion picture are concentrated by the lens projector and then focussed, after passing through the discs, on to the light-sensitive cell.

The operations at the receiving end are very similar. When the transmitted currents are received they vary the illumination of the lamp in the same manner, and these variations on being sent through a rotating disc at the receiving end trace out a series of parallel lines on the receiving screen.

The synchronism of the transmitting and receiving apparatus is effected by means of synchronous motors driven from the alternating current supply-mains, both receiver and transmitter motor being supplied from the same A.C. generator. Jenkins did not synchronise by wireless or by telephone, but took advantage of the A.C. power supply to automatically synchronise his machines.

Dr. Alexanderson's Experiments.

Dr. E. F. W. Alexanderson, of the General Electric Company, Schenectady, New York, has, besides inventing phototelegraphy apparatus, also turned his attention to experiments in television, but lays no claim to having solved the problem.

Early last year (1927), Dr. Alexanderson read a paper before the American Institute of Electrical Engineers, in which he described his television system without, however, giving a demonstration. He explores the image by means of a rotating mirror polyhedron, each mirror being set at a slightly different angle from that proceeding it, the image being cast upon the revolving polyhedron by a lens and reflected upon a light-sensitive cell. To obtain more detail, Alexanderson suggested dividing his image into zones and using a plurality of cells with a corresponding plurality of light sources at the receiver.

His receiving station consisted of a similar optical arrangement to that at his transmitter, light sources replacing cells and a screen replacing the object being transmitted. His fluctuating light he proposed producing by means of high speed shutters controlled by oscillographs.

Synchronism was to be obtained by the use of synchronous motors.

The transmitting apparatus consists of a revolving wheel or drum about $2\frac{1}{2}$ feet in diameter which carries twenty-four mirrors 8 by 4 inches on its 10-inch flanged periphery. The drum is direct-coupled to a motor running at a high speed. There is a lens for focussing the image and a brilliant source of light from which seven beams of light radiate and consequently seven spots of light traverse the picture or scene to be transmitted. In addition, at the transmitting end, seven photo-electric cells are required. The arrangement of screen, revolving mirrors with lenses, and light sources is shown in Fig. 62.

Alexanderson is experimenting with this apparatus by the use of seven distinct wave-lengths, each carrying a crude image. Each of these crude images will be blended at the receiving end into one perfect image of good definition, *i.e.* no blurring of outline.

The reason for using seven light beams instead of one is because of Dr. Alexanderson's contention that it is impossible sufficiently to illuminate a large screen with a single spot of light in the small space of time required in television work. The mechanical inertia of the apparatus militates against the attainment of such an object. The gain in the amount of illumination thus varies directly as the square of the number of light sources, since with seven light sources and seven spots of light 49 times as much illumination will be obtained in the same interval of time. Moreover, there will be 168 light-spot traversals over the screen for every revolution of the mirror drum, since there

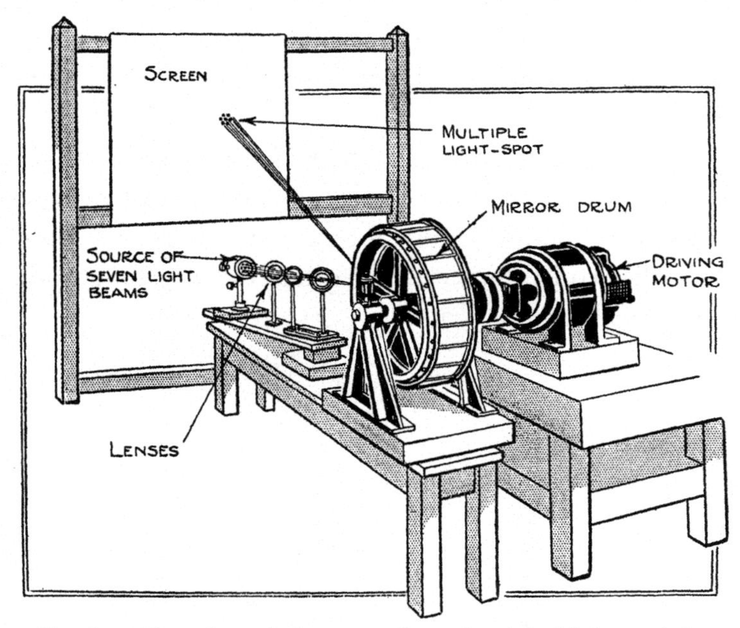

FIG. 62.—Alexanderson's Apparatus (reproduced by kind permission of the *Wireless World*).

are seven light sources and 24 mirrors. A further advantage lies in the fact that seven spots of light will traverse the whole of the screen in one-seventh of the time that it takes one spot to do so.

As the mirrors revolve these seven beams trace seven lines of light simultaneously, at, say, the top section of the screen and then perform the same operation over the next section, and so on until the whole of the screen has been traversed. Each of the seven light-spots traces its own picture and has to be controlled independently from the transmitting end. It is for this reason that the seven photo-electric cells are required. Hence a multiplex radio system of transmission sending out seven different carrier waves should be capable of transmitting seven crude but differing pictures that can be blended into one picture at the receiving end.

The suggested distance apart of these carrier waves is 100 kilocycles, so that with a wave-band of 700 kilocycles a radio channel for television purposes could be utilised on a short wave-length in the neighbourhood of 20 metres.

CHAPTER VI

RESEARCHES WITH THE CATHODE-RAYS

Historical.

THERE are many secrets of nature that have been unlocked by modern science, especially in the realm of the infinitely small, such as those revealed by investigation into the structure of the atom and the electron, the detection of invisible rays, and other recent discoveries. The enlistment of some of these hitherto little-known phenomena into the service of television has been attempted on more than one occasion. For instance, the use of the Cathode-rays in Belin's, Dauvillier's, and Campbell Swinton's forms of apparatus already mentioned. The employment of Cathode-rays together with other aids is exceedingly interesting as the unique characteristics of these rays suggest to the inventive mind many possibilities. A brief glance at their properties and marvellous behaviour should therefore prove of service.

Cathode-rays were first brought to notice when men began to experiment with vacuum tubes and the properties of vacuous spaces generally and in particular the phenomena associated with electric discharges in a glass tube from which air had been evacuated. The advances made in this direction have been due largely to the researches in high vacua by

Gaede in Germany, Langmuir and Dushman in America, Knudsen in Denmark, and Dewer and N. R. Campbell in this country.

The Phenomena of the Discharge Tube.

It is well known that if a closed glass tube with a metal disc electrode sealed into each end, *i.e.* a Cathode plate and an Anode plate (Fig. 63), be connected to a high vacuum pump, after the first reduction of pressure a form of a " spark " discharge will pass which only requires for its production a small applied P.D. between the electrodes. The discharge consists of

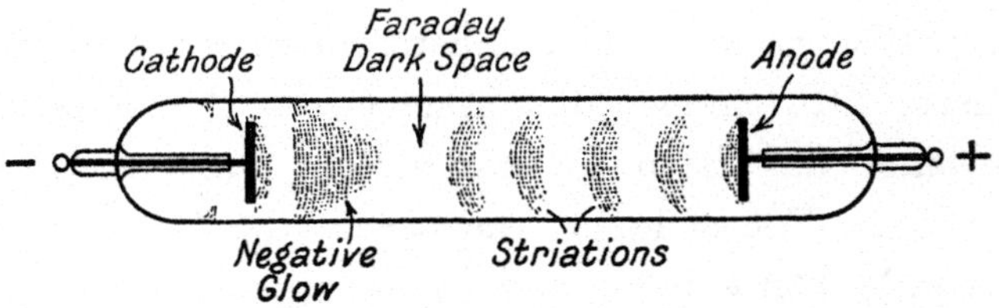

FIG. 63.—Electric Discharge in an Exhausted Tube.

negative charges of electricity—electrons. This discharge soon changes when the pressure of the gas is reduced. Exhaustion of the discharge tube beyond 1/10000th or 100-millionths of an atmosphere causes rapid changes in the discharge phenomena. It is then possible to detect luminous streamers which appear to proceed in a direction from the Cathode and penetrate a short distance along the length of the tube. Still further reductions of pressure approaching 10-millionths of an atmosphere make the walls of the tube show a bright green fluorescence. A suitable object interposed in that part known as the Crookes' dark space causes a sharp shadow of it to appear on the walls of the tube. Goldstein gave this phenomenon or stream of radiation from the Cathode the name " Cathode-rays."

Properties and Characteristics of Cathode-rays.

These Cathode-rays are deflected by a magnetic field, the usual method of applying the field being by means of a short solenoid coil of about 6 inches

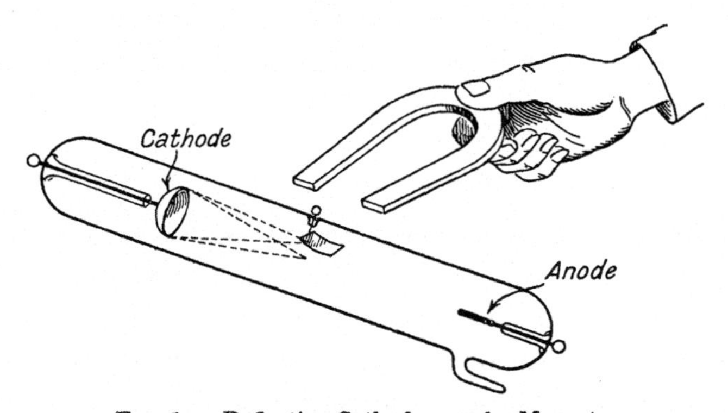

FIG. 64.—Deflecting Cathode-rays by Magnet.

diameter, a fact first discovered by Dr. Fleming. A horse-shoe magnet will produce the same effect.

An interesting experiment may be performed by making use of a concave Cathode which focusses the rays, thus converging them to a point. Now Cathode-rays when focussed on any material substance produce great heat—glass is melted, platinum foil rendered red-hot, etc.—due to the impact of electrons. If therefore a piece of platinum foil is hung inside the tube, but not too near the focus of the rays, an ordinary horse-shoe magnet will deflect the rays so as to bring the focus on to the foil and raise it to a bright red heat (Fig. 64).

The rays are produced in the form of a thin, pencil-like discharge and this pencil of rays can be moved in any direction, either magnetically or electrically. It has no weight and therefore no inertia, and there is no limit to the speed at which it can travel. When this pencil of rays strikes a plate of fluorescent material, a brilliant spot of light is produced so that by using the Cathode-ray in conjunction with a fluorescent screen we can get a receiving device capable of following almost any speed.

The Cathode-ray Oscillograph.

Crookes, Braun, and others improved on the original form of vacuum tube and the Cathode-ray oscillograph tube was evolved from it. The Cathode-ray oscillograph, the outcome of vacuum tube research, is unsurpassed as an inertia-free recorder, having been used for studying the wave-form and purity of high frequency currents like those obtained in line telephony and in wireless practice with conspicuous success.

The likelihood of Cathode-rays serving a useful purpose in experimental television may be appreciated perhaps more readily by glancing at the construction of a typical Cathode-ray tube. Fig. 65 gives a general idea of the form and construction of such a tube. It differs a little from the early forms of tube in that the source of electrons is a hot filament instead of a gas discharge. The inside of the tube at its large

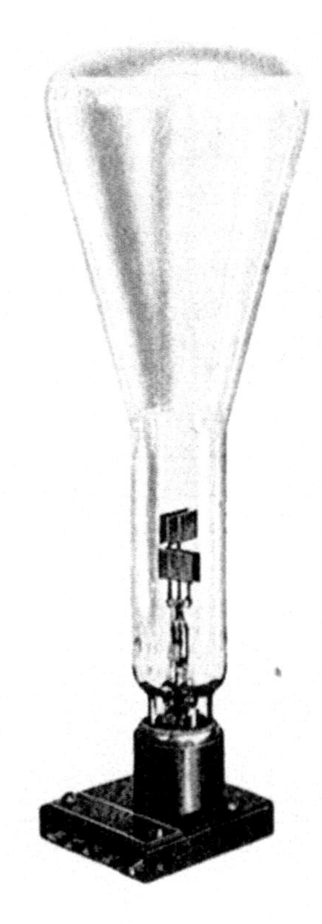

FIG. 65.—The Cathode-ray Tube or Oscillograph.

[*To face page* 112.

end is coated with a fluorescent substance—a mixture of calcium, tungsten, and zinc silicate. The spot where the electrons strike is thus rendered bright and luminous for visual observation. The focussing of the ray on the fluorescent screen is brought about merely by adjustment of the filament current, usually about 1·3 to 1·5 amps. being consumed by the filament when the beam of electrons is correctly adjusted. The filament is heated by a 4- or 6-volt accumulator.

During the last few years there has been extensive development both in the construction and technique of application of the Cathode-ray tube. With present-day apparatus, a single spot of light can be seen by the eye traversing the fluorescent screen at a rate of between 200 and 300 miles an hour.

Subsequent to the discovery of Crookes' tube, Braun's tube, and their modern development—the Cathode-ray Oscillograph—several inventors conceived the idea of utilising the Cathode-ray by deflecting it a great number of times per second for the purpose of producing a series of dots on a screen so many times a second, persistence of vision being relied on to give a defined image. MM. Belin and Holweck's and M. Dauvillier's apparatus and their application of the Cathode-ray in attempting to solve the problem of television have already been outlined in the last chapter; in the present chapter a few other types of apparatus employing the Cathode-rays are now given.

8

Rosing's Attempts with the Cathode-ray.

Acting on the ideas inspired by such a promising ally in field of television research as the Cathode-ray, Boris Rosing, a Russian professor, in 1907 brought out a device of a novel and interesting character. His transmitting arrangements were similar in principle to the others, but his receiving device was very original,

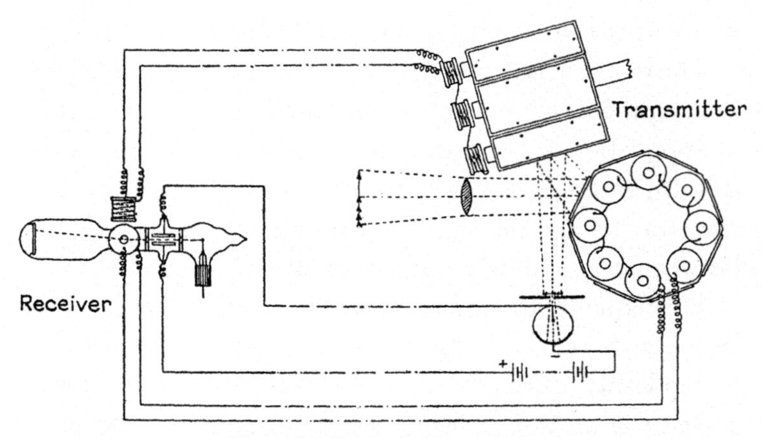

FIG. 66.—Rosing's Apparatus.

as he dispensed altogether with mechanical parts and used instead the Cathode-ray. Rosing used as his transmitter two mirror polyhedrons revolving at right angles to each other, their combined motion causing an image of the object transmitted to pass over a light-sensitive cell (Fig. 66). The varying current from the cell was transmitted to the receiver, and here it passed through a magnet which deflected the Cathode-ray away from an aperture placed in its path, the amount of the ray which passed through being proportional to the current passing through the magnet cell. This ray was caused to traverse a fluorescent screen by currents sent out from coils joined to the mirror polyhedrons. By this means he abolished mechanical inertia at his receiver but not at his transmitter.

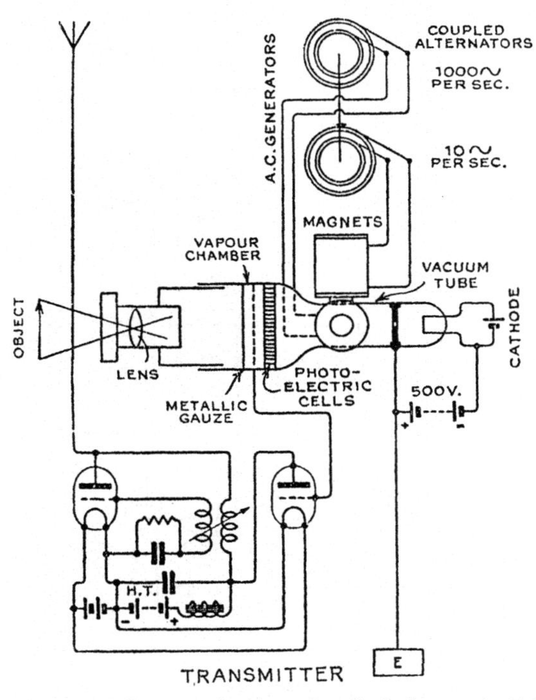

FIG. 67.—Proposed Construction for using Cathode-rays in Television. Transmitting Circuit.

Mr. Campbell Swinton's Suggestion.

In this connection it should be noted that Mr. Campbell Swinton published a letter in *Nature* prior to the publication of Rosing's device, and in this letter he suggested a design of apparatus in which Cathode-rays could be both at the transmitting and receiving ends. And, further, in an address given before the Rontgen Society in 1911, he pointed out that by employing an imponderable agent of extreme tenuity like the Cathode-ray, the difficulty of securing the essential feature of extremely rapid and accurate

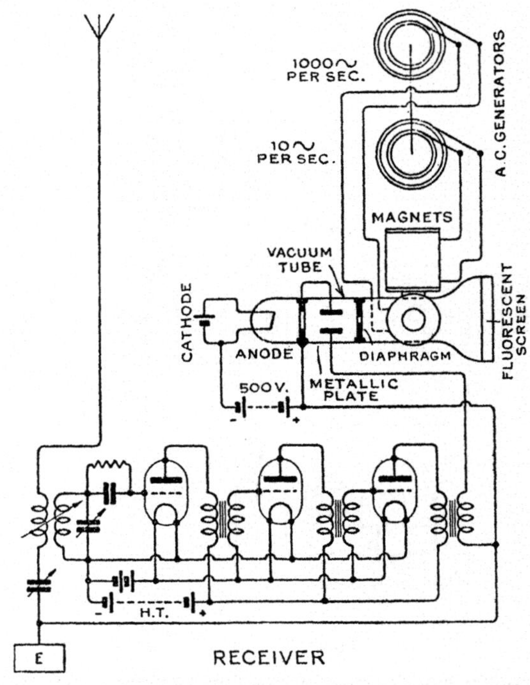

FIG. 68.—Proposed Construction for using Cathode-rays in Television. Receiving Circuit.

motion required with mechanical parts could be removed.

Mr. Campbell Swinton proposed to use at the transmitting and receiving ends of his apparatus a Cathode-ray tube having the Cathode itself heated by one or two cells (Figs. 67 and 68). At both ends the Cathode-rays impinge on screens. At the transmitting end the screen consists of a number of small cubes of potassium (or rubidium) surrounded with insulating material, thereby forming a mosaic. On the one side of this screen the rays impinge, and on its other side is a chamber filled with sodium vapour or any other gas that conducts negative electricity (electrons) more readily under the influence of light than in the dark. The metallic cubes are made of potassium, rubidium, or other strongly active photo-electric element because such elements readily discharge electrons when exposed to light. Upon this mosaic can be projected an image of the scene to be transmitted, the mosaic to be traversed by a Cathode-ray, each cube discharging in turn as the ray travels across it, the discharge being proportional to the amount of illumination received. In the case of cubes on which no light is projected, no action takes place, but in the case of those cubes that are brightly illuminated by the projected image, the negative charge imparted to them by the Cathode-rays passes through the ionised sodium vapour along the line of the illuminating beam of light until it reaches the screen, whence the charge travels to the plate of the receiver. This plate on becoming charged acts on the Cathode-rays in the receiver. The fluctuating current thus produced controls the intensity of the Cathode-ray at the receiver, this Cathode-ray traversing a fluorescent screen in synchronism with the ray at the transmitter, mechanical exploring devices being thus dispensed with at both receiver and transmitter. Cathode-rays can be bent by the action of a magnet, and therefore by synchronously deflecting the two beams of Cathode-rays, one at the transmitting end and the other at the receiving end, the varying magnetic fields of two electro-magnets are set up. The electro-magnets are placed at right angles to each other and supplied with alternating current of widely different frequencies. This causes the two beams to sweep over the picture surface in a tenth of a second (the maximum limit for visual persistence). At the receiving end the screen on which the beam impinges must be a sensitive fluorescent substance, so that variations in the intensity of the rays cast on the screen produces a replica of the picture transmitted.

CHAPTER VII

IMAGES AND THEIR FORMATION

THE results of modern research of radio seeing seem to indicate quite definitely that with the present means at our disposal an image of the scene or object must be formed before the transmission can be attained with any degree of success, and it is the perfecting of the means and devices for transmitting and receiving this image that enables television to be successfully accomplished.

The study of Optics assists very considerably in following out the ideas and principles applied, and therefore a few cursory remarks on the theory of light and its transmission may prove helpful.

It is proposed to recount the principles on which images are formed preparatory to a description of the latest developed system of television so that the practical application may be understood more readily.

Light-Sources and Illuminants.

Before we can see a body, it is necessary for it to be either self-luminous, that is to say, the minute particles of which it is composed must themselves set up the vibrations of waves that affect the eye, or the body itself must be capable of reflecting light from some self-luminous source. Thus the sun is a self-luminous body, but the moon is rendered luminous by reflecting the rays of light that stream upon it from the sun. In television we are mainly concerned with the latter kind of luminosity. Whether we transmit a daylight scene or a scene in a darkened room, it is the rays of light thrown on to the scene, either those proceeding from the sun or from an artificial source of light, that render it possible to transmit the scene.

The sun, a lighted candle, an electric glow lamp, a fire-fly, are objects which are said to emit light, *i.e.* are self-luminous. These self-luminous bodies are not only seen directly by the eye by virtue of the rays proceeding from them, but they render visible all surrounding objects that are not self-luminous, such as the articles of furniture in a room when an electric lamp is switched on. The rays of the latter being transmitted in all directions fall on non-luminous bodies like the walls and furniture and the rays are thrown back or reflected so as to reach the eye.

How Light Travels.

Light, or to speak more broadly radiation, is, according to modern ideas, believed to consist of electro-magnetic waves set up by vibrations of the minute entities that compose its source and transmitted by wave motion through the all-pervading medium known as the ether. Light is definitely known to travel at an inconceivable speed, faster than that of any moving material body. Calculations made during astronomical observations and measurements carried out with optical apparatus confirm this statement. The accepted figure for the speed of light is 186,000 miles per second in all directions. A ray of light from the sun (92,700,000 miles distant) thus takes about eight minutes to reach the earth. Its speed is equivalent to that of a body going round the earth seven and a half times in one second.

The question arises—How does this energy travel so vast a distance? We can only imagine two methods :—

(1) By movement of matter through space (Corpuscular Theory).

(2) By a handing on of energy from point to point (Wave Theory).

Hertz' Experiments in Electro-magnetic Radiation.

The mode by which radiant light and electricity, which are now regarded as one, travel is not known to us for a certainty. Hertz, however, in his classical experiments was the first to show that they are identical when oscillations of electric current are set up across an air-gap; such oscillations produce waves in the neighbouring ether just as a tuning-fork sets up waves of sound in air. Hertz detected these waves by suitable apparatus, investigated their properties, and in particular clearly demonstrated that they are propagated with a velocity of 3×10^{10} cm. per second. This value is the same as that obtained for the velocity of light. Hence electro-magnetic radiation, or the setting up of electro-magnetic waves, has all the properties of light, with the only difference that it is on an enormously greater scale. Therefore we may conclude that light is electro-magnetic radiation set up by the vibrations of the infinitely small particles that constitute the source, and on account of their minuteness send out very much shorter waves and accordingly waves of far higher frequency than the Hertzian waves.

From a certain phenomena in connection with photo-electricity it seems very probable that there is a kind of corpuscular theory of light.

The wave-theory of light, however, has hitherto held the day, and all practical experiment and calculations have so far demonstrated that the hypothesis is a workable one; it serves as a foundation on which the superstructure of light in all its practical applications can rest.

Light Waves.

The wave theory of light perhaps gives rise to some difficulty in accepting the view that light is due to wave-motion. Our everyday ideas are that wave motion will travel round corners, especially in the case of sound, whereas light when it meets an opaque obstacle casts a sharp shadow. This is explained by the difference in length between a sound wave and a light wave. Sound waves are ordinarily a metre or so long, light waves only one twenty-thousandth of a centimetre. If we could only have the objects enormously large in the case of sound waves and extremely small in the case of light waves, a more accurate idea and comparison could be conceived. For example, when a church bell is sounded and a house is the obstacle, a very marked sound shadow is formed, the sound being much less intense if the hearer stands with the house between him and the bell. This may be observed in a still greater degree if a shadow is formed by a hill to the sound waves from a big explosion. On the other hand, if the shadow thrown by a small sharp-edged object placed in front of a brilliant source of light be examined by a fairly high power eye-glass, it will be found that a little light does get behind the edge.

The shadow of a needle or a hair when light from a single point or a single narrow slit is incident upon it is found to be, not a fine black shadow, but, on the contrary, a shadow with curiously fringed edges and with a line of light right throughout the very middle of the shadow, which is caused by the light waves passing by it, spreading into the space behind, and meeting there.

This is the chief feature about light-waves, namely, their very small length; but even the shortest of them differ in size. In consequence of this latter feature, some are so small that the human eye cannot detect them—the mechanism of the eye is not adapted for their frequency of vibration to effect the sensitivity of the optic nerve and thus convey any impression to the brain without artificial aid. Hence there are visible and also invisible rays. This will be referred to more fully later on.

The difference in size of the waves is called their wave-length, that is, the length from the crest of one wave to the crest of the next (Fig. 69), and since the creation of short waves involves a greater frequency of vibration or movement than the creation of long waves, it is now becoming the fashion to measure all physical phenomena involving wave motion in frequencies. The velocity or speed of light, however, whether the waves are short or long, remains the same,

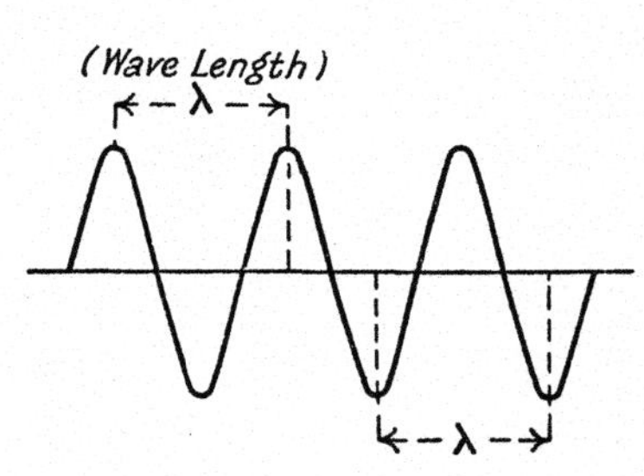

FIG. 69.—Wave-length.

and the equation—

$$\text{frequency} = \text{velocity/wave-length}$$

holds good in all cases.

Lights and Shadows.

If an object intercepts the rays of the light proceeding from any source, we obtain what is known as a shadow

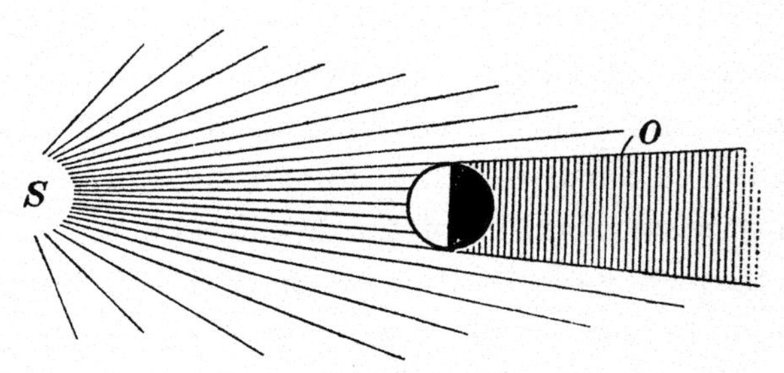

FIG. 70.—A Shadow Cone.

of that object. This result is not desired in true television. For example, let *S* denote a source of light and let an opaque body be placed in the path of the rays of light proceeding from *S* (Fig. 70). Since the rays that fall directly on the opaque body are blocked and those that just pass it are not bent, it follows that a *darkened* cone, *O*, extends beyond the opaque body called a shadow cone and any point within this cone receives no light from *S*. A screen placed at right angles to the axis of the shadow cone will show a well-defined shadow of the object. Very sharp shadows are formed when a naked arc light is the source.

If the source of light is large in comparison with its

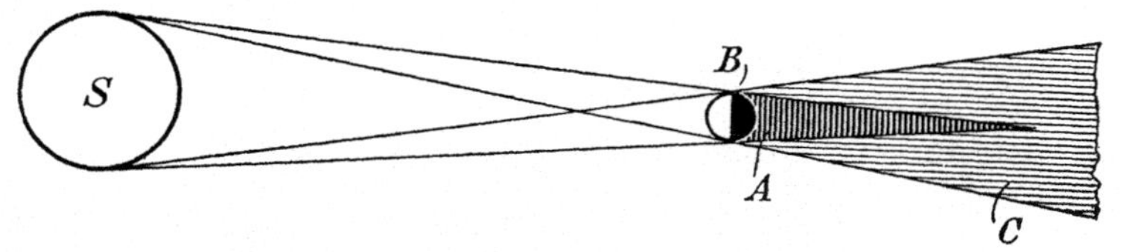

FIG. 71.—Umbral and Penumbral Shadow Cones. (A, Umbral; B, Opaque Sphere; C, Penumbral.)

distance from the object, the rays from every point on the source go to form a separate shadow cone from the object, and it is only the space common to all these shadow cones which is free from light. Compare the sections of the shadow cones *A* and *C*, termed the *umbral* and *penumbral*, respectively, thrown from an opaque sphere, *B*, by light from opposite points of an extended source, *S* (Fig. 71).

The foregoing accounts of Lights and Shadows, although perhaps given in rather an elementary form for the benefit of readers unacquainted with optics, will enable us to see clearly the difference between true television and the systems that have already been demonstrated as such. For true television, we need to have an *image* formed of the scene to be transmitted. The light from some source must be reflected from all points on the scene and be brought by means of a lens to form an image. True television means the transmission of the image of an object with all gradations of light, shade, and detail, so that it is seen on the receiving screen as it appears to the eye of an actual observer. Images may be formed either by one or more lenses or by means of mirrors, as we shall see in the following paragraph.

Refraction of Light Rays.

Light is assumed to travel in straight lines in any medium of uniform density, such as air, glass, water, etc., but in travelling from one medium to another it suffers deviation on account of its refraction at the surface of the separation of the two media. Let *AO* (Fig. 72) represent a ray of light passing through air and incident at *O* on the surface of a piece of glass, and let *OB* represent the refracted ray. Then the angle *AON* is called the angle of incidence and *BON′* the angle of refraction, where the straight line *NON′* represents the normal to the surface at *O*. It is for this reason that an image can be formed by a lens.

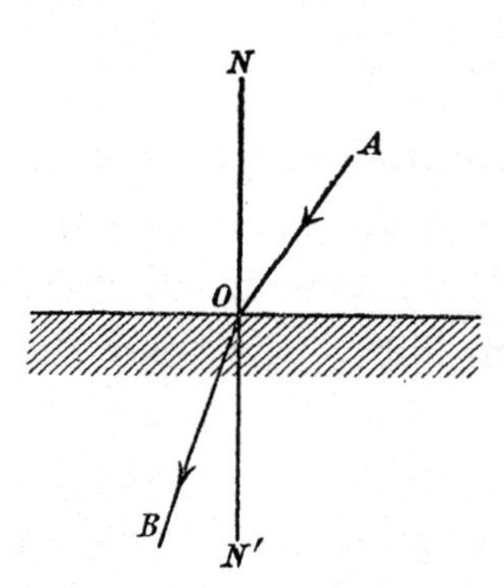

Fig. 72.—Incident and Refracted Ray. (In Refracted Light the Angle of Incidence is equal to the Angle of Reflection.)

IMAGES AND THEIR FORMATION

Images.

When an object or body is seen by the eye, what is known as an image of the object is thrown on the retina where the nerve-centres convey the impression to the brain. It is this image or concentration of all the light rays proceeding from the object that produces vision. Hence in television before a scene or object can be transmitted any distance, an image of it must be formed, otherwise we get a shadow of the object and the successful results achieved in television are due to this transmission of images, not shadows, of the scene or object. Shadows are easily transmitted, as we saw in Chapter V. The best method of forming an image for television purposes is to use either a mirror or a lens.

Mirrors and Lenses.

If we have a mirror with a surface that bulges out, it is called a convex mirror (Fig. 73). If it is hollowed

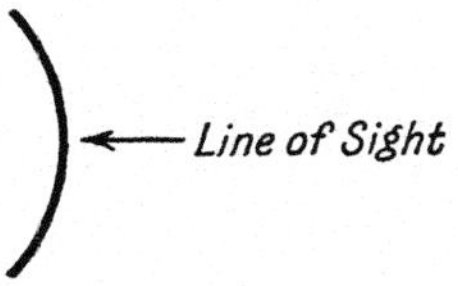

Line of Sight

FIG. 73.—A Convex Mirror.

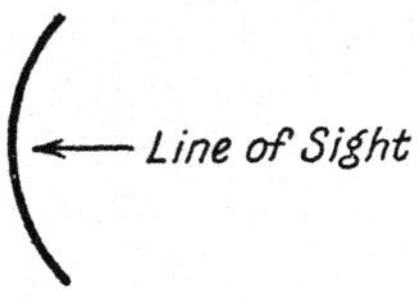

Line of Sight

FIG. 74.—A Concave Mirror.

out, it is called a concave mirror (Fig. 74). A convex mirror will cause a divergence of the rays; they appear to come from a virtual focus from behind the mirror

(Fig. 75), but no real image is formed. If a concave silvered mirror be placed in a beam of light from any

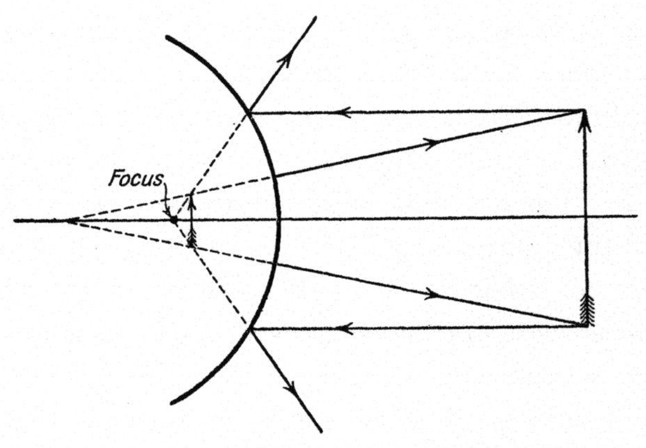

FIG. 75.—Image formed by Convex Mirror.

source, it will cause the rays to be focussed at a point in mid-air (Fig. 76), forming a real image. With a

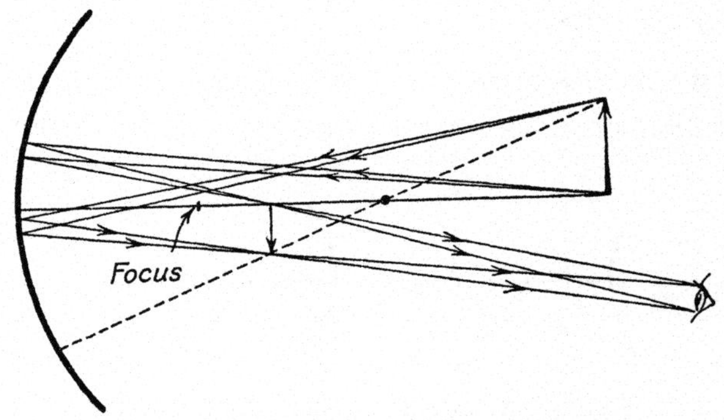

FIG. 76.—Image formed by Concave Mirror.

piece of glass that is thicker in the middle than at the edge—a convex lens—the waves or rays converge to a focus and a real image can be formed. With a piece of glass that is thinner in the middle than at the edges—a concave lens—the effect is just the opposite, the wave will emerge as a bulging wave as if diverging from some virtual focus and no real image is formed.

A lens may thus be regarded as a combination of two refracting surfaces either of which may be convex, plane, or concave.

If we consider light to be transmitted by means of wave-motion, it is quite easy to understand many of the laws and phenomena that are connected with it. For example, the reflection of light from a plane or curved surface that is polished is simply a re-bound of the waves that impinge on the surface. Light travels more slowly in glass than in air, so that in consequence it follows quite simply that if the waves strike obliquely against the surface of a glass, that part of the wave-front that strikes the glass first will go more slowly after entry and the other part which is going on a little longer time in air gains on the part that entered first, so that the direction of the wave-front is changed and the line of march is also changed. This may be made clear by the aid of Fig. 77.

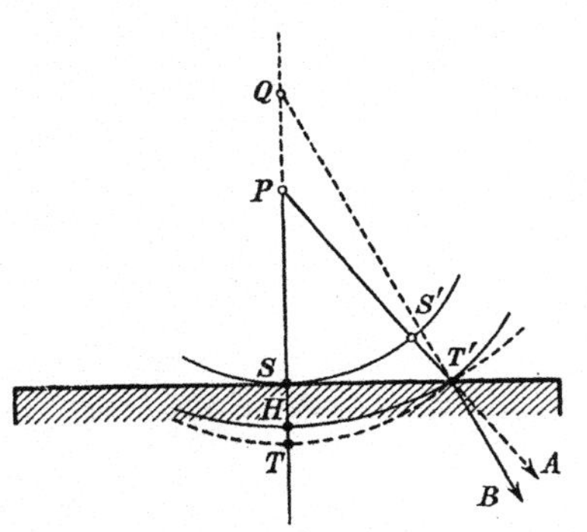

FIG. 77.—Illustrating Refraction of Light.

If waves of light from *P* strike against the surface of a thick glass plate a wave reaches *SS'*. A little later

9

it would (in air) reach TT'; but it has struck a denser medium (glass) and the part of the wave that enters first will only reach H, or two-thirds of the distance.

A set of arcs can be described by means of compasses to represent the various wavelets, the arc in each case being made only two-thirds of the distance that the wave of light would have had to go if after passing the surface it could have gone on to TT'. The overlapping wavelets build up the new wave-front HT',

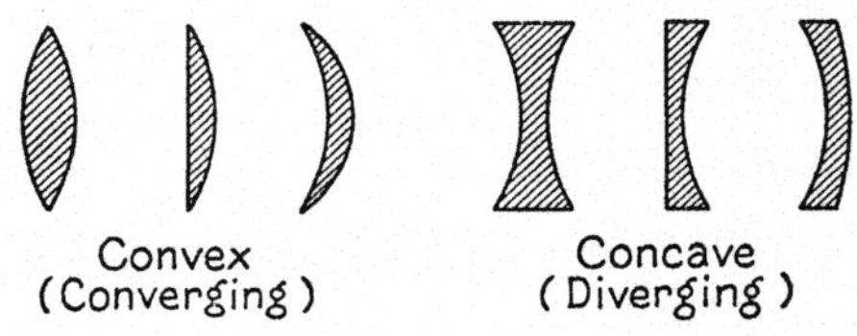

FIG. 78.—Various Types of Lenses.

which is a much flatter curve having its centre at Q. The wave therefore will appear to proceed from this point. Again, the wave-front that has travelled in the direction PT' has its direction changed—instead of going along $T'A$ in a straight line it takes the course $T'B$ as if it had come from Q. This sudden change in its path caused by a ray of light entering a denser medium is known by the name of *refraction*.

This property of refraction can be used to converge light to a focus. For example, if we take a piece of glass that is thicker in the middle than at the edges then we can use it to form a real image.

Glass is the medium used in focussing the object and

forming an image for television purposes, and when made in the form of a solid with spherical surfaces, *i.e.* either a convex or concave lens, is capable of converging or diverging the rays of light passed through it. In television, since the rays have to be converged, convex lenses only are necessary. Fig. 78 shows sectional views of the forms that lenses may take.

CHAPTER VIII

THE BAIRD TELEVISOR

THE systems described so far have in a few cases enabled shadows to be transmitted, but all have failed to produce true television, which was first attained by the apparatus invented by Mr. J. L. Baird, and is now in practical operation. The apparatus designed for making use of the Cathode-rays is perhaps the most interesting, and while a system of television based on the application of Cathode-rays is most fascinating in its possibilities, up to the present time no practical success has been achieved by it, the difficulty of obtaining a sufficient response from the light-sensitive cell in this, as in other systems, preventing true television from being devised. Mr. Baird, using a purely mechanical exploring mechanism, succeeded some three years ago in publicly demonstrating the transmission of outline images by wireless between two separate machines, and subsequently, in January 1926, he demonstrated true television to members of the Royal Institution, real images with gradation of light and detail being transmitted. This was the first demonstration of true television ever given.

Outline of Principles Employed in Baird's System.

The principle followed by other inventors has been

adopted in this system, namely, that of rapidly traversing an image of the object or scene to be transmitted over a light-sensitive cell in a series of closely-drawn parallel paths. The picture reproduced is therefore one made up of fine parallel lines.

At the transmitting end, the light-sensitive cell used is the principal agent in effecting the transmission of the picture. The light proceeding from a brilliant source is reflected from the picture surface and focussed by means of the projection lens and through revolving discs on to the light-sensitive cell. The special type of light-sensitive cell used by Mr. Baird gives an instantaneous effect, and it is possible with such a cell to send a figure, a picture, or a series of moving pictures, like a scene made up of moving people, in rapid succession. The finely-drawn lines of light are swept across the picture by means of revolving discs having lenses, apertures, etc., which will be described in detail later on in this chapter. The varying gradations of light and shade of the picture, object, or scene alter the intensity of the rays reflected from it, which are focussed on to the light-sensitive cell, thereby causing electrical current variations to be given out by the cell which vary in strength in accordance with the light variations and are by this means transmitted to the receiving apparatus.

The variations in light intensity from the picture as they fall on the cell produce variations in electric current just as variations in speech uttered into a telephone transmitter produce variations in current along the line wire and actuate the diaphragm of the receiver at the distant end. In the case of television, the varying currents set up by the light-sensitive cell are sent to that portion of the transmitting set which throws them into the ether. These currents on being received by the receiving set vary and control the light from a lamp placed behind an arrangement of revolving discs similar to that at the transmitting end. The

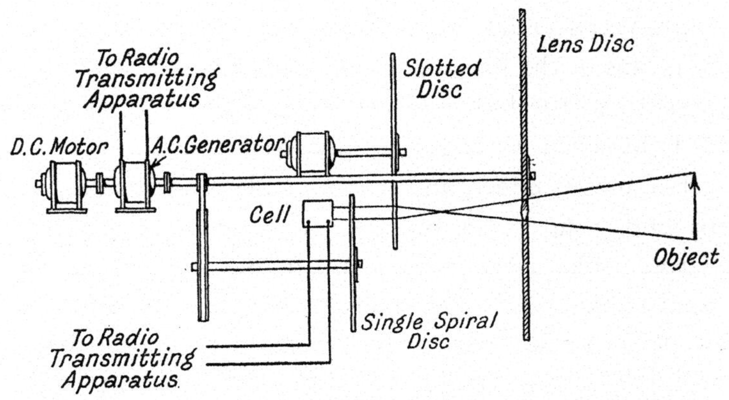

FIG. 79.—Diagrammatic Sketch of Transmitting Arrangements.

principle as applied in actual working at the transmitting end is shown in diagrammatic form in Fig. 79.

Original Apparatus—(a) Transmitting End.

The apparatus used in the first of these demonstrations is now in the South Kensington Science Museum, and consists at the transmitting end of a roughly constructed disc of cardboard containing 32 lenses in staggered formation (see Fig. 80) mounted on a shaft. Behind this and mounted on the same shaft are two additional discs, one with a large number of radial slots

(Fig. 81), and behind it another disc with a single spiral slot (Fig. 82). A sketch showing how these discs are mounted was given in Fig. 29, Chap. II.

The essential parts at the transmitting end are therefore :—

(1) A revolving lens disc.
(2) A revolving slotted disc.
(3) A revolving disc with single spiral slot.
(4) A light-sensitive cell.
(5) Object or scene.
(6) Source of light.
(7) Radio transmitting set.

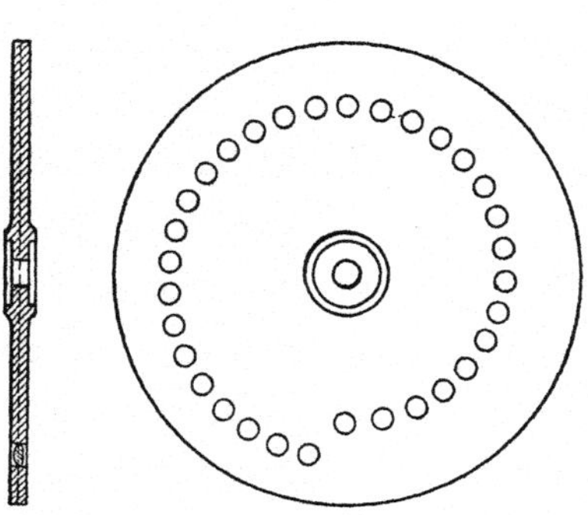

Fig. 80.—The Lens Disc (32 Lens).

The revolving lens disc (Fig. 80), in front of which is placed the object, is the principal item of apparatus. This disc, which is provided with a single spiral of 32 convex lenses, is rotated at a high speed—800 revs. per minute, thus causing a series of images of the object or scene to pass across the aperture to the light-

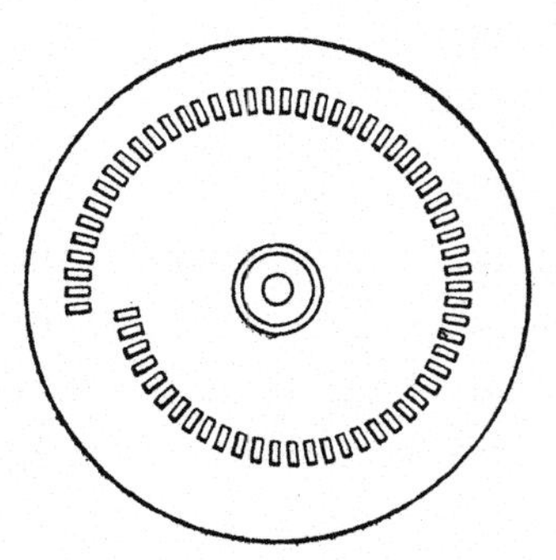

Fig. 81.—Disc with Radial Slots (64 Slots).

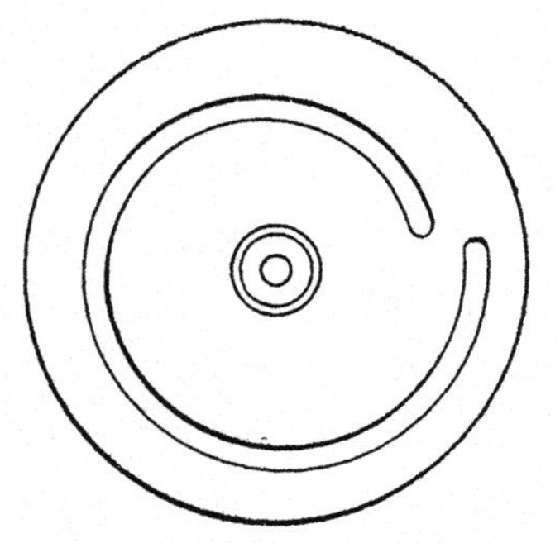

Fig. 82.—Disc with Spiral Slot.

sensitive cell. Before reaching this aperture the light is broken up by the slots in the second disc, which revolves at 1000 revs. per minute. The effect of the disc having the spiral slot is to give a backwards and forwards motion to the slot admitting light to the cell, and thus divides the image into a greater number of strips. Without the use of the disc with the spiral slot there would be only one strip for each lens. By using this disc any required number of strips may be obtained by the use of only a few lenses.

The two terminals of the photo-electric cell are connected by insulated leads to the radio transmitting apparatus; alternatively they can be connected to an ordinary twin pair telephone line.

(*b*) *Receiving End.*

The principal parts of the receiving end are almost identical, and Fig. 83 gives a diagrammatic sketch of this arrangement. They are :—

(1) A revolving lens disc.
(2) A revolving disc with spiral slot.
(3) A glow discharge lamp.
(4) A ground-glass screen.
(5) A radio receiving set.

It will be observed that behind the second disc is the lamp that is lighted by the receiving current, its position corresponding with that of the light-sensitive cell at the sending end. The variations of light intensity fall on a ground-glass screen, showing a reproduced image of the object or scene transmitted.

There is no slotted disc at the receiving end; its presence is quite unnecessary.

To obtain synchronism, two motors are employed—a direct current motor which supplies the driving power and an alternating-current generator running at 500 cycles per second which sends out a synchronising signal. The alternating current from this generator and the fluctuating current from the cell are superimposed upon a carrier wave sent out and transmitted

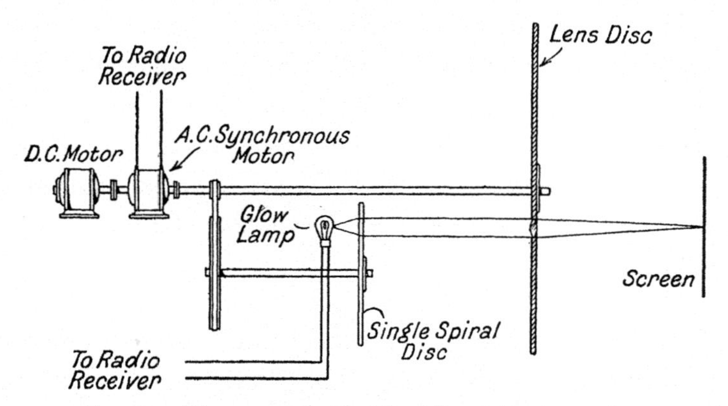

FIG. 83.—Diagrammatic Sketch of Receiving Apparatus.

to the receiver. At the receiver the two currents are filtered out, the alternating current after amplification is used to control the speed of a synchronous motor directly coupled to the shaft of the D.C. motor driving the receiving apparatus. Synchronism is obtained approximately by adjusting the D.C. motor, the A.C. motor being used to prevent hunting. At Mr. Baird's demonstration two separate transmitters and receivers were used for simplicity, one set for synchronising and one for Television.

The fluctuating current, after amplification, controls the light of the lamp, which is a glow discharge lamp of the Neon type (Fig. 84). The light from this lamp is caused by the rotation of the spirally slotted disc, and the lens disc, to traverse the screen exactly in step with the traversal of the image over the cell at the transmitter.

In criticising this original form of Mr. Baird's apparatus, the first point which appears obvious is that a limit would arise in endeavouring to obtain a large finely-grained image. Mechanical considerations would prevent the discs from revolving beyond a fixed speed of possibly 3000 revs. per minute as a maximum, whereas the use of the Cathode-ray gives us an exploring device without mechanical limits of any sort.

Mr. Baird, however, in his patent No. 265640 gives a method whereby this mechanical limitation may be overcome. His method consists in using what is described as an optical lever. He uses a succession of exploring devices, each device exploring the moving image of the one preceding it, so that the speed of traversal of the image is doubled with each operation without increasing the mechanical speed. It is an application to television of the principle of relative motion. Fig. 85 is a drawing reproduced from the specification which indicates one of the methods of applying this principle.

The two systems of lensed discs revolve in opposite directions, so that the motions they give to the image

FIG. 84.—Neon Lamp.

[*To face page* 138.

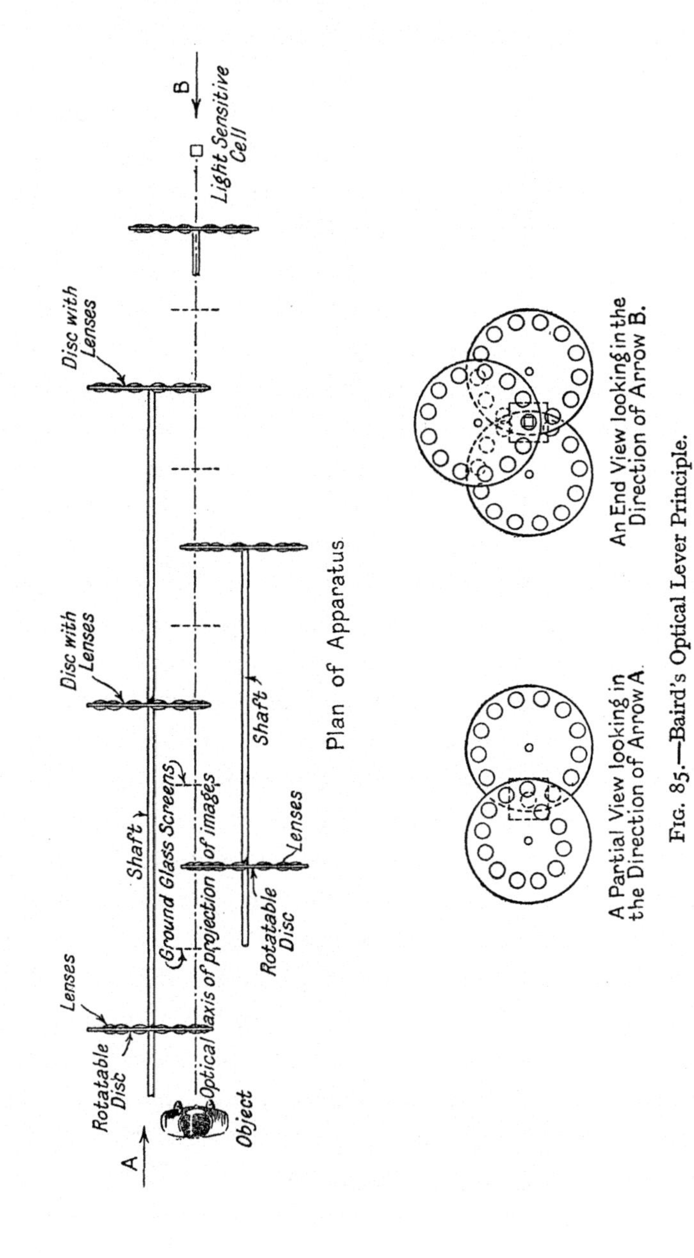

FIG. 85.—Baird's Optical Lever Principle.

are additive, lateral motion being given by the final lens disc.

A further point arises. The image is reproduced by a moving point of light traversing a screen. This point of light covers the whole screen, so that, if the point is small and the screen large, immense intrinsic brilliancy is necessary if adequate illumination is to be obtained. Such brilliancy as that obtainable from lamps of the glow discharge type would be sufficient to cover only a very small screen, and, in fact, even the most intense source of light obtainable, which is the arc lamp, would be inadequate to cover a screen of large dimensions. The image given by Mr. Baird's machine, while quite sufficiently brilliant, measured only about 2 by 3 inches in his first machine, although latterly by increasing the brilliancy of his glow lamps he has succeeded in bringing this measurement up to 8 by 12 inches. To cover a screen equivalent to the modern cinema screen and with equal brilliance is, however, a different matter, but while it seems that this could not be done with the use of a single moving light spot, there is no reason why, as indicated in Mr. Baird's patent specification No. 266591, a plurality of such points should not be used.

Accordingly, two or more photo-electric cells or other light-sensitive cells may be used as indicated in Fig. 86 with the arrangement of spiral-lensed disc, slotted disc, and other parts already mentioned for the purpose of causing the image to traverse the light-sensitive cell in a series of strips. The light waves from the view or scene by this means impinge intermittently on the separate light-sensitive cells, each cell dealing with its own wave-band of the view scene, or image transmitted and controlling its own light source. It is necessary to have a different frequency of intermittence for the light waves incident on each of the cells, so that each cell sets up a current having a frequency differing from that of any of the other cells joined in circuit as light-sensitive devices. The signals of different frequencies so obtained are to be transmitted on the carrier wave sent out by the transmitting station and received at the distance end, where they are separated out by means of filter circuits or sent on separate wave lengths and reproduce the image.

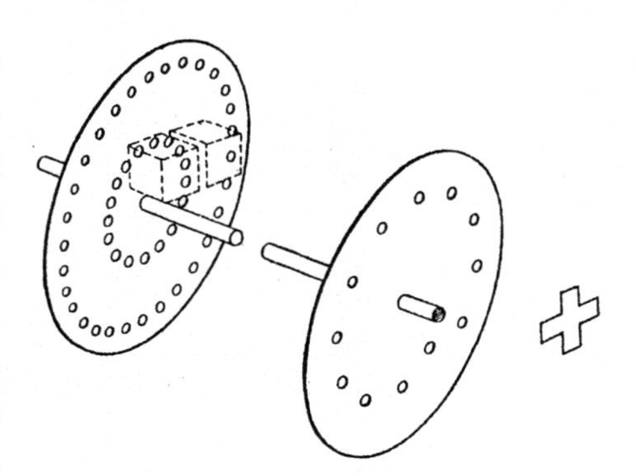

FIG. 86.—Arrangement with two or more Photo-electric Cells.

A diagrammatic view of the arrangement of parts and connections is shown in Fig. 87 and should be self-explanatory, as the descriptions of the various parts have been added on the drawing. The lensed disc has 12 lenses set in spiral contour and is mounted on a shaft with which it is rotatable. The double slotted disc is a rotatable disc having two sets of radially arranged holes or slots, the number of holes in the outer set being double the number of those in the inner set.

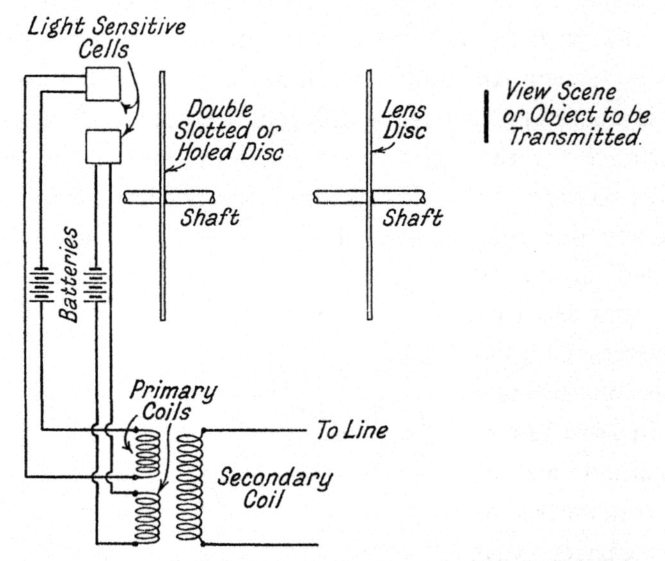

FIG. 87.—Diagrammatic View of Arrangement in Fig. 86.

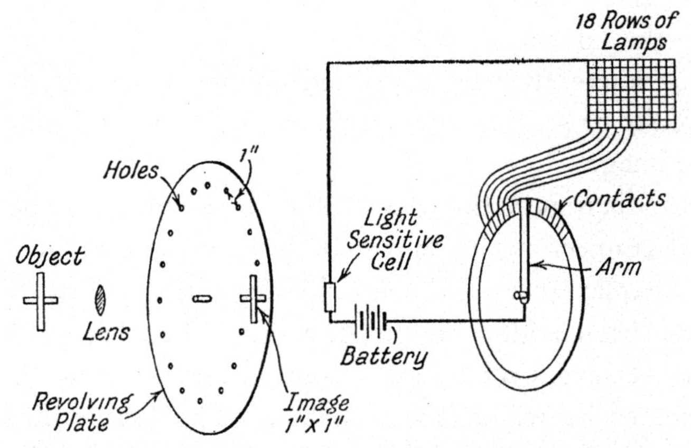

FIG. 88.—Baird's Method of employing Plurality of Light Sources.

This latter disc revolves at a speed so that 1000 interruptions (or more) per second of the light waves from the inner set of holes are given. The outer set of holes may be arranged to give, say, 2000 or more interruptions per second of the light waves. The light-sensitive cells are connected to ordinary H.T. batteries and to a transformer as shown. Each light-cell, however, has a primary coil of its own, but the two primary coils are connected to a secondary coil which is common to both.

Another altogether different method is indicated in Mr. Baird's Patent No. 222604 (Fig. 88). Here, in place of a single light source, a plurality of light sources disposed to form a screen is used. These are fed in succession by a commutator revolving in synchronism with the transmitter, each lamp being thus connected in turn with the transmitter, so that a moving light spot traverses the screen. With this system there is no limit to the size of the screen nor to its brilliancy, but the complexity of the apparatus involved would appear a considerable barrier to its practical use.

As in previous types of similar apparatus, a light-sensitive cell is placed behind an exploring disc at the transmitter and the variations of light falling on it cause the current set up in the cell to vary. A valve amplifier strengthens this varying current so that it can be transmitted, when connected to a wireless transmitter set, to the receiving station.

At the receiving end a brush arrangement fitted at the end of an arm (see Fig. 88) revolves in exact synchronism with the transmitting disc already referred to. As this brush revolves it passes over a series of contacts marked in the drawing, each of which is connected to a small electric glow lamp. There is an indefinite number of these lamps; the more lamps there are the more perfect in detail will be the reproduced image, since these lamps constitute the screen on which the picture can be viewed.

Each hole in the transmitting disc as it revolves sweeps out a strip of the image, and the arm, which revolves in exact synchronism with the disc, sweeps over the contacts connected with the first row of lamps, thus lighting each lamp in turn as it touches the corresponding contact. Each hole in the disc has therefore its corresponding row of lamps. If at the moment a bright part of the image is traversed by the light that passes through a hole, the appropriate lamp is lit brightly; if, on the other hand, light from a dim part of the image passes through a hole, the corresponding lamp will be dull.

A disc of 18 holes should have 18 rows of such lamps, and each row may have any number of lamps in it, as already mentioned, to give better definition. Since there must be a contact for each lamp, 18 rows of 20 lamps in each row would require 360 contacts.

From what has already been said on the general theory of working television apparatus, it will be evident that the varying brightness of the numerous lamps that form the screen will reproduce the image, and that visual persistence will blend the rapid successive variations into one whole image.

Another interesting patent which is worthy of description indicates the method of obtaining intense illumination of the object to be transmitted without the

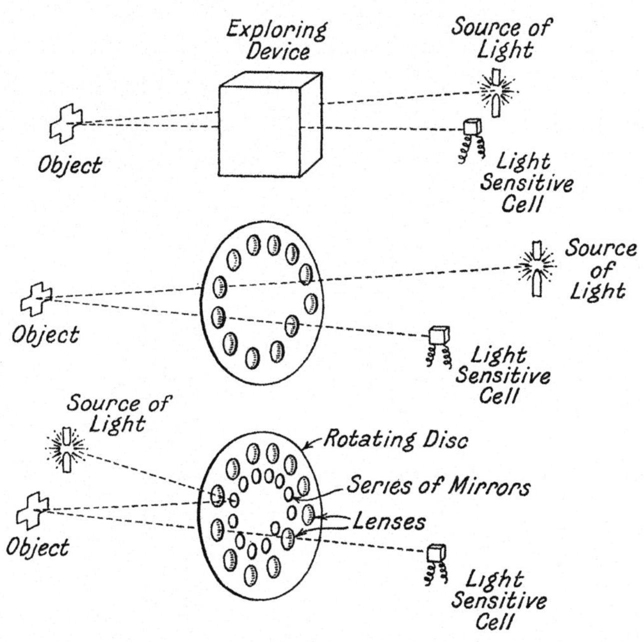

FIG. 89.—Baird's system of Exploring the Object to be Transmitted by a Single Point of Intense Light.

disadvantage of brilliant flood lighting. The system is described in Mr. Baird's Patent No. 269658 and consists of exploring the object to be transmitted by a single point of intense light. This pencil of light is caused to traverse the object and is used in conjunction with a stationary photo-electric cell. As the point of light is continually moving, it may be made very

IO

intense without inconvenience to the sitter. The patent describes methods of using a moving light spot alone, and also in conjunction with a device causing the image to traverse the cell simultaneously with the traversal of the light spot over the object.

This combination gives the advantage that the maximum light available at any instant is concentrated on the cell by the action of the second exploring device. The drawing (Fig. 89) indicates purely diagrammatically the various features of the invention.

CHAPTER IX

TELEVISION TECHNIQUE

Present State of the Art.

REVOLUTIONARY advances in technique have been made during the past two or three years, notably those brought about by Mr. Baird, the details of which were dealt with in the preceding chapter. Therefore leaving the development of the art as already outlined, we now come to an examination of present-day practice of reproducing scenes that are ordinarily out of range of human vision, a matter that will acquire greater importance in succeeding years.

Present-day methods of achieving true television, that is, transmitting actual scenes as distinct from cinematelegraphy or phototelegraphy, differ from the devices that made Korn's and Belin's early successes in phototelegraphy epoch-marking events in the past. Their crude but encouraging results were only achieved in those pioneering days of the art by relying on a line or metallic conductor to carry the transmission over two different points, and when at a later stage transmission without wires was essayed, the employment of syntonic wireless with carborundum detector (Fig. 90) was the only available means of bridging space. To-day more ingenious devices and apparatus and improved methods of transmission can be incorporated into the electrical circuit arrangements, and the operation of transmitting sight without the aid of any material link whatever may be quite accurately described as "wireless seeing," in just the same way that speech across countries, oceans, and continents may be termed "wireless telephoning."

Modern Requisites and Procedure.

It will be obvious from the few details already given of how various investigators have endeavoured to

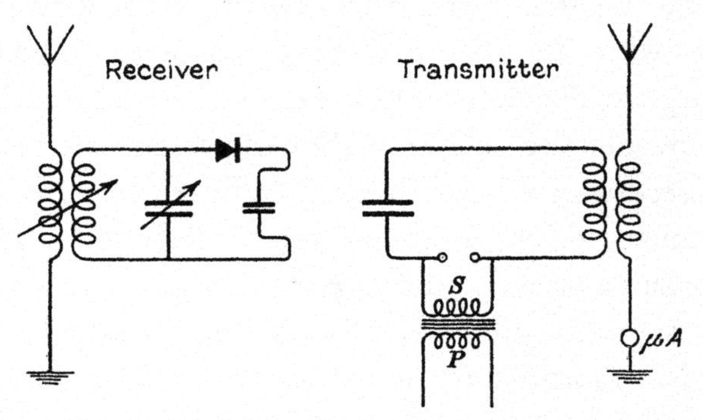

FIG. 90.—Syntonic Wireless Circuit with Carborundum Detector.

design apparatus for seeing by wireless that certain requisites are necessary and that certain well-defined principles must be followed. These may be briefly summarised as follows :—

(1) A source of light is necessary to illuminate the scene, object, or objects to be transmitted. This must be, so far as experiment has taught us, a strong and brilliant illuminant, because it can be used only after reflection and consequent attenuation.

(2) It is necessary to focus the object or scene so as to form an image. Some means must be adopted, such as a system of lenses or mirrors for image formation, both at the transmitting and receiving ends.

(3) This image, by means of a device at the transmitting end, must be split up into strips which are received synchronously in a similar fashion and by a similar device at the receiving end and rapidly thrown on a screen. The only way in which the whole image can be rendered visible in the ordinary way is to show it on the screen at the same rate and in the same manner as pictures are shown at the cinema, namely, about sixteen whole pictures in succession every second, the picture thus appearing as a continuous one, due to visual persistence.

(4) During this process the image with its variations or graduations of light and shade must be cast on to a device that will convert the graduation of light and shade into variations of electric current at the transmitting end. This is the electric "eye." The various forms of light-sensitive cell that are used as electric eyes have already been fully described, the photo-electric cell at present holding and will continue to hold the field until outclassed by a better device.

(5) Feeble currents only can be expected to be set up by this means, since no material substance is known that will give out currents strong enough for the purpose required and hence a magnification or an amplifier of the current impulses is necessary. The best known amplifiers of current are the ordinary valve amplifiers.

(6) An illuminant—a glow discharge lamp—is necessary at the receiving end which has its intensity of illumination controlled by the varying received current. The image thus formed has therefore variations of light and shade similar to the transmitted image when it is thrown on to a receiving screen.

Illuminants.

The difficulty of providing a suitable source of light for use with either a selenium cell or any form of photo-electric cell has always been one not easily overcome. The source of illumination must be one of steady and uniform intensity, but withal of sufficient power and brilliance, so that the light reflected from the object may be of maximum intensity.

Originally Mr. Baird used a metal filament projection lamp of 1000 candle-power with his apparatus. This was quite suitable for inanimate objects, but much too bright to be comfortable for a human face, in fact even a 500 candle-power lamp at a short distance has a most unpleasant effect upon the eyes. A bank of 20 ordinary 40 watt lamps at about 2 feet from the sitter was subsequently adopted. The brilliance of these lamps was controlled by a resistance so as to give ample illumination without distressing the person whose image was being transmitted. Fig. 91 shows the bank of lamps employed by Mr. Baird.

Synchronism.

A simple method of effecting the synchronism of the transmitting and receiving apparatus is that adopted

FIG. 91.—Bank of Lamps.

To face page 150.

in the Baird system of television. At the transmitting end the apparatus, that is to say, the rotating discs for focussing and subdividing the image, are driven by a shaft connected to two motors, one a D.C. motor which supplies the driving power to the shaft, and the other an A.C. generator having a frequency of 500 cycles per second which sends out a synchronising signal. The alternating current from this generator and the fluctuating current from the light-sensitive cell are sent on separate wave-lengths or superimposed on a carrier wave and sent out to the receiving end, where the two currents are filtered out. Like the carrier wave in radio telephony transmission, this carrier wave is modulated or moulded to conform to the frequencies imposed on it. The alternating current after amplification is used to control the speed of an A.C. synchronous motor directly coupled to the shaft of the D.C. motor driving the receiving apparatus. Synchronism is obtained approximately by adjusting the D.C. motor, the A.C. motor being used to prevent hunting. By this means isochronism is obtained. To obtain synchronism, the motor of the receiving machine is rotated about its spindle until the received picture is correctly framed.

A modification of this device was used by the American Telephone and Telegraph Company in the recent experiments in America. Instead of using a D.C. motor to give the drive, a low frequency A.C. motor was used, which has the advantage of making the initial process of getting the two machines in step simpler. It has, however, the very serious disadvantage of requiring another synchronising line or wave-length. To obtain synchronism, the American Telephone and Telegraph Company use the system described by Mr. Baird of rotating the driving motor about its spindle.

Television Radio Equipment.

The equipment of an ordinary short-distance television wireless circuit so far described has been chiefly that comprising the purely mechanical parts and their operation. It may be of interest therefore to outline very briefly the electrical equipment and circuit connections.

As with the transmission of music and speech without the aid of wires, so with the transmission of sight, the careful balancing of an aerial circuit by means of suitable inductance and capacity is an essential. We have the same conditions to observe and consequently the same wireless apparatus parts to join in circuit.

As already mentioned, when initial efforts at transmission were made experimenters were very much handicapped because syntonic wireless with carborundum detector was the only known means of bridging space without a wire conductor. To-day we have in conjunction with continuous wave transmission, the thermionic valve, which is a great advance in the method of effecting communication without wires between any two points. Again, a quarter of a century ago, the development of television was hindered for the want of suitable energising apparatus, but there is no need for it to languish on that account to-day, since the thermionic valve can be utilised both for controlling and amplifying television currents in the various sections of the path between one observer and another. Further, the use of transformers and those individual combinations of inductance and capacity known as filters for selecting and rejecting undesired frequencies are now found to be a very important means towards perfecting results.

Short Wave Wireless Television.

When it became evident that radio television was an accomplished fact, it followed that the next stage was to design and instal equipment necessary to ensure successful commercial operation. General experience in radio transmission points to the fact that fading of signals and the occurrence of atmospherics were the difficulties to be encountered in transmitting vision over considerable distances. The preliminary data already on hand showed what minimum amount of power was necessary for particular or given transmission, the necessary receiving arrangements and connections and the best wave-length on which to work being determined by trial and experiment based on current radio practice.

The extremely rapid manner in which short wave stations have sprung up in recent years, hundreds of transmitting stations working on wave-lengths between 150 m. and 15 m. having been set up by both commercial companies and Government administrations all over the world, decided the Baird Television Development Company in experimenting with systematic long-distance trials on the short wave for television purposes. One important advantage of the short wave is that a comparatively small expenditure of time and money is required to set up, for example, a 2 kw. 50 m. transmitting set complete with a small effective aerial 30 or 40 feet high for a range covering thousands of miles.

Within the last three or four years it has been discovered that waves below 100 m. (approximately) display phenomena which are not met with in long waves. The latter travel in the form of a direct or earth-bound wave and over a great distance may be attenuated by absorption to an almost negligible intensity. On the other hand, short waves are absorbed but very little in the upper regions of the atmosphere, although they may be in the ground.

The technique of producing and detecting short waves is admittedly difficult, but, beyond that there is, in addition, the drawback of their remarkable inconstancy. Hence, scenes televised in daylight quite successfully on a certain wave-length may be a complete failure in the dark at night, and even when sent over different tracks success may be confined to one track and not to the others.

The Special Aerials.

Having decided that short-wave working would prove on the whole more efficient and satisfactory for television transmission than the employment of long waves, the next step was to devise a special form of aerial suitable for the purpose. Since the power may more or less be concentrated in the one *direction in which it can be utilised*, beam radiation, as practised by Hertz years ago, has been tried in many instances by exciting a plurality of spaced aerials—but, generally

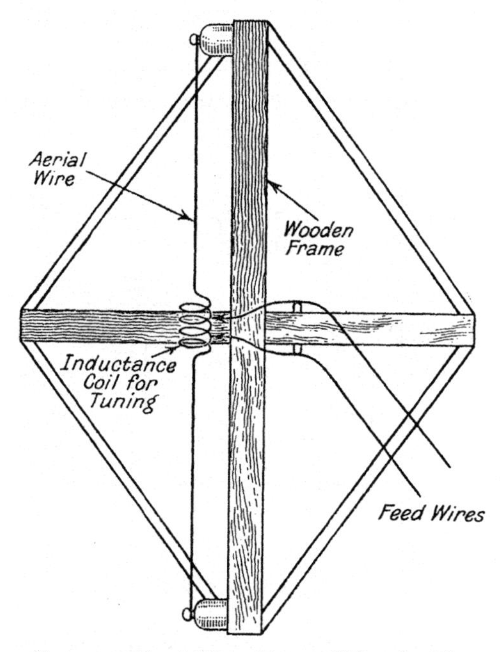

FIG. 92.—Short Wave Transmitting Aerial.

speaking, a single vertical aerial without mirror or other similar device and of the type shown in Fig. 92 has been found sufficient to cause radiation most strongly in a direction inclined to the horizontal when it is excited with an appropriate harmonic of its fundamental frequency.

The Transmitting and Receiving Circuits.

In conjunction with the design of special aerials, the question of design of circuits for the transmitting and receiving sets connected to the purely mechanical portion of the apparatus is of prime importance. As at present much of the experimental work must be treated as confidential, it is not possible to give the

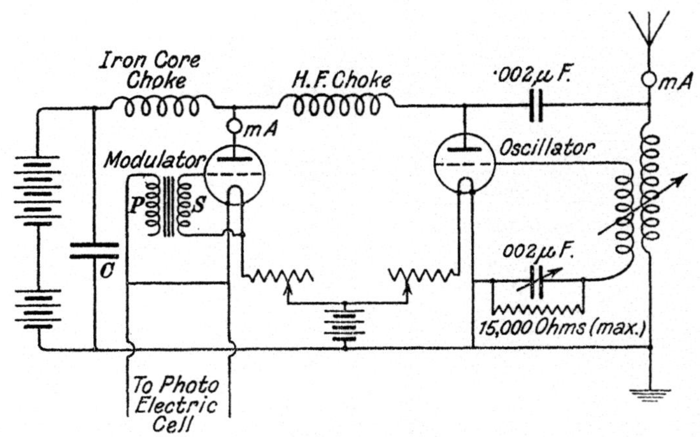

Fig. 93.—Transmitting Radio-circuit.

diagrams of the actual circuits now in use, but Fig. 93 gives a suitable circuit for working on the 200 metre wave-length that is normally employed in television. Here again, acting on the ideas inspired by wireless practice in cognate fields, the general type of circuit having what is known as choke control has been adopted.

The reduction of the wave-length to 40 metres has been effected in certain instances, and it is found that for transmitting images over very short distances good results have been obtained by employing a circuit with grid control.

The receiving circuit may be any one of the well-known types of short-wave receiving circuit.

Transatlantic Television.

Although the public mind is no doubt surfeited with successive wonders in scientific discovery, there is little doubt that another has yet to be added, in that seeing across the ocean and continents is only a matter of time.

Assuming a London–New York Television Circuit possible, it should be remembered that, so far as our present knowledge extends, the conditions governing the use of such a circuit worked on a short wave-length would be very different from a circuit over which long waves are sent. The signal strength on a short wave-length would vary from hour to hour, sometimes from minute to minute, and while atmospherics would be not nearly so troublesome as on long-wave transmission, they would constitute a factor to be taken into account.

Experiments are being conducted at the present time with a view to establish a television circuit across the Atlantic, and probably results of a sensational character will be available for publication shortly. While adopting an attitude of reserve concerning the success of such an endeavour, it should be remembered that not many years ago Marconi in his first efforts at ocean telegraphy merely obtained a series of three dots. To-day messages can be sent and received quite satisfactorily between Europe and America. In like manner, seeing across the ocean may become an established fact.

Its Possibilities.

At the time of writing, there are several television stations in the country all working on the system invented by Mr. Baird. The stations between which operations are carried out by the Baird Television Development Company, Ltd., their distances apart, official call signs and wave-lengths, and opportunities for receiving scenes by means of Televisors will form the subject of a public announcement very shortly.

Now there is no reason from a technical standpoint why these distances should not be greatly increased. Instead of the old tools being relied on, new ones in the shape of oscillographs, radio valves, the thermionic tube, the photo-electric cell, as well as many other ingenious mechanical contrivances, not to mention the short wave beam system of transmission, are available.

Television over distances of small range is quite an accomplished fact and it only remains to perfect the details of the scenes transmitted. Since, moreover, television is worked on the same principle as radio telegraphy and radio telephony, there is no reason why seeing events that are happening in America cannot be just as easy of accomplishment as ocean talk to America. The same theoretical and mathematical considerations are applicable. Hence the aim at the present time is to extend the range of transmission.

In this respect the utilisation of the short wave or

beam system is found to be very much better as a working and commercial proposition than the adoption of long waves. Primarily short wave transmission has two advantages, (1) its directional character and (2) the comparatively small transmitting power required to ensure successful results. The theoretical considerations respecting wireless equipment and the use of short waves for television circuits, however, it is intended to discuss more fully at some future time in another book, when television practice has become more advanced.

CHAPTER X

RECENT DEVELOPMENTS; VISION IN DARKNESS; THE NOCTOVISOR; THE PHONOVISOR; LONG DISTANCE TRANSMISSION

ONE of the most remarkable developments of television has been achieved by Mr. Baird's successful application of the infra-red ray to his televisor.

The observed facts as revealed by experiments which have taken place during the last hundred years show us that waves of varying length are transmitted through the ether. These waves vary in length from thousands of metres down to lengths so small that even present-day apparatus cannot measure them. In recent years, the longer waves have been made quite familiar to us on account of their use in wireless transmission of sound and speech. The waves that measure only a fraction of a millimetre, however, are known to us only as light waves. They vary in length between the limits of 0·00076 mm. and 0·00039 mm. Ordinary white light manifests itself to us when the whole group of the waves between these limits affect the optic nerve. If, however, we look at the same light through an atmosphere of fog it appears red, due to the shorter waves being absorbed and the longer waves only—those near the 0·00076 mm. limit—being able to penetrate the fog and affect our eyes.

Red rays are relatively long when compared with violet rays, but assuming a hair to be $\frac{1}{1000}$th of an inch thick, its diameter is still nearly forty times greater than the length of a wave of red light.

Invisible Rays.

Scheele, the discoverer of oxygen, was the first to give a lead towards the discovery of other rays than those which produce ordinary white light. Acting on the knowledge that white light could be split up into the primary colours—red, orange, yellow, green, blue, indigo, violet (the spectrum)—by means of a prism and also that sunlight changed the colour of silver chloride from white to purple, he sought by experiment to find out which of these seven colours of sunlight produced the maximum effect. He found that the maximum effect was obtained when the silver chloride was exposed to the rays at the extreme violet end of the spectrum.

This was a step forward. Herschel was the next to throw further light on the subject while investigating the heating properties of the visible rays produced by the sun in order to ascertain which of them (the red, yellow, blue, green, or violet) had the least heating power. He split up the white light from the sun into the coloured spectrum produced by a prism and then tested each colour by allowing it to fall on the bulb of a very delicate thermometer. By trial and experiment in this manner he found that he could get a maximum heating effect when the thermometer was in a position beyond the red, namely, in the path of an invisible ray. He pursued his experiments further and established beyond any doubt that the sun gives out invisible rays whose wave-lengths are greater than the longest red rays (0·00076 mm.). These rays are known as infra-red rays (Fig. 94).

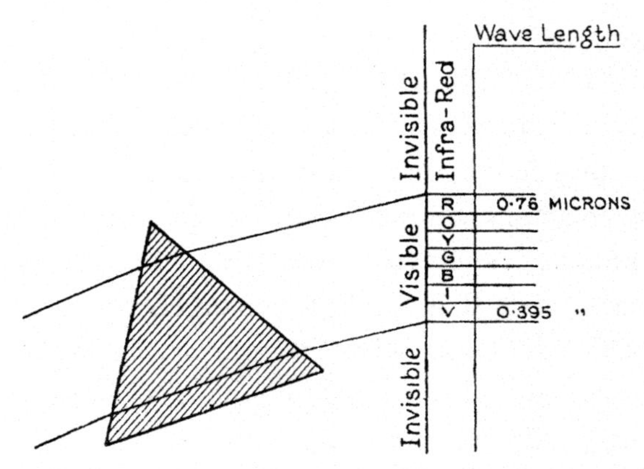

FIG. 94.—The visible Colours produced by a Prism—a part of the great Electro-magnetic Spectrum.

Properties of Infra-red Rays.

The infra-red rays make themselves apparent by the heat they produce. If they are allowed to fall on very fine wire, the wire increases in temperature sufficiently to observe a change in its resistance. This change of value in the electrical resistance of a body is very noticeable in their effect on a Selenium cell. If a beam of infra-red rays be allowed to fall on a Selenium cell then the resistance of the cell is altered during the time the rays fall on it. Such a cell, therefore, can be used as a detector of infra-red rays.

Use of the infra-red rays to affect a Selenium cell

for practical signalling purposes was made by Ruhmer over twenty-five years ago, who gave a demonstration before the Electrical Society of Berlin on March 19th, 1902.

If a thin sheet of ebonite or bakelite be placed in the path of the sun's rays, the infra-red rays will pass through with very little diminution in intensity, although the visible rays are completely absorbed.

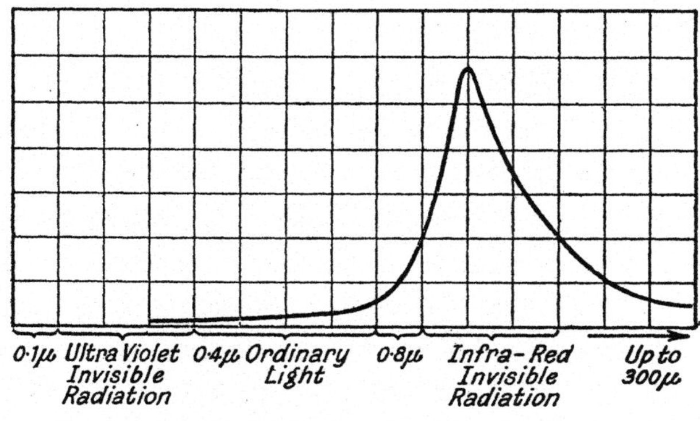

FIG. 95.—Radiation Energy Curve for the Spectrum.

This fact was noted by Ernest Ruhmer over twenty years ago.

The photo-electric cell is sensitive to rays of the spectrum beyond the range of vision both in region of the ultra-violet and of the infra-red. Such cells as the potassium cell possess their greatest sensitivity in the ultra-violet region, others, like the rubidium cell, have their greatest sensitivity in the yellow band, but these cells as a class are the least sensitive to the infra-red (Fig. 95). The following figures taken from Thompson's "Discharge of Electricity through Gases" give a good indication of the characteristics of the usual types of photo-electric cells :—

Type of Cell.	SENSITIVITY.			
	Blue.	Yellow.	Orange.	Red.
Rubidium .	0·16	0.64	0·33	0·039
Potassium .	0·57	0·07	0·04	0·002
Sodium . .	0·37	0·36	0·14	0·009

Sensitivity to white light being taken as unity.

Vision in Darkness.

While the sensitivity of the eye is limited to the wave-lengths from 0·4 to 0·8, the photo-electric cell is sensitive to much wider bands. By making use of this fact and using the invisible infra-red rays in place of light, Mr. Baird was enabled to demonstrate vision in total darkness, his first open demonstration being to members of the Royal Institution on December 31st, 1926.

It would seem at first surprising that these rays should have proved successful rather than the ultra-violet rays, which have a much superior photo-electric effect. The ultra-violet rays, however, have a very small penetrative power. Ordinary glass offers a resistance to them and they are dispersed by the atmosphere much more rapidly than visible light is dispersed. They also have a dangerous action on the eyes and skin, and these reasons probably led Mr. Baird to choose the infra-red rays in preference to the ultra-violet. As already stated, the presence of these infra-red rays was first detected by Sir William Herschel when investigating the spectrum by means of a thermometer. He found that the great heating effect took place at a region beyond the red end of the spectrum and that these rays also affect photo-electric cells, although to a much less extent than the rays of the visible spectrum.

While they are quite invisible to the eye, they are otherwise identical in their properties to light. They can be reflected and refracted and can also affect a photographic plate, although the effect is extremely slight and exceedingly long exposures are necessary.

The Noctovisor.

One of the most surprising developments of television is unquestionably the recent demonstration of Vision in total Darkness. In these demonstrations, which were given by Mr. Baird in December, persons sitting in total darkness were seen and recognised by observers on the screen of a modified form of television apparatus which Mr. Baird has termed a " Noctovisor."

This remarkable result is achieved by using in place of light the invisible rays beyond the red end of the spectrum. Photographs were taken many years ago in the early days of photography by means of these rays by Abney and others. The rays have also been used for signalling as they affect Selenium cells and also, although in a much lesser degree, photo-electric cells. It remained, however, for Mr. Baird successfully to apply them to television and thus render actual direct vision in darkness possible.

Mr. Baird's " Noctovisor " (Fig. 96), as he terms

it, consists essentially of a Television Transmitter and Receiver directly coupled together. The eye of the Television Transmitter scans the scene which, although in total darkness as far as the human eye is concerned, is flooded with infra-red rays. These rays affect the cell in precisely the same fashion as light rays, so that an image appears upon the screen of the directly-coupled Receiver.

The nature of the image thus received is somewhat distorted, red appearing as white and blue appearing as black, a further peculiar effect is that smoke and vapour are semi-transparent, so that the device to some extent renders vision through fog possible.

The fog-penetrative powers of the infra-red rays are, of course, no new discovery, it being well known that the penetrative power of light varies as the fourth power of its wave-length, red light penetrating fog some sixteen times better than blue light.

Advantage of this phenomenon is now being taken in aerodromes, where neon tubes with their deep red glow are used to guide the airmen, on account of the fog-penetrating powers of these red rays.

A short-range demonstration of the fog-penetrating powers of these rays was given by Mr. Baird on the screen of his " Noctovisor " recently, a dummy's head in a room filled with a fog perfectly opaque to ordinary vision being clearly seen on the receiving screen. This demonstration, however, showed only the penetrative power over a short distance, and it will be interesting to see if the effect is available under the conditions which prevail, say, in a sea fog, where the

FIG. 96.—Mr. Baird testing the Fog-penetrating Power of his " Noctovisor."

[*To face page* 166.

penetration to be commercially effective must have a range of miles rather than yards.

Advantage has been taken of these rays for signalling purposes during war time, the rays being used in place of visible light to actuate photo-electric relays.

Mr. Baird by successfully using these rays in conjunction with a coupled television receiver and transmitter has rendered actual vision possible in total darkness. The appearance of coloured objects when received by the infra-red rays differ considerably from their appearance under normal illumination, due to red colouring appearing as vivid white, and the use of these rays is much more difficult than the use of normal illumination, so that there is no advantage in their use for television as such.

The ability, however, to see without the use of light has obvious uses in warfare, and a further possible application for the " Noctovisor " is disclosed by the fog-penetrative qualities of the rays used. Since these rays penetrate some sixteen times farther through fog than does visible light, it is reasonable to expect that this may open up a field of utility for the " Noctovisor " in increasing the range of vision through fog. Any increase of visibility in foggy weather would be of immediate advantage to the mariner and the aviator.

The "Phonovisor."

In transmitting the image of any object by television, the traversal of the image over the cell causes the production of a fluctuating electric current, the character of the fluctuations being determined by the shape and appearance of the object or scene being transmitted. If the fluctuating electric current is received on a telephone in place of a televisor, a noise is heard, this noise having a different character for every object, so that every scene may be said to have its corresponding "image sound." By listening intently it is quite possible to distinguish two different human faces by their equivalent sounds, and the difference between a hand and a face is very marked. Every movement is audible, opening the mouth, for example, causes an immediate change in the note.

By recording these sounds on a phonograph, a permanent record can be taken, and if these records are played again into a microphone connected to a televisor working in synchronism with the phonograph the original image is reproduced, thus turning a scene into a fluctuating electrical current, then into a sound, then indentations on wax, then reversing the whole series of processes and recovering the scene from the wax disc, so that we have a means of storing living images upon phonograph records. Mr. Baird has given the name of "Phonovisor" to this device, which has interesting possibilities.

It is perhaps not too wild a flight of fancy to say that by a development of the "Phonovisor" the blind may one day learn to know their friends by the sound of their faces. Whether this method will in any way displace the cinematograph is doubtful, but it certainly is a noteworthy scientific achievement and had the cinematograph not already been in existence would have been of the first importance.

RECENT DEVELOPMENTS

Long Distance Television.

In April 1927 television was demonstrated in the U.S.A. by the American Telephone and Telegraph Company. Their system consisted in exploring the scene at the transmitter by means of a point of very intense light, and by this means obtaining an illumination of immense brilliance without dazzling the person being transmitted. The arrangement used consisted of a disc with a spiral of holes revolving in front of a power arc lamp as shown in Fig. 97. By this means a spot of light from the arc was caused to traverse the face of the person being transmitted. The light reflected back from the face was caught by three large photo-electric cells, and the fluctuating current from these cells transmitted after amplification to the receiving station. By means of a commutator it was fed to a bank of small neon lamps, arranged in the form of a screen, each lamp being fed in succession so that a spot of light in effect traversed the screen.

A small machine was also shown with a single neon tube behind a disc with holes in spiral formation.

Synchronism was obtained by means of synchronous motors, separate synchronising signals being transmitted between receiver and transmitter. A large low frequency synchronous motor being used to obtain approximate synchronism and a small high frequency synchronous motor used to prevent hunting.

While the question of distance is not of primary importance, considerable distances have now been spanned. The American Telephone and Telegraph

Company have transmitted between New York and Washington, a distance of 200 miles, and Baird in this country has transmitted between London and Glasgow, a distance of 400 miles. In both cases, telephone lines were used, and it is remarkable that successful results were achieved in spite of the capacity effects inseparable from long-distance lines.

The apparatus used appears to have been comparatively simple in the London to Glasgow demonstration, as a total of three operators only were required, the receiving instrument being contained in a box which as shown in the photograph (Fig. 97) is little bigger than a large suit-case.

The American Telephone and Telegraph Company employed a great number of operators in their demonstrations, nearly 1000 men having been engaged in the New York to Washington demonstration. Their apparatus involved no fewer than four separate channels, two for synchronising, one for the television impulses, and one for speech, whereas Baird used two only, one for television and one for speech.

At the time of writing television sets are not available to the public, but their advent cannot be long delayed. The first sets will, we may anticipate, show only the most simple of scenes, a head-and-shoulders view of the person speaking, or possibly a simple scene, such as a few figures on a stage with little detail; but with the rapid development of wireless broadcasting in view we may reasonably expect that the development of television will continue steadily until results rival the perfection of the present-day cinema, and with the

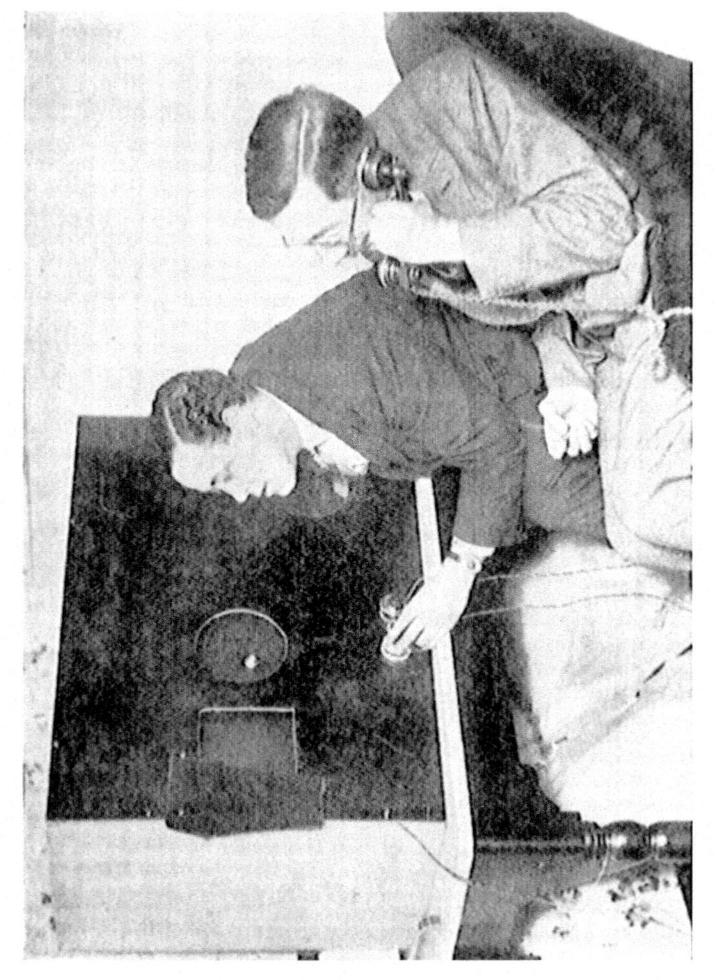

FIG. 97.—Television between London and Glasgow. Professor Taylor Jones in Glasgow sees and speaks to Mr. Baird in London.

[*To face page 170.*

perfection of television, allied arts, such as Noctovision, will develop along parallel lines, so that perhaps we may conclude by quoting Professor Ayrton's remarkable prophecy when, in 1880, he stated: "The day will come when we are all dead and forgotten and our electric cables have all rotted away. In these days a man who wishes to speak to a friend will call him with a world-embracing electric voice and his friend will reply, perhaps from the slopes of the Andes, perhaps from a ship in the midst of the ocean, or if there is no reply, he will know that his friend is dead." Ayrton's world-encompassing electric voice is with us now, within less than fifty years of the utterance of what must have seemed at that time a wild flight of fancy.

Television will give to us electrical vision and an annihilation of space which far exceeds Ayrton's prophecy.

APPENDIX

A London–New York Television Circuit.

At the time of going to press Transatlantic Television, which was referred to in Chap. IX as being in the experimental stage, has become an accomplished fact.

In the early morning of February 9th a demonstration was given in Hartsdale, a suburb of New York, to the Press, when faces in London were seen on the screen of a small Television receiver in New York. The first face transmitted was that of Mr. Baird, and this was followed by the face of a Press representative (Mr. Fox), and later by the face of a lady (Mrs. Howe). The faces of Mr. Baird and the Press representative were recognised; the face of the lady was, however, indistinct, the features not being clear.

The transmission took place on 45 metres from a 2-kilowatt station situated at Purley, fifteen miles out of London, the Television transmitter being at the head-quarters of Mr. Baird's Company in Long Acre, London. The Television impulses were transmitted over telephone wires to Purley, and after amplification at Purley were used to modulate the carrier wave of the 45-metre transmitter. Reception at Hartsdale took place on a one-valve receiver using reaction. The received signals were subsequently amplified by a four-valve low-frequency amplifier, and then used to control the light of the glow discharge lamp in the receiving Televisor.

Communication between Hartsdale and London was carried out by Morse signals from a small wireless transmitter working on 37 metres from Hartsdale.

The picture received was 2″ by 3″ in size, the over-all dimensions of the receiving apparatus being 2′ by 3′ by 8″. In the experiment only one operator was required to operate a receiver of the simplest possible form, and it is probably owing to the simplicity of the apparatus that success was possible.

Experiments are being continued in the Company's laboratories in Long Acre, and apparatus is in use experimentally which produces a life-size picture. A larger power wireless transmitter is also in course of erection, and with increased power and perfected apparatus much better results are anticipated.

Owing to the achievement of this Transatlantic Television the distances mentioned on p. 169 have naturally been considerably increased.

INDEX

INDEX

Printed in Great Britain by
Richard Clay & Sons, Limited,
bungay, suffolk.

Television apparatus will require even more exacting standards of quality and performance from its components and accessories than Radio apparatus does. ¶ This is particularly true of Valves, which must be perfectly matched and efficient to the last degree. ¶ Six-Sixty Valves perfectly fulfil these requirements. Manufacturers of Radio apparatus standardise on Six-Sixty, and Six-Sixty Valves will also be standardised for Television. ¶ We shall be glad to hear from anyone interested in the question of Valves for Television. It is a matter to which we are devoting particular attention

First in Wireless in 1919
First in Television in 1928

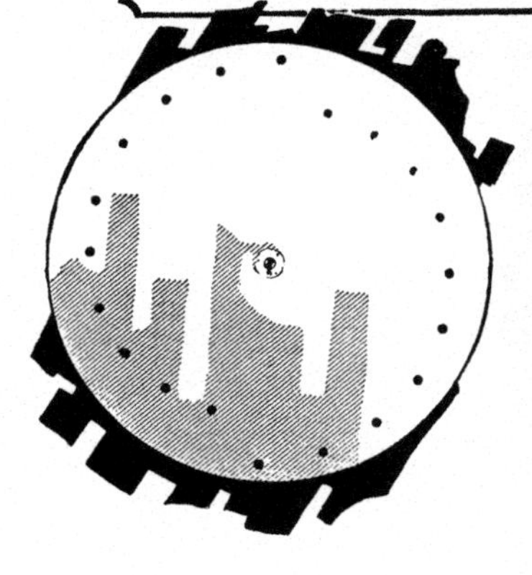

BACK in 1919 (before Broadcasting began) Peto-Scott Co., Ltd., were the first to supply the needs of the early Wireless experimenting public. Now in 1928 (after nine successful years) Peto-Scott Co., Ltd., are the first to supply all the needs of the early experimenter in Television. To the latter, as well as to the former, Peto-Scott offer the same standard of quality and the same unrivalled service. We have always in stock all the components required to construct a complete Televisor.

Here is what you require :—

4-volt Electric Motors, 21/-.

6-in genuine Mangin Mirrors as specified, 12/-. Mounted in adjustable metal stands, 19/6.

Osglim Lamps, 3/3.

Aluminium Discs, sold in pairs, 20″ and 14″ dia., accurately punched to specification and fitted with brass flanges for attaching to motor spindle, 12/6 per pair.

Selenium Cells, Dust and Moisture proof. Mounted, £1.

Complete parts for Amplifier, £5 : 6 : 3 including free Blueprint and wiring diagram.

We can supply 1/60th horsepower Electric Motors to work 200/250 D.C. or A.C. Speed, 3,000 r.p.m. Price £1 : 10 : 0. (These will take little current from the mains and will save cost of heavy accumulators.)

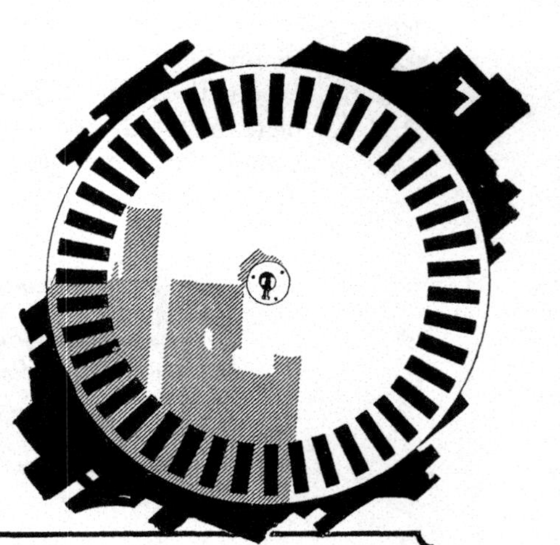

PETO-SCOTT CO., LTD.

77, City Road, London, E.C. 1.

***Branches:* 62, High Holborn, London, W.C.1 & 4, Manchester St., Liverpool.**

2021

THE BEST CONDENSERS IN THE WORLD

J.B., S.L.F. Slow Motion.
(J.B. True Tuning S.L.F.)

Look at the workmanship, the accuracy in every detail, the perfect finish, and you'll be convinced cf the efficiency of J.B. Condensers. Go a step further, and substitute J.B. in any Receiver and the resulting improvement will immeasurably strengthen that conviction.

J.B. Log. Slow Motion

Every modern improvement, every detail that makes for sharper and more efficient tuning is incorporated in J.B. Condensers. TELEVISION—the "BIG THING" of the future—is relying on that essential accuracy and sharpness in tuning which J.B. Condensers ensure.

Look ahead and be in a position to take advantage of coming developments by making sure that your radio receiver is up to date.

J.B. Log. Plain

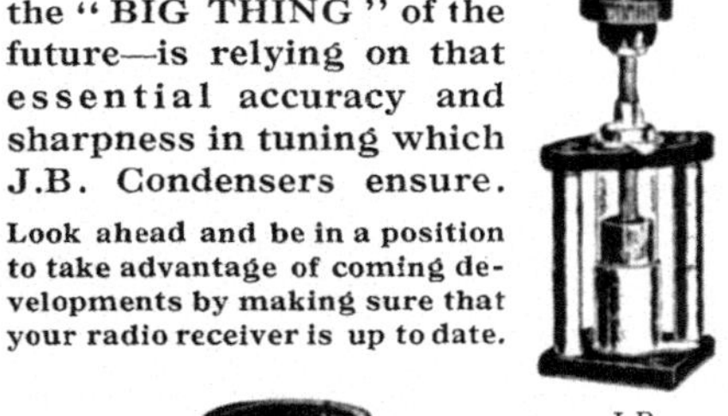

J.B. Neutralising Condenser

J.B., S.L.F. Plain

JACKSON BROS
8, POLAND ST.—OXFORD ST.
LONDON - W.1
Telephone:— GERRARD 7414

11 K.W. (Geneva Rating) Standard Broadcasting Transmitter

Standard
Broadcasting Equipment

The table below outlines the general characteristics of the complete range of Standard Broadcasting Equipments available.

Telephony Rating			Approx. Weight of Transmitter	Floor Space Required	Water-cooled Valves Employed	
Geneva	Unmodulated Input Power to Aerial	Total Power Consumption	Lbs.		Rectifier	Amplifier
Medium Power						
1½ K.W. ...	1 K.W.	11 K.W.	3,200	170 sq. ft.	—	1
3 K.W. ...	2 K.W.	16 K.W.	3,500	170 ,,	—	2
7½ K.W. ...	5 K.W.	25 K.W.	15,000	780 ,,	3	2
11 K.W. ...	7½ K.W.	30 K.W.	20,000	780 ,,	3	2
15 K.W. ...	10 K.W.	60 K.W.	26,000	1,040 ,,	3	5
High Power						
22½ K.W. ...	15 K.W.	70 K.W.	46,000	3,200 ,,	3	7
30 K.W. ...	20 K.W.	90 K.W.	54,000	3,200 ,,	6	10
45 K.W. ...	30 K.W.	135 K.W.	61,100	3,200 ,,	6	14

Illustrated literature will be gladly sent upon request.

Illustrated literature will be gladly sent upon request.

Speech Input Equipment Control Desk

Advertisement of Standard Telephones and Cables Ltd., Connaught House, Aldwych, London, W.C.2 (Central 7345. 10 lines).

Vol. 3 OCTOBER 1930 No. 32 SIXPENCE MONTHLY

TELEVISION

A
VISION
WIRELESS
RECEIVER
See Inside

THE WORLD'S FIRST TELEVISION JOURNAL.

COMMERCIAL TELEVISION

When may we expect it?

By The Editor.

DURING the last two years or so we have read in the general press a great deal about television, what it is, and how it is accomplished. Many descriptions have been published of various successful demonstrations given in this country and in the United States of America, and the Man in the Street is no doubt very anxious to know just when television will become a commercial possibility.

At the end of last year Mr. J. L. Baird, the leading inventor in the field, made a prophecy that he considered it possible that televisors of a crude form might be available to the general public before the end of 1927. This prophecy was, however, made in the most guarded language, and was, of course, merely a prophecy. Time has shown that it was somewhat premature.

As and when the inventions reach the marketing stage the Baird Company will no doubt proceed to license manufacturers to construct and market this apparatus, and furthermore will arrange to receive royalties, probably on a basis similar to that on which the B.B.C. commenced its operations. We do not find, however, that any definite statement or implication appeared in the prospectus of the Baird Company that these televisors were to be marketed within any specified time.

On inquiring into the facts we find that although the Baird Television Development Company has, in accordance with legal precedent, taken to itself the widest possible powers under its Memorandum of Association, including the broadcasting of programmes and the manufacture of televisors, it is obvious from the very name of the Company that it exists primarily for the purpose of developing Mr. Baird's inventions and acquiring Patent and other protection for them.

Let us examine the position carefully in the light of past experience in the matter of great scientific achievements.

Take flying, for example. When the brothers Wright made their first historic experiments, optimistic prophecies were made to the effect that in a few years' time there would be no more railways or steamships; we should travel by air instead. Experience has shown that many years elapsed before even the first air service was established on a commercial basis, and railways and steamships are still with us to this day.

Again, let us consider wireless. Marconi first spanned the Atlantic from Poldhu to Newfoundland on the 12th of December, 1901, and it was confidently predicted by him and by others that a transatlantic wireless telegraph service, to rival the cables, would soon be a reality. But year after year passed, and the long-promised commercial service did not materialise, with the result that the Marconi Company came in for a considerable amount of criticism. Actually it was October, 1907, before the first transatlantic wireless service was opened on a commercial basis.

Consider the development of radio-telephony. From the date of Fessenden's original crude experiments, nearly twenty years elapsed before wireless telephony could be said to be a practical commercial possibility; yet modern broadcasting owes its existence to those original experiments.

Wireless, or radio, as we know it to-day, really means wireless reception. Millions of us listen-in daily; yet scarcely one in a million knows anything at all about the transmitting side of the question, and the transmitting side, if not more important, is at least as important as the receiving side. Wireless receivers were in existence long before the advent of broadcasting and there were many amateur enthusiasts to operate them. But all they had to listen to was Morse code from ships and other commercial transmitting stations. Until the prototype of the modern broadcast transmitter was developed to a reasonable degree of perfection there was no speech or music to be heard.

In the light of such experience we may reasonably expect a similar process of development to take place in television. There is no doubt about it that television is an accomplished fact, in spite of the verbose statements to the contrary which have been made in certain quarters, and in spite of the pseudo-scientific arguments, designed to prove the impossibility of its achievement, which have been loudly voiced by carping critics. Mr. Baird has publicly demonstrated it in this country innumerable times, and according to a recent newspaper announcement has lately transmitted between London and New York.

But, as we have already seen, a period of time must always elapse between experimental achievement and commercial exploitation. In these days of enormously accelerated scientific progress, however, we may perhaps be forgiven for expecting the advent of commercial television within a shorter space of time than it has taken other inventions to appear upon the market.

In the meantime we, from our own knowledge of this new science, would strongly advise the public to take no heed of the irresponsible remarks of ignorant critics; for it must not be overlooked that the actual knowledge of these same carping critics is but little greater than that of the Man in the Street, and their practical achievements in the field of television, nil.

The Baird Company, secure in the knowledge of the results which it has already obtained, will, we feel sure, treat such worthless criticisms with the supreme contempt which they deserve, and continue to work steadily towards the realisation of the plans which it has made.

Seeing Across the Atlantic!

AT the beginning of this year the record distance over which television had been publicly demonstrated was between London and Glasgow, by Mr. J. L. Baird. To transmit vision over such a distance—435 miles—seemed at the time to be a most phenomenal achievement; yet, just after we had gone to press with this issue there burst upon the world the startling news that the Atlantic had been spanned by television! again by Mr. Baird!

Just what does this mean? It means that recognisable images of human beings seated in the heart of London were seen in New York, over 3,500 miles away!

This public demonstration, carried out in the early hours of the morning of February 9th, turns out to be the culmination of months of secret experimenting.

On the night of the demonstration there assembled at the offices of the Baird Company, in Long Acre, a small party made up of Press representatives and privileged guests. The transmissions commenced at midnight, London time, or 7 p.m., New York time.

In order to give the watchers at the New York end an opportunity to adjust the receiving apparatus, the image of a ventriloquist's doll was first transmitted. The image sound produced by this doll, which sounded for all the world like the drone of a huge bee, was then sent over a telephone line to the company's private experimental wireless station at Coulsdon. From this station the image sound was then flashed across the Atlantic on a wave-length of 45 metres.

On the American side, the signal was picked up by an amateur receiving station at Hartsdale, a suburb of New York. After amplification the signal was then applied to the receiving televisor, upon the ground glass screen of which the image appeared. This screen measured about two inches by three inches.

Four watchers were anxiously gathered round the apparatus. These were Capt. O. G. Hutchinson, the Joint Managing Director of the Baird Company, who had gone to New York specially to conduct the experiments; Mr. Clapp, one of the company's engineers; Mr. Hart, the owner of the amateur wireless station at Hartsdale, and Reuter's press representative.

When the image of the doll's head had been satisfactorily tuned in, Mr. Hart started up his transmitter, called a receiving station operator at Purley, near London, and asked that Mr. Baird should take his place before the transmitter instead of the doll. This message was telephoned from the receiving station to the laboratories at Long Acre.

For half an hour Mr. Baird sat before the transmitter, moving his head this way and that, until the message came through from New York that his image had come through clearly. Mr. Fox, a Press representative, then took Mr. Baird's place, and continued to sit before the televisor until word came through that his image was coming through excellently. It appeared that Mr. Fox's features were particularly striking, from a television point of view, and transmitted better than those of other sitters.

Mrs. Howe, the wife of another journalist present, was then transmitted, and, although her features were not recognisable at the American end there was no mistaking the fact that a woman was seated before the transmitter.

Those assembled at the London end were able to see, on a check receiver, a pilot image of what was being transmitted. This image, which was full size, showed the head of the sitter, the complete details of the features showing in black relief on an orange-coloured background. By means of this pilot image the transmitting operator was enabled to check the outgoing transmission and correct any irregularities.

Atmospherics and other interference, and also fading of signals marred the image as received at the New York end at times, but in spite of these disabilities, reception was, on the whole, very good. The demonstration proved quite conclusively that if a much higher powered wireless transmitter had been employed, the image would have been received in New York entirely free from atmospheric and other disturbances. An important feature is that only two operators were required to attend to the television transmission, one at each end of the circuit.

By special arrangement with the Baird Company, we are being afforded special facilities and information which will enable us, in our next issue, to give our readers a more technical and illustrated account of how this latest wonderful dream of science has been achieved.

Clairvoyant: "Might just as well shut up shop now she has a Televisor."

Transatlantic Television.

*[It is regrettable that so many people in this country should find it necessary to rush into print either to " damn with faint praise " or adversely to criticise and belittle the pioneer work of Mr. J. L. Baird. It is refreshing, therefore, to read the whole-hearted admiration of the American Press, some extracts from which we reproduce below. Truly, " A prophet hath no honour in his own country."—*ED.]

The *New York Times*, Feb. 11th (Editorial): " Baird was the first to achieve television at all, over any distance. Now he must be credited with having been the first to disembody the human form optically and electrically, flash it piecemeal at incredible speed across the ocean, and then reassemble it for American eyes.

" His success deserves to rank with Marconi's sending of the letter " S " across the Atlantic—the first intelligible signal ever transmitted from shore to shore in the development of trans-oceanic radio telegraphy. As a communication Marconi's " S " was negligible; as a milestone on the onward sweep of radio, of epochal importance. And so it is with Mr. Baird's first successful effort in transatlantic television. His images were crude; they were scarcely recognisable; they faded and re-appeared, as the atmospheric conditions varied; but they were the beginnings of a new branch of engineering. . . .

" All the more remarkable is Baird's achievement because . . . he matches his inventive wits against the pooled ability and the vast resources of the great corporation physicists and engineers, thus far with dramatic success. Whatever may be the future of television, to Baird belongs the success of having been a leader in its early development."

The *New York Herald-Tribune*, Feb. 12th: " Baird has been experimenting a long time with television, and it has been his ambition to be the first across the ocean, in the well-known Lindberghian manner. He has succeeded, for, if the images that were received on the televisor in New York were crude, they were pictures, nevertheless. . . . If it be appreciated also that Baird is an experimenter of the most classic type, and that he has been struggling along for years with the crudest of equipment, built in the skimpiest shop, his recent stunt is nothing short of marvellous . . .

" When engineers in New York successfully demonstrated television on a telephone line about 200 miles long, between New York and Washington, Baird showed he could do the same thing by screening pictures in Glasgow of persons in London, a distance of 438 miles. It is said that probably one thousand engineers and laboratory men were involved in the American tests. Only a dozen worked with Baird."

The *Sun Telegraph* (Pittsburgh), Feb. 9th, referring to the received images, says: " They were comparable to the visions brought in at the A.T. and T. demonstration by air, from no farther away than New Jersey. The vision of the dummy, in fact, was clearer than those, but the moving faces were not so strong."

Television in mid-Atlantic.

A Passenger's Story.

By A. J. DENNIS.

WITH the normal wireless work of the ship going on in the usual way, a notable television triumph was accomplished on the Cunard liner *Berengaria* in mid-Atlantic a few weeks ago.

A little group of people (of whom I was privileged to be one) crowded together in a small reception room amid a maze of wires, batteries and tubes, and saw, projected on a screen in front of us, images of people sitting at the time in front of a transmitter in the laboratory of the Baird Television Development Company in Long Acre, W.C.

True, the images we saw were sharply defined only momentarily, and at times it was impossible to obtain any results at all, as, for instance, when a morse station got to work on the wave-length being used. But one image was sufficiently clear for Mr. Stanley Brown, Chief Wireless Operator of the *Berengaria*, to recognise his fiancée, Miss D. Selvey. All of us could pick out the face of a girl; we could see the way in which her hair was done; we could see her head turning slowly from side to side.

Previously we had seen the image of the head of a bald man with deep-set eyes clearly and sharply defined, resembling a photographic negative held up to the light, with only the outline of the chin a little indistinct.

This was a dummy which was used when the possibilities of transatlantic television were being investigated a short time ago.

It was not so much the result of the experiment—decided upon only the day before the ship sailed and performed under obvious and exceptional difficulties—that impressed those who, like myself, know nothing of the technical mysteries and intricacies of television. Rather it was the potentialities of the discovery which were driven home.

It did not seem difficult to imagine that in a few years' time passengers on liners like the *Berengaria* might sit at their ease capturing all the excitement of, say, the F.A. Cup final on the television screen in front of them as the match was being played at Wembley.

We could visualise the possibilities, and a little thought showed us that television will bring people in the most out-of-the-way regions into close and constant touch with scenes and events which they now see perhaps only once or twice in a lifetime.

Mr. S. W. Brown's Story.

In a special interview with a representative of TELEVISION, Mr. S. W. Brown, Chief Wireless Operator of the *Berengaria*, stated that when the image of Miss Dora Selvey, his fiancée, appeared on the screen, he had no difficulty at all in recognising her, first by her characteristic style of hairdressing, and later by her profile. " It was a wonderful experience," said Mr. Brown, " to be able to see Miss Selvey like that in mid-Atlantic, and the achievement clearly demonstrates the enormous progress which has been made in television."

Baird Televisors for America First

BRITISH INVENTOR'S TRIUMPH

TELEVISION BROADCASTING STARTS IN AMERICA.

In view of the remarks made by Capt. O. G. Hutchinson, of the Baird Television Development Co., during the course of an interview which we reproduce below, we were not altogether surprised when, just as we were going to press, we received the following cable from our New York correspondent:—

"As from Saturday, May 12th, Station WGY will broadcast television programmes three days per week."

For the information of our readers, Station WGY is one of America's greatest broadcasting stations. It has a power of 50 kw., or twice the reputed maximum power of Daventry, 5XX.

IT was with somewhat mixed feelings that we read in the daily press recently that the Baird Television Development Co. had entered into a contract with a powerful American syndicate, as a result of which the Americans acquired certain patent rights in the Baird apparatus. Still more disturbing was the statement of one of the Americans concerned, that within a few months of the return of himself and his colleagues to the States televisors would be on the market over there.

This news seemed to us to be of such importance to readers of TELEVISION that we immediately sought an interview with Capt. O. G. Hutchinson, Joint Managing Director of the Baird Company, whom we found quite ready to discuss the history and meaning of the deal.

"Early this year," explained Captain Hutchinson, "I went to New York to superintend the final details of some experiments which culminated on February 9th in the reception in New York on the 'Televisor' screen of crude but recognisable images of a number of persons sitting in London.

"The demonstration aroused intense interest in New York, and indeed throughout the United States, and I was approached by a group controlling several broadcasting stations who were anxious to be the first in the world to broadcast television. So anxious indeed were they that they were willing to pay well for the privilege.

"Other commercial groups came to see me with a view to acquiring the patent rights in the invention for America. One of the most powerful of these groups was so keen on 'getting in on television' that it immediately commissioned two of its technical experts to investigate the television situation and report on its commercial possibilities. After these experts had seen all there was to be seen relating to television in America they crossed over to Europe and finally came to our laboratories in Long Acre. Their arrival here coincided with the preliminary tests of Mr. Baird's latest 'Televisor.'

"They waited a few days until the tests were complete, and then we gave them a demonstration by wireless and by wire.

Sir Charles Higham, the British advertising expert, photographed on the "Leviathan" prior to sailing for New York recently.

"The result of this demonstration was that they cancelled the reservations already made on the *Mauretania* to return to New York and cabled their principals that they had found real commercial television, and that in their opinion the Baird Television Company had the basic patents of the only practicable system in existence.

A Powerful Group.

"After this the principals of the group entered the scene and we made a deal with them which I think will work out satisfactorily to all concerned.

"The group concerned is extremely wealthy and powerful in the field of American radio. It controls a vast chain of radio stores extending throughout the United States, Canada, and Mexico. It also controls several radio broadcasting stations, and is arranging to purchase others which will immediately be put into service as television broadcasting stations.

"The manufacture and sale of receiving 'Televisors' for home use will be commenced at once.

"A special radio-television station is being erected on Long Island, near New York, for the purpose of co-operating with our trans-Atlantic station over here which is now ready, thus bringing two-way television between London and New York a step nearer.

"Under the terms of the contract we retain a 50 per cent. interest in the profits of the American, Canadian, and Mexican companies which are being formed, and we get the benefit of any improvements which may be acquired or developed by the American companies, while they on their side get the benefit of any improvements made by Mr. Baird in this country."

In answer to our inquiry as to the amount of money involved in the

TO BE THE WORLD'S FIRST MOBILE TELEVISION STATION.
The "Leviathan" leaving New York. (Photo by courtesy of U.S. Lines.

transaction, Captain Hutchinson replied that he could only say it was a large amount and considered very satisfactory by the directors of the Television Company.

Sir Charles Higham, the well-known publicity genius, left for America in company with the American representatives, and we understand he is handling for them certain publicity matters on the other side. The party sailed on the United States liner *Leviathan*, which ship has been selected to be the first mobile television station. Negotiations are in hand with the object of equipping the *Leviathan* with both a transmitting and receiving apparatus.

Captain Hutchinson concluded by stating that " television should now go ahead by leaps and bounds. Our American friends are real live men, and as there is no broadcasting monopoly over there I would not be a bit surprised to hear within the next month or so that regular television programmes were being broadcast; and I believe that once they

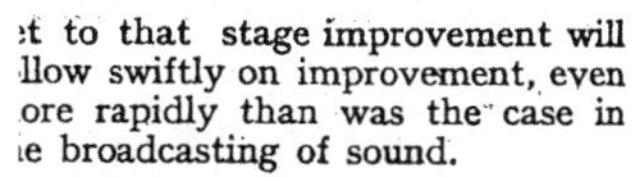

:t to that stage improvement will
llow swiftly on improvement, even
ore rapidly than was the case in
ιe broadcasting of sound.

" I do not wish to overstate the
ιse in any way, but our American
ιusins have certainly got a go-
ıeadness about them that we would,
am sure, do well to emulate."

The Wireless Room of the "Leviathan," which will be closely linked with television in the near future.

America Leaves Us Behind Again

Television for the Home—but not OUR Home

By R. F. TILTMAN, F.R.S.A., A.M.I.R.E., A.Rad.A.

Author of "Television for the Home," "Wireless Without Worry," Etc., Etc.

FROM brief notes which appeared recently in the press—in some papers only in the financial columns—many people know that the Baird Television Development Company has sold to an American syndicate certain rights of the Baird television system.

The general public, however, certainly does not realise to the full exactly what this fact implies, and the object of this Article is to bring to light without delay the regrettable states of affairs which will be realised to the full by the "man in the street" before many weeks have passed.

To put the matter briefly, within a few weeks now the organisation which has acquired the American rights of the Baird Television system will be selling home "televisors" throughout the U.S.A., and will commence a regular television broadcast service from a number of radio stations.

Thus television will enter American homes—but what about British homes?

Britain's Prestige.

It is a universally recognised fact that we have for years led the world in television research and development. The American press, in common with that of all other countries, has freely recognised Mr. J. L. Baird, the British inventor, as the first man in the world to demonstrate true television, and at no time in the past three years has the supremacy of his system been seriously challenged.

Furthermore, those who have had opportunities of witnessing the gradual improvements and developments of Mr. Baird's apparatus in the past two years have looked forward with confidence to the day when the system would be sufficiently advanced to permit the marvels of broadcast sight to enter into the home just as the broadcast voice has brought entertainment and interest to millions.

It has been confidently predicted that very shortly a television broadcast system would be started—for it is not usual to hold back a development of this sort from the public until perfection is reached; as witness the early and crude state in which inventions such as the phonograph, bicycle, motor-car, aeroplane, etc., were put into general use; or in more modern times the introduction of the broadcasting of sound, and of "listening in" by means of crystal sets.

Therefore, in view of the fact that television is an all-British invention, people here have confidently awaited the commencement of a broadcast television service in Great Britain immediately the Baird system was sufficiently developed to permit its introduction to the public.

The Power of American Gold.

Now that an immensely wealthy and powerful American organisation has seen fit to purchase for America (after their technical and patent experts had reported upon it after witnessing demonstrations) certain rights of the Baird system with a view to commencing an immediate service for the public it is perfectly obvious that the system *is* sufficiently advanced for that purpose.

As the Baird system *is* sufficiently advanced to warrant its being introduced in an early form in the home, why is no television service being started in this country, the country of origin of the invention?

That is the question which the public will ask—which the public is entitled to ask—and when the facts are more fully realised the voice of the public on this point will be loudly heard!

Is Mr. Baird to blame for this state of affairs?

I am perfectly sure that this is not the case. As a matter of fact, I happen to know that it is due to Mr. Baird's patriotism that the invention was developed so far in this country, for, early in 1927, representatives of an American Corporation tried hard to take him and his world rights back to America in order to develop the whole system from that side of the Atlantic.

No. The fault lies with conditions over which the inventor has no control, conditions which one thing alone will change—public opinion.

The American organisation which is starting a television service controls a number of radio broadcast stations in that country.

There is the explanation. To develop a television service in its quickest and most satisfactory manner it is really necessary for it to be allied with existing broadcast telephony, and there is no difficulty about that in America where there is healthy competition in broadcasting instead of a monopoly.

A demonstration of television was carried out on board the "Berengaria"—one of our largest liners—on March 7th. But owing to the broadcasting restrictions over here it has not been possible to follow this up.

"Leviathan" First.

Thus we find that the "Leviathan," of the United States Lines, is to be the first ship in the world to be fitted with a permanent Television installation. This is again made possible by the fact that, in America, the control of broadcast stations is in the hands of private enterprise.

How absurd for Britain—the birthplace of television—to sit calmly by while other countries reap all the benefits of a television system devised by British brains.

Seeing Round the World*

What Television will mean to YOU

By SHAW DESMOND

"SEEING at a distance" is going to touch the lives of each one of us; take us out of our ruts; bring us fortune and perhaps, sometimes, fortune's maladroit sister "Misfortune" · and. generally, turn our
down. It is going to change the outer lives of men and women as much as the telegraph or telephone. It is going to do more—it is going to make both those instruments of human advance and human torture (perhaps one and the same thing!) infinitely more potent whether for good or evil.

From the moment that you, John Brown, or you, Mrs. John Brown, get out of bed in the morning until you pass into your beauty sleep at night, television is going to haunt you. Nor, indeed, will either of you, separately or together, be able to escape the multitudinous activities of the televisor even when you have retired to your room! The televisor is the eye that sees everywhere . . . *but*, and it is a very big "but," NOT *unless you wish it*! You need not answer the telephone by your bedside unless you wish. Similarly, you need not be seen by the televisor unless you wish. **But the insistence of the televisor will undoubtedly be as great as that of the telephone.**

Hard Facts of the Future.

Here I shall indulge in no airy prognostication. I shall confine myself strictly to facts, or to things about to become facts. When I speak of buying a beefsteak by the televisor. When I speak of buying a Paris "creation" by the televisor. When I speak of addressing thousands of your constituents to be when running for that incredible talking-shop known as the House of Commons—in all these I shall be dealing with hard facts about to mature or already matured.

You are a business man, Mr. Brown. You telegraph. You telephone. But now you are going to "televise" (patent applied for). You are going to be "televised" (additional patent secured).

You will find that fellow Smith at the other end of your wire—or rather in life. No use for Smith, whom you loathe, to disguise himself behind a mere "voice," as he does now over the telephone. You will be able to mark each flicker of Smith's eyelashes ("confound the fellow! why can't he smile?"), note each gesture, draw your own conclusions. For you are "televising" Smith.

Mr. SHAW DESMOND, the well-known novelist and publicist, tells our readers, in his usual inimitable and amusing style, how he thinks television is going to affect our everyday life.

Smith's usually oily voice on the 'phone, with Smith's plausibility, but minus Smith's expression, may mislead you to-day when you telephone. It can't mislead you to-morrow. Smith's face has got to "come across with the goods." Smith has to speak something like the truth because Smith's face is there to give the voice the lie, if necessary.

The telephone is probably the world's biggest time-waster! That statement I make without fear of effective contradiction. The last banker who in his time steered America through one of her banking crises—and he admitted the essential truth of it, after consideration.

"Business is Business."

For to-day you, Mr. John Brown, probably send three times as many messages over the 'phone as you need to send . . . just because the 'phone is at your elbow. If you were an American you would send five times as many. I sat once in a New Jersey house where for some hours I listened to forty-eight messages on the 'phone (a most reprehensible proceeding!), and afterwards put it to my host that of these forty-eight messages, thirty-seven need never have been sent. He denied. Thought. Admitted!

But with the televisor automatically put into action with the lifting of the telephone receiver you will think twice before you send a useless message. It is one thing "to talk to a voice." It is another thing "to talk to a man." That is the difference.

With a voice, you can avoid responsibility. You are under no compunction to "watch out." You are not being observed. But when you talk to a man, not in the spirit but in the concrete flesh, you have to watch your step. You will be apt to think twice before you make connection. That is human nature.

Or a contract has been signed in New York whilst you sit in your London office. It purports to be signed by Jones. It has been "wirelessed" that it contains certain clauses. You want to see that signature. You want to see those clauses. You don't want any "funny business," because you are not quite sure of Jones.

So you ask the New York end kindly to hold up the contract before the televisor. To "televise" the contract. You see exactly what is being signed. You know where you are.

But all this drops into nothingness when we come to what I will call "television-advertising." It is going to shake the world to its publicity centres. It is going to revolutionise the sales of everything from chewing-gum, that modern devitamised food, to white elephants and cures for corns.

As I have already dealt with what I have called "television broadcasting" fairly fully in the columns of various London newspapers, I propose here to leave out the minuter details of the enormous possibilities of wire-less-broadcasting if ever the "B.B.C." pass out of the stodgy official stage to be taken over by private enterprise. Yet I will venture to set out the conclusions to which I have arrived after discussing it with various editors of dailies, publicity specialists, and big business leaders.

The Programme of To-morrow.

My conclusions, after these conversations, set out in order of importance are :—

1. That television-advertising will probably be a commercial proposition within two to three years and in common use within three to five years.

2. That used with broadcasting and compared with the present use only in advertising of the printed word, in which *sight* alone is used, it will be as much more effective as would be the comparative effects of a play on a man sitting in a theatre with eyes open, and a man sitting in the same theatre blindfolded. When *seeing* is added to the hearing we have a pretty perfect combination.

3. That the day is now fairly close when the great dailies will use "television-broadcasting" as part of their daily work, in a way to be shown later in this article, and that instead of hurting them this new advertising medium will enormously increase their circulations and power. *For nothing can completely replace the printed word.*

4. That the great stores, etc., will be able to reach their millions by the "advertising-televisor" where to-day they reach their thousands.

5. That as television is a British invention, and by virtue of the international strength of the Baird patents, if England be alive to the possibilities of the new medium, it will give her an initial impetus over her competitors similar to the initial impetus which steam gave her a century ago in the industrial world.

Compressed, we shall see (for steps are now being taken indeed to make it effective) our great dailies broadcast to millions of readers and "listeners-in" the front page of their edition of the *morrow* with, printed across it, the words: "This space has been reserved in tomorrow's edition for Messrs. Blank, the great popular emporium, to advertise their motor-cars," or dresses or hats, as the case may be.

> **Mr. Shaw Desmond says on this page :—**
> **"There will be scarcely a house in England tomorrow which will not have its own television screen... entertainment both visible and audible... a pretty perfect combination."**

But you will ask: "How shall I be able to see this?"

As Sherlock Holmes would say to his dear Watson: "That is elementary."

There will scarcely be a house in England to-morrow which will not have its own television screen. You will sit in your armchair after you come home from "the city," put a pipe on, switch on your wireless and look at the screen which will be hanging like a picture before your eyes. On that screen you will see perhaps a curtain as in a theatre. The curtain will roll up. And behind the curtain you will see the front page of your favourite daily, to-morrow's issue, with to-morrow's date, and underneath the announcement I have mentioned. *C'est tout!* After a minute or so, when you have properly tuned in your televisor, the entertainment, both visible and audible, will commence.

It is possible that, if the editor of TELEVISION permits, I may return to this "television-advertising" in a later article. I will now content myself with stating, first, that an invention is now maturing which will certainly make this "home-screen" possible; secondly, that it is considered sufficiently "brass-tacks" to have interested leading business men and newspaper men; and, lastly, that as Edison recently said to me in his laboratories in New Jersey, **"there is practically no limit to the possibilities of power-projection," whether that projection takes the form of wireless motors or wireless broadcasting, or television-advertising.**

We are only at the beginning. And it is now, as always, the first step, already taken, which costs.

So much for the business side of television and Mr. John Brown. What about John Brown's wife?

"Take Politics."

Woman has come out of her shell —commercially, socially, artistically. But the shell has not yet quite dropped from her. Bits of it are clinging to her still!

She is finding all sorts of difficulties in her new freedom. She finds that "the brutal male" has still to be reckoned with; that sex-war is fact, not fiction; and that she must fight for what she wants, and that she can only hold what she wants so long as she fights!

Television may be described as the midwife of the new society—of that society in which woman will play her part equally with man.

Take politics!

Do you realise that there is actually in existence a televisor which can stand on your table and which takes up little more room than a typewriter? Do you realise that this handy little instrument is about to be used by Mrs. Phillipson, Lady Astor and Mr. Lloyd George, also by Messrs. Winston Churchill and Ramsay MacDonald? They may not know it yet—but they will either have to use it or lose their seats! This is not hyperbole, it is stern fact. **The vote is to be "televised."**

No more meetings in little stuffy halls when the candidate, his halo newly polished, seeks the suffrages of the intelligent electorate! No more tortuous train journeys! No more travelling a hundred miles by outraged constituents to the lobby of the House to call the sitting member over the coals!

Not on your life! (I regret this American slang, but would plead that everybody is doing it, from peers to postmen!)

Think of the awful advantage of a beautiful woman would-be M.P. over a plain common or garden male when she takes out her powder puff, does something diabolic with a lip-stick and pencil, and puts on a devastating hat!—to appeal over the televisor to her constituents to be! Think of Jix or Winston in the same position. Where does the mere male come in?

He doesn't!

"An Awful Prospect!"

Nor is this any joke. It is going soon to be hard brute fact for the wretched male politician. For the televisor will be used in every election from Land's End to John o' Groats.

It might even mean a female majority in the House. Awful prospect!

A female Prime Minister. A female War Minister. A female Minister of Morals! The mind baulks. For all these things the televisor may make possible.

But coming from suffrage to sausages.

Has any woman reading these lines known what it is to go out under a broiling sun to see Mr. Diehard's legs of mutton or steaks? Has any unhappy housewife (and we men simply don't know what women go through in the house!), following me thus far, ever had to trust to luck and Mr. Greenheart's honesty for plums or apples or oranges or Brussels sprouts? Why, every daughter of Eve of them all has been through it.

Enter the televisor!

You "televise" Mr. Diehard, the butcher. You tell Mr. Diehard that you don't want any aged meat, nor are you primarily concerned with Mr. Diehard's bank balance—but rather with your own digestion and that of your husband. Has Mr. Diehard any *really tender* beefsteak?

No use Diehard saying he *has*! You want to see what you are buying. Won't he please hold up the piece of meat for your inspection through the televisor?

Mr. Diehard sighs and does so. No chance now to send round a bit of tender and a bit of tough to help each other out. He's got to deliver the goods—the identical goods you have inspected through the televisor. Result: happiness, digestively and maritally. More human happiness turns upon human digestion than most of us are ready to concede!

Or you, Mrs. John Brown, are living in the wilds. Down there in the heart of Devonshire or up in the north-west Highlands from which I have just come after listening to the complaints of all sorts and conditions of women who "won't buy a London dress because they can't first see what they are getting."

The Mannequin Parade.

You want a London, a Paris, or a Viennese dress. You don't order in the dark or "in the blind" as the idiom goes on the continent. You wireless Selfridge or Peter Robinson's or Wörth, tell them what you want, and ask them kindly to place the dress before the "exhibition screen." You have the morning of your life enjoying the latest creations a thousand miles away (perhaps to-morrow three thousand miles away when you see and speak with New York as easily as with Balham or Blackfriars) without having to stir a foot from your own home, without exertion, and without that irritation which I gather from my women friends is inseparable from "buying a dress," which is worse than buying a horse or buying a gun.

"Oh! but," you say, "what about the colour?"

Well, what about it?

Did you not know that we are now well on the high road to solving colour transmission by television, as it has already been solved on the films? That is but a detail—an important and even difficult detail—but a detail now on the point of solution.

You will be able to see the delicate jade in a Paquin dress as clearly as though you were in their Paris show-rooms. The costly exquisiteness of the "absolutely simple" dress of the really great designer will be as plain to your eyes as though the mannequin were strutting before you in your own room—*as she will*. For **I do not doubt that ultimately either a genius like Baird, or another, will bring in the stereoscopic effect into the televisor** which will let those to whom you speak walk, as they certainly will talk, *right out of the screen*.

A Word to the Wise.

You, Mrs. Brown, will be able to indulge your *penchant* for fashionable "At Homes" without going out. Your friends will be able to come into your room without *their* going out. You can have a "Televisor—At Home" hour fixed beforehand when all your friends will touch the switch, let the screen roll up, and find themselves speaking together in that screen, as though in the flesh.

And, of course, I have only touched the fringe of the commercial and social and political potentialities of television. And, of course, as certainly, all sorts of clever—clever people will say: "But that's all in the future. It's not going to be in our lifetime."

For the comfort or dismay of those critics (and no intelligent man objects to intelligent criticism) may the writer, who does not own a television share, and as novelist and publicist has in this no axe to grind, state the following unchallengeable facts:—

First, that **a man has sat in a chair in London and his image has been "televised" to New York, his features and personality being recognised there.** Secondly, that cases like that of the wireless operator of the Cunarder *Berengaria* in which, when one thousand miles out at sea, he recognised his *fiancée* in a televisor screen, will become each day more common and soon will no more call for comment than a long distance telephone conversation. Lastly, that television in the United States at least is already recognised as one of the most formidable business propositions on that live-wire continent, with already a formidable organisation and revenue of its own, and

The author on this page has a few remarks to say about the many applications of television to the ordinary events of everyday life—politics, buying that new frock, interviewing the butcher, and the "at home," new style.

that what America is doing to-day England will be doing to-morrow (would that it were the other way about !).

If the doubting Thomas or Thomasine (for the feminine doubter does exist) want further hard unassailable facts then let them digest these :—

The fact brought to my notice when I was engaged recently in completing the scenario of one of my novels for a leading producer, that **no film of a certain type is being written to-day without consideration of its future television possibilities.** The fact that, as Mr. M. A. Wetherell, the internationally known producer of the " Robinson Crusoe," the " Victory," " Somme," and other films recently stated to me : " Television is going one day to revolutionise the world of pictures from roof-tree to foundation." And that final, hard-headed fact, for " money talks," that Hollywood, which I visited not very long ago, as also the world of the legitimate theatre, are being deeply exercised in their minds by the possible effects of television upon their entirely separate worlds.

If this consensus of evidence will not convince you, Mr. John Brown, and your charming partner, Mrs. John, that television is not something " in the air " but something already with its tentacles set deep into solid earth, then I am afraid

The recording turntable of the phonovision apparatus in the Baird Laboratories.

nothing will convince you. And if the England for which you both stand will not be convinced—so much the worse for England ! But the men who run England are, many of them, already half-way on the road to conviction—to that strange compelling conviction—that we are about "*to see round the world* ! "

Television and Broadcasting

By The Technical Editor

Our readers have, no doubt, noticed the absence from our columns recently of the Technical Editor's "Technical Notes." This omission has been due to the fact that the Technical Editor has been absent on a visit to the United States. In the following article he outlines some of the impressions which he has gained.

LONDON is the television centre of the world. Whether or not it is rash to make such a statement, it is my considered opinion after travelling in America for four months with eyes and ears wide open. The opinion was not formed until after I reached home, for I was out of touch with all the recent developments here, and in fact must admit that I did not even see the recent numbers of this magazine until I reached London again.

I always feel rather sorry for the American general public, because they are not as well served as we are in this country in the way of general news. During my stay there I only once heard any mention of Baird television, in spite of the fact that television is almost a general subject of conversation. The reason is a very simple one

Patriotic Publicity.

Publicity in America always tends to be patriotic—if one may debase the meaning of this word—and the exploits of Americans are applauded to the complete exclusion of similar or better exploits by foreigners. This was brought home very clearly, when I read in the leading New York newspaper an editorial article devoted solely to explaining that Lindbergh was not the first man to fly across the Atlantic!

Although this kind of publicity is rather obnoxious to us in this country one really wonders whether it would not be sound policy for us as a nation to adopt it, at least to some extent.

There is no need to enlarge on the general thesis of the part which publicity and advertising plays in American life, but it was interesting to note how it affected the whole question of broadcasting. The listener over there pays no fees directly for his programmes, for they are all supplied as indirect publicity for someone or other. One result is that the programmes are much more "popular" than they are here, and hence more entertaining, and I think the B.B.C would be well advised to lower its standard somewhat, so as to approximate to the American level in the interests of the general listener.

IT is very interesting to note that our Technical Editor, after an extensive tour of the States, returns with the conviction that "London is the television centre of the world."

One very important factor in connection with the American broadcast system which has impressed him is that it is "ready for television."

In contrast to this, our Technical Editor admits dismay at the attitude towards television which has been adopted by the B.B.C. in this country.

The matter about which I really wish to write here is not a criticism of British broadcasting, but the possibilities of broadcasting television. As matters stand at present there is a grave risk that the broadcasting of television (not transmission to a few isolated observers) might easily take place in America before it happens in this country, for the simple reason that the American broadcasting system offers such opportunities, and in fact invites experimental novelties. In a word, the American broadcast system is *ready* for television, and it is clear that as soon as any television apparatus is available in America appropriate broadcasting will commence, with the inevitable result that the whole science will progress more rapidly.

The recently published developments of the Baird system in which daylight illumination has been used, and colour television has been transmitted, show that this country is far ahead of America at the present time, but it is quite reasonable to speculate as to whether the lead can be maintained when the whole onus of development is thrown upon a single company, or even upon a single inventor.

The Baird Television Development Company have promised that apparatus will be on sale within the next few weeks, and that they will broadcast television; all honour is due to the enterprise of those in control of the company, but I should like to see some further support accorded to them.

Attitude of B.B.C.

I have been rather dismayed at the attitude which appears to have been taken up by the B.B.C. on the question of television. The B.B.C. is no longer a private corporation; it is now a national organisation, and I quite seriously ask that those in control of it should take a very broad national view of all questions which come within their scope. Undoubtedly television is one of these questions, for it can be so closely allied to broadcasting, and the least that they should do is to provide facilities for the broadcasting of television.

I have not recently seen any reasoned statement of the views of the B.B.C. on television, but the *obiter dicta* which one comes across appear to suggest that they are not only inadequately informed, but also uninterested, and it is that attitude which is so much to be deplored. I venture to suggest that it is the duty of the B.B.C. to be interested in television upon purely national considerations if no others. We do not want the history of the synthetic dye industry to be repeated in the television industry.

TELEVISION AND THE FILMS

A Talk with Leading Film Producers

By SHAW DESMOND

Author of "Passion," "Bodies and Souls," "Gods," etc. etc.

In the following article, specially written for "Television," the famous Novelist, Journalist, Lecturer, and Politician Mr. Shaw Desmond gives us a glimpse into the film of the future and draws a vivid pen-picture of the wonders and almost unlimited possibilities of Television.

The above picture shows a group of modern cinematographers at work. In time to come (probably not so far distant, either) we shall see men with a television transmitter of, probably, somewhat similar bulk busy "shooting" scenes direct on to thousands of home and public televisor screens all over the country.

TELEVISION is the magician of the future. Television is going to give us picture-magic. Television is now to make the dead bones of the films "rise up and walk." It is about to give us new worlds for old!

The pictures are on the threshold of a revolution quite unguessed at by the picture fan.

Thomas Edison said to me a little time ago in his laboratory in West Orange that the world was on the eve of momentous happenings through the newly-invented "power-projection" through the air, from the control of pilotless aeroplanes from the ground and the explosion of ammunition dumps at any distance to the finer and more delicate projections of "wireless," especially as regards the living picture. It is in the fairyland of the pictures of the future that wireless plus television will play a decisive rôle.

"To this power of projection there are no limits," said the famous inventor, who is notoriously conservative in his statements.

Already by wireless waves we are able to see the man with whom we speak upon the new "television-telephone." And, as we know, there is now in existence a method by which the dummy figures of the film become alive through the spoken word, their movements and speech being made as visible and as audible to the audience as those of the ordinary actor, synchronising.

I have reason to think, also, that we shall soon see the figures "come out of the screen," as in the old-fashioned stereoscope. When that day comes the theatre proper will be directly challenged.

Personally, I do not believe that, ultimately, the dummy of the picture, even when he walks and talks like a living man, can ever be other than a sublimated robot. The theatre proper will always have its arena, one in which it cannot be challenged. No robot can compete finally with flesh and blood. But what about flesh and blood taken *direct from life* and seen through the televisor?

Films "from Life."

There is no evidence yet of a "revolt against the robot," as seen in the films. But there *is* evidence of a revolt against the "dummy"—that is, a demand for the "character" film as opposed to the film of the pretty face. The present movie stars of the glassy smile and lips painted like pieces of butcher's meat may before long, like Othello, find their occupation gone. The woman "character" actor will remain. People like Wallace Beery, Vera Gordon, Jean Hersholt and Farrell McDonald are to be exalted, whilst that great Austrian, Emil

Jannings, if he be not destroyed by the American process of "capitalising talent," may be taken as the type of the future. All this is paving the way for the television-film "taken from life" itself.

There are, through this new trend, to be many radical changes in the studios, many heartburnings and many jealousies, though that day is not yet. And Hollywood is going!

Two of the greatest living producers, including Mr. M. A. Wetherell, are in agreement with me that, with the coming of the "character" film, even before the advent of the television-film, of the Charles Dickens type, we are going to see the novelist proper pressed into the service of the pictures to write the "captions." As novelist, I am interested naturally in this, and know of others who, like myself, have been approached by film companies for this purpose.

The Trend of Things.

"Captions, hitherto, have often been crude and ill-informed," said a prominent London producer to me recently in the intervals of making a new war film, "sometimes shocking the onlooker by their exaggeration and lack of continuity. The trained novelist, and especially the coming of television-films, will change all that.

Producers of the first rank, with princely salaries (some of them with salaries varying from £5,000 to £15,000 per annum) who are, I understand, in agreement with the view that economy in word and action is the thing at which the future film will aim.

One first rank producer, Mr. Alfred Hitchcock (who is only 27), says: "What we must strive for at once is the way to use these film nouns and verbs as cunningly as do the great novelist and the great dramatist, to achieve certain effects on the audience." The "direct life-film" will demand this.

This trend away from the "pretty-pretty" and the saccharine offerings of the worst of the American films is going to make a road for both the "shadow-pantomime" and for the educational film.

The Shadow Film.

The famous German film, perhaps one of the half-dozen greatest films ever shown, called "Warning Shadows," which showed at the London Tivoli, is but the pioneer of a series of similar films. Shadow-sending by television is remarkably easy.

"The shadow-pantomime," as I have called it, that is the whole film going forward in a series of shadow-pictures without words, will always have to depend chiefly upon a "selective" audience, but the Germans are now preparing a series of such pictures, though, of course, without the televisor.

We reproduce above a photograph of the Author whose contributions to this magazine are very popular features. In the article printed here Mr. Shaw Desmond's remarks on the film are succinct and pithy.

The grown-up often loves shadow-pictures. We are all like children in this. The kiddies usually prefer a shadow to a "straight" picture. For the "shadow" offers possibilities of "suggestion," and by "leaving it to the imagination" gets effects not possible to the straight film. A television shadow-film, *showing the play as it was acted*, would fill any theatre!

Now we come to what is the most important of all film evolution—the Kiddies' Picture-Theatre.

I am convinced that the education of the future will use the picture rather than the book. The picture which I saw of the ways and habits of the lobster, taken in its natural surroundings, will teach a child more in one hour about natural history than half a year of natural history books.

Experiments recently made showed that of children taught a certain subject by the book, less than 30 per cent. passed an examination in that subject. Of children of the same age taught the same subject by "the movies," 90 per cent. got through.

We shall see, in the not distant future, educational picture-theatres established in the great cities, seating anything from two to five thousand children, and showing lobsters, lions and men as they are living "at the moment of showing." The London County Council may one day use such theatres for its extension lectures, as may other similar institutions. I pioneered this suggestion long years before the war, before television was heard of, but it may not take even another five years before we see it in action, *plus* television.

Momentous Events.

The educational side of the future film will be much helped by the coming of the music-picture. Here, we are on the edge of momentous events.

Already, as I have said, by "wireless waves," we are able to see the man with whom we speak upon the new "television-telephone" by utilising the translation of light into sight. Now we are about to see the translation of light into sound.

"The film of the future will carry its own incidental music *in the film itself*," said the men who made the "Robinson Crusoe" picture, to me. This will avoid those painful "cuttings off" when the music has not kept pace with the screen. The music will be actually in the screen, the light waves being converted into sound, and I believe that some such method has already been demonstrated in London.

"Music in Colours."

We have now gone a step further—that is, *the conversion of sound into colour*.

This I recently saw demonstrated in Philadelphia by the inventor, Thomas Wilfred, in combination with the famous Philadelphia Orchestra. It was just "music in colours."

Leopold Stokowski, one of the first living conductors, in helping at this demonstration, said: "These music-pictures may in time lead to new combinations of form and rhythm which have hitherto only been the dreams of a few great spirits."

(*Concluded col. 3, opposite page.*)

(*Concluded from opposite page.*)

To watch upon the screen, as the orchestra played, the folding and unfolding of fairy scenes of colour and the painting by fairy hands of pictures actually expressing the music, was a new revealing of the possibilities of the "pictures."

The "automatic film" has already passed from the realm of phantasy to that of fact. Here, indeed, is an unexploited field for the educational film, helped by television.

Peeps at Mars and Venus.

Such films will be placed in the midst of the jungle to register the ways of the Wild Things that live therein. They will be "set" under water to tell us the life story of the man-eating shark and the great whales and the terrible decapod, with its ten waving tentacles, which haunts the sea-abysses. A way to overcome water-pressure and darkness is already being experimented with.

What the film of the future will do is to destroy the mystery of old earth by making all parts of it accessible to everybody—some day, as one thinks, without leaving one's own fireside. But this very passing of earthly mystery will drive man, the inquisitive animal, to realms outside the world. It will, I venture to prophesy, soon begin to be suspected that if "etheric wireless" ever passes the stage of experiment we may be able to picture the doings upon Mars or Venus and even to decide once and for all whether they are inhabited, as many astronomers contend.

But this is something yet to be realised. Almost all the other developments suggested are already passing or have passed from experiment to fruition, partly thanks to television.

"The pictures" are yet but babies learning to take their first steps.

Moving Shadowgraph Experiments in America

In America an enormous amount of interest is being taken in Television, and many experimenters are now devoting their attention to the subject. In the following article a description is given of a moving shadowgraph apparatus which has been developed by Mr. C. Francis Jenkins, who has interested himself in the subject for many years. The apparatus described, however, does not permit of the achievement of Television; it transmits and receives only the shadows of simple moving objects.

ONE of the best known experimenters in America who has been striving for some years to achieve television is Mr. C. Francis Jenkins, of Washington, D.C., who claims to be the inventor of the prototype of the modern cinema projector.

Before attempting to solve the television problem Mr. Jenkins experimented over a considerable period with photo-telegraphy apparatus, in conjunction with Mr. D. MacFarlane Moore, who has done a great deal of work in connection with neon tubes.

In the course of his experiments Mr. Jenkins produced an entirely new contribution to optical science, which is now known as the Jenkins Prismatic Disc. This disc consists of a circular plate of glass, the edge of which is ground to the shape of a prism, the angle of which varies continuously and gradually round the circumference of the disc. By using combinations of two or four of these discs, and rotating them by means of electric motors, Jenkins was able to explore a picture by means of a beam of light, which was caused to bend to and fro by virtue of the combined action of the rotating prismatic discs.

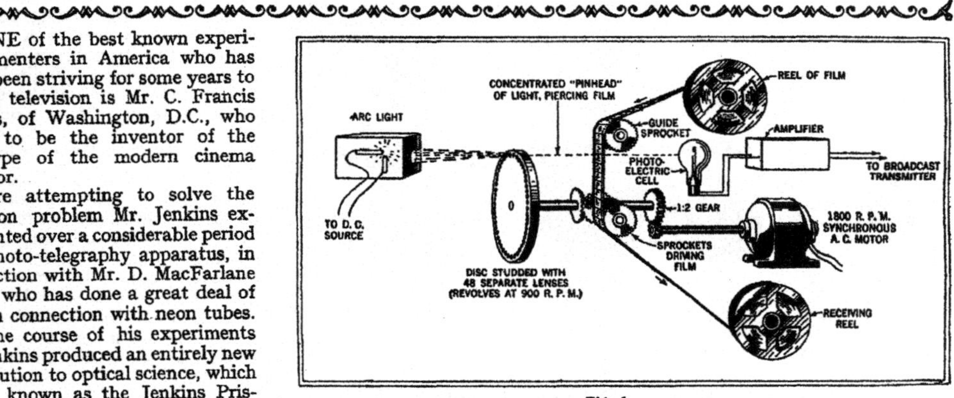

Fig. 1.
The general arrangement of the transmitter.

Fifteen per Second.

Finding this apparatus unsuitable for television purposes, Jenkins turned his attention to lens discs, and in the latest type of Jenkins machine, particulars of which are available, the lens disc carries forty-eight lenses, and the purpose of the apparatus is avowedly to transmit and receive, not television, but special cinematograph films. Using a wavelength of 300 metres, Mr. Jenkins has recently been broadcasting these special shadow films from his laboratory in Washington. There is no detail in the films, only a plain black and white silhouette of simple scenes such as a little girl bouncing a ball.

The films are reeled off through the transmitter at the rate of fifteen pictures per second—one less than the ordinary cinematograph. They are reproduced at the receiver at the same rate, and the received images, viewed through a magnifying glass, appear to be about six inches square. According to witnesses' reports, the received shadowgraphs are very clear, and the illusion of motion is excellent.

The Mechanical System.

The general layout of the new Jenkins animated shadowgraph transmitter is shown in Fig. 1. The film reels are mounted on a simple framework, one above the other, in such a manner that the film is pulled downwards by a set of sprockets which are driven by an electric motor. One end of the shaft which drives the sprockets is fitted with a gear pinion which meshes with another smaller one. The small pinion is mounted directly on the shaft of the electric motor, which is a synchronous A.C. motor capable of running at 1800 R.P.M. Because of the speed reducing action of the gears, the pictures are pulled past the sprockets, or any fixed point next to the film, at the rate of 900 per minute, or 15 per second.

The Function of the Lenses.

At the other end of the shaft which drives the sprockets is a heavy metal disc, about 15 inches in diameter and about one inch thick. The edge of this disc is studded with forty-eight separate little lenses, each of which has an "optical speed" of *f.* 3.5. These lenses are designed to concentrate the light from a powerful arc lamp into an intensely brilliant "pinhead" beam, which is caused to pierce the film as the latter travels down past the back of the disc.

Immediately behind the film is mounted a photo-electric cell, which is so placed that the "pinhead" beam of light, after passing through the film, impinges upon it. The cell is connected to a three-stage resistance-coupled amplifier, the output of which is connected to a further amplifier of similar design but having eight stages. Both amplifiers are very heavily screened by double copper shields.

A close study of Fig. 1 will make the operation of the apparatus clear. The lens dics revolve at 900 R.P.M., or fifteen times per second. The separation between the centres of the lenses is just equal to the width of the film. The latter moves steadily downwards at the rate of fifteen pictures per second. Its action is not jerky, as in a cinematograph projector.

Exploring the Picture.

The arc lamp on the left projects a powerful converging beam of light through one of the lenses of the disc, which lens further converges the beam to a point which "scans" or travels across the film from one side to the other due to the rotary motion of the disc. When the next lens picks up the beam from the arc lamp the film has moved downwards slightly, so that the second beam travels across the film on a parallel but slightly higher path. Succeeding lenses of the disc trace further parallel paths across the film, until each picture has been explored by the forty-eight lenses. As soon as the beam of light from one lens runs off the film, the beam from the succeeding lens starts to pierce the film on the other side. This movement is continuous during the operation of the mechanism.

Thus forty-eight separate beams of light travel across each individual picture in one fifteenth of a second. At the start of each fifteenth of a second period, a fresh picture slides into position and another series of forty-eight light beams start to pierce it.

While this movement is taking place the light beams, after shining through the film, fall on the photo-electric cell with degrees of intensity which depend upon the density of the parts of the film through which they shine. The electrical output of the photo-electric cell varies in proportion to the variation of the light intensity, and, after amplification, this varying current is caused to modulate the radio broadcast transmitter.

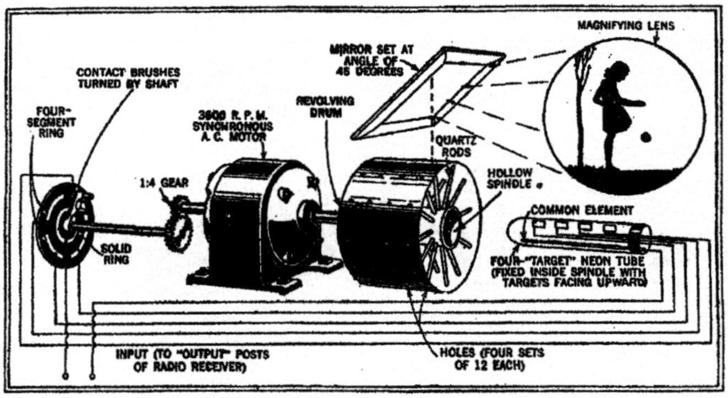

The Light Conductors.

The apparatus for receiving these transmissions has many points of difference from any other form of receiver previously described in this magazine. It is illustrated in Fig. 2, and consists of six essential parts. The heaviest unit is a 3,600 R.P.M. synchronous A.C. Motor, to the shaft of which is attached a hollow metal drum about seven inches in diameter and about five inches wide. The centre of this drum is a hollow spindle with a thin wall.

In corresponding places on the drum and the spindle are four spiral rows of tiny holes, twelve holes to a row. A short piece of quartz rod between the outside and inside connects each pair of corresponding holes. The purpose of the forty-eight little quartz rods is to conduct light from the inner spindle to the holes in the outer drum with as little loss as possible.

Fixed inside the hollow spindle, with the flat little plates facing directly outward, is a special neon tube having four discharge electrodes. This tube is about four inches long and one inch in diameter, the little discharge electrodes, or plates, being about one quarter of an inch square. A straight wire running near the four electrodes acts as a common element. In Fig. 2 this tube is shown withdrawn from the spindle, in order to illustrate it; in actual use it fits inside the latter without touching it.

Fig. 2.
The arrangement of the receiving apparatus. The neon tube is shown withdrawn from the drum for clearness. In operation it fits inside the hollow shaft of the drum.

The Output Connections.

The other end of the motor shaft is fitted with 1.4 reducing gear which drives a revolving switch. The revolving element is simply a pair of contact brushes connected together. One brush effects continuous electrical connection with a solid brass ring embedded in an insulating disc, while the other makes a wiping contact over the four sections of a split ring.

The four segments are connected to the four discharge electrodes of the neon tube, while the solid ring is connected to one of the output terminals of the radio receiver. The common element of the neon tube is connected to the other output terminal of the wireless receiver.

Picking-up the Signals.

All the receiving apparatus described so far is contained in a wooden box measuring about two feet long and a foot square at the end. Directly above the top of the revolving drum is a square opening in the top of the cabinet, and over this opening an ordinary mirror is mounted at an angle of 45 degrees to the top. About a foot in front of the mirror, and standing upright, is a magnifying lens about ten inches in diameter.

Following the action of the receiver, the modulated picture signals from the distant transmitter are picked up by an ordinary wireless receiver, amplified, and fed to the moving picture receiver. Assuming that the contact brushes have just made contact with the upper right-hand ring, as shown in Fig. 2, and that one of the quartz rods in the first, or outermost circle, is pointing straight up, this condition corresponds with the start of a picture in the transmitter, when the light spot is just commencing to sweep across the film.

Building-up the Image.

As the contact brushes have just closed the circuit to the neon tube electrode at the extreme right, this electrode lights up immediately and fluctuates in brilliancy exactly in accordance with the modulation of the signal. The fluctuations of light are carried up the quartz rod and projected through the holes in the outer drum upon the mirror. The light thus reflected from the mirror follows the shading of the images on the original film, so that a picture is built up in the mirror. This picture may then be observed through the magnifying glass.

A complete picture of forty-eight lines (corresponding to the rate of transmission) is built up on the mirror with every four revolutions of the drum At the beginning of the second revolution, the contact brushes turn to the next segment of the switching ring (because of the gearing) and the second electrode of the neon tube becomes operative The third and fourth quarters of the picture are similarly built up from the third and fourth electrodes, and the cycle then commences again with the first electrode.

During one second the drum revolves 60 times. Since four revolutions create one picture, 60 revolutions create 15 pictures, which gives the speed of 15 pictures per second mentioned when the action of the transmitter was being discharged.

It is, of course, necessary for the transmitting and receiving mechanisms to run absolutely in synchronism, but Jenkins does not provide any special synchronising method. For the purpose of the demonstrations given in Washington, the transmitter and receivers were driven by synchronous A.C. motors, and as the motors took their power from the same power line, it is reported that little difficulty was experienced in keeping the pictures steady.

Mr. C Francis Jenkins with (over his left hand) the "prismatic disc" which he invented. This disc has an outer rim ground in the form of a prism of gradually and continuously varying angle. By r[illegible]ed to explore a picture.

Witnesses describe the pictures, as viewed at a distance of about ten feet from the magnifying lens, as being clean-cut silhouettes against the characteristic reddish background provided by a neon tube.

Synchronising Difficulties.

This apparatus is interesting and distinctly novel, but it should be clearly understood that it does not produce television. It transmits and reproduces only moving shadowgraphs of simple figures and scenes which have previously been specially prepared and recorded on a standard cinematograph film; and as there are no refinements of shading and detail to be handled, the problem of designing suitable apparatus is vastly simpler of solution than is the problem of designing apparatus which will enable true television to be accomplished.

Also the problem of synchronism has been entirely side-stepped, for although several similar synchronous motors fed from the same power supply will maintain a fairly even average speed, there is still a difficulty known as "phase swinging" to be contended with. Phase swinging will not cause the received image to be distorted, but will cause it to move slowly up and down, or from side to side, on the receiving screen. To correct this fault manually, and to correct also minute changes in motor speed by any manual method, requires considerable skill, and even then the operation partakes of the nature of a juggling feat.

In the following powerful and well-reasoned article Dr. Robinson makes a striking appeal for facilities to broadcast television. As a former member of the Imperial Communications Committee—the ultimate authority on wireless communication in this country—his remarks on the subject will be read with great interest by our readers. Next month Dr. Robinson will explode the erroneous idea, voiced by some critics, that television broadcasting will require a very wide waveband.

TELEVISION—AN APPEAL FOR BROADCASTING FACILITIES

By J. ROBINSON, M.B.E., D.Sc., Ph.D., M.I.E.E., F.Inst.P.

Dr. J. ROBINSON.

BROADCASTING is a comparatively new service, and it has become firmly established as a necessity of life. Wireless has performed many wonders so far, and we are by no means near the end of its possibilities. Not the least of its achievements has been the broadcasting of speech and music, and the average man in the street is of the opinion that this is the chief application of wireless.

When a new development takes place in any walk of life there is a tendency to wish to employ the method of broadcasting in connection with it. This is particularly the case when the new development has some association with the press or with the communication services.

We have in the last few months heard many times about wireless being employed to transmit pictures, and in fact our newspapers employ this very rapid method to obtain pictures of events happening at a considerable distance. This new development enables the editors of newspapers to obtain a picture of an event at a distance of five hundred miles in probably less than one hour, whereas the system employed until quite recently—that of taking a photograph and sending it by post—required probably one day. Of course this new method of transmitting still pictures can be accomplished either by wireless or by cable.

Control of Broadcasting.

We have also heard at some length about **television, which consists of the instantaneous transmission of events actually in progress, thus enabling actual vision of motion to be accomplished.**

It is not surprising that our minds should be directed to the facilities of broadcasting for these new services, for wireless, though not essential, is most useful for them. The control of broadcasting of speech and music in this country (though not that of wireless generally) is in the hands of the British Broadcasting Corporation.

The problems introduced by the possibilities of optical broadcasting are so vast that the general public expects that no mistakes will be made by the responsible authorities. I have not heard that the control of optical broadcasting has been given to the B.B.C., and before such a step is taken very careful consideration must be given to the subject by the Government or their advisers on this subject, the Imperial Communications Committee.

Grave Responsibilities.

However, there are rumours that the B.B.C. is attempting to absorb these new functions, and that they are contemplating the transmission of still pictures. Very grave responsibilities rest on them in this connection, and one of the first things to be done before any decision is made to give broadcasting facilities to one system and not to the others is to have a thorough examination of the whole situation. In this country we have several systems for transmitting still pictures and one system of television. Any decision by a body like the B.B.C. will have very far-reaching effects, and it is expected that they will have taken every possible step to seek out information about these systems. It is not sufficient for them to wait for things to be submitted to them. **Their responsibilities to the people of this country demand more than a passive waiting attitude, and they must be actively employed in searching for and investigating any alternatives to any system which may have been submitted to them by any one active personality.**

Fields of Utility.

The two types of optical transmission—that of the still picture when some minutes are required to complete one picture, and that of television which transmits a reproduction of actual events in motion—both have their different forms of utility at the present time. The still picture is now being employed in very excellent manner for newspapers, and further for what is called facsimile transmission by the communication companies, both cable and wireless.

More Scope in Television.

In fact we might say that the still picture has reached a stage where any further developments will be along detailed lines. It is not, however, necessarily the case with television that further developments will be only on minor details, for this is a very much more difficult problem than that of the still picture, it is of much more recent growth, and above all its potentialities are unlimited, particularly in comparison with the still picture. **In fact the position is such that the transmission of still pictures is merely a phase of development, and when television has proceeded somewhat further, still pictures will be completely abandoned, just as the development of the cinematograph caused the abandonment of the magic lantern except for special purposes.** We could write at considerable length on the possibilities of television, of actually looking at happenings and people thousands of miles distant, but I shall leave this to the imagination of the reader.

"A Hindrance."

Whether broadcasting of television will become as general as that of sounds is presumably a matter of opinion at the moment, but it is difficult to understand why there should be any doubt about it There may be some reasonable difference of opinion as to how long it will take for television to be widely used in the home, but it is absolutely certain that this state of affairs is coming.

Can we look on the broadcasting of still pictures as a step in this direction? This question appears to be troubling the B.B.C., but in my opinion a very emphatic answer can be given. **The broadcasting of still pictures will be a hindrance rather than a help to the development of the vastly important subject of television.**

Insufficient Interest in Still Pictures.

In this connection it must be clearly understood that the word "Broadcasting" is used here, and not "Wireless," for there is no doubt that in the laboratory the methods of the still picture have been of use in the television field. In broadcasting, however, we are bringing the subject right into the homes of the people, and there must be some justification for doing so.

In the present stage of wireless, where there is a congested ether, and when the ether is guarded jealously by those who allocate a new wavelength for a new service, we must be absolutely certain that there are very good reasons for employing the wavelengths and times allocated to broadcasting for a new service. **It is exceedingly difficult to imagine what reasons, if any, can be put forward to justify the employment of broadcasting facilities for still pictures. There cannot be sufficient interest in any form of still pictures to justify broadcasting of this nature.** The B.B.C. will be well advised to leave this subject to be applied in its legitimate fields—that of the newspapers, and that of facsimile transmission for the telegraph companies.

Broadcasting Weather Maps.

There has even been a rumour that it would be useful for the B B C. to broadcast weather charts, but it cannot be seriously considered by anyone that the general public will wish to undertake expenditure and spend some minutes every day merely to receive a chart which very few will understand, particularly when they can receive without any trouble the present excellent B.B.C. weather reports. Again, let it be clearly understood that we are here referring to broadcasting as at present widely known, for there is something to be said for the transmission of such weather charts to sailors, but surely this application of wireless is not one for the B.B.C.

Provide Facilities for Broadcasting Television.

Many more aspects of the broadcasting of optical effects could be brought forward, but on each of these already discussed there appears to be only one legitimate course open, for in every case, that of the interest of the public, that of the responsibility in allocating wavelengths and time, that of future potentialities, and that of the putting before the public, or that portion of it which is interested in development work, of

(P. & A. Photo.)

Dr. FRANK CONRAD, Research Engineer of the Westinghouse Electric and Manufacturing Co., and the apparatus he has constructed for transmitting cinema films. Its resemblance to C. Francis Jenkins' apparatus, described in our last issue, will be apparent.

something to work on, the only course to follow is to provide facilities for the broadcasting of television.

This of course involves the question whether television has reached a sufficiently advanced stage. There is one company in this country and one only—the Baird Television Company—which has carried television to a fairly advanced condition. In fact Mr. Baird personally is the real pioneer of this subject, and **those who have already witnessed what he has accomplished are almost invariably enthusiastic about the results.** Naturally, perfection has not yet been achieved, "*but actual television is being carried out.*" It is a subject of tremendous possibilities and also of tremendous difficulties, but what is possible at present is very suitable for broadcasting.

Present Possibilities.

It is possible to give small scenes with very good definition, and scenes of the magnitude which is at present possible can be transmitted by wireless means without an undue absorption of ether space. It is, for instance, possible to transmit a single individual so that no imagination is required to recognise him or her. It is possible in such circumstances to notice the movements made by such an individual.

The question may be asked whether it is worth while broadcasting scenes of such small dimensions even though they can be transmitted and reproduced. There is no doubt that it is useful to commence broadcasting with such material at hand, and one way in which this can be used is to show a picture of an actual performer who is broadcasting for telephony purposes. In this case one would have a combined wireless telephone receiver and television receiver, and, if necessary, both sound and vision can be obtained simultaneously; or again, we might concentrate on the acoustical part and occasionally switch over to look actually at the performer.

The Baird International Television Company's stand at the Radio Exhibition at Rotterdam, where actual demonstrations are now in progress.

Value to Advertisers.

It is thus possible to give comparatively small scenes with excellent definition, and it merely remains for those skilled in the art of what is suitable matter for broadcasting, to take up the subject. Many subjects could be chosen by such experts which would interest the public, and the public would be interested because they are viewing "action," or something in motion.

Quite apart from this aspect, there are various applications of television, such as its use by advertisers. In some other countries, particularly the United States, advertisers are allowed to use the broadcasting services, and they combine in this way the two functions of advertisement and the provision of amusement and interest for the public. This form of using broadcasting might with advantage be considered by the B.B.C. for sounds, and there is no doubt that for television purposes at present it will be particularly appropriate.

There is one feature of this subject which may form an imaginary stumbling block, and that is that television may absorb a large slice of the ether, or in more academic language a wide frequency band. I shall deal with this subject in a later issue of this journal, but it is very important to have it clearly in our minds that at the present time and in the present stage of development, a single service of television will not absorb as much of the ether as a present-day telephony service.

A very important feature is that television is a new subject, and one with enormous potentialities and, further, with many problems still to be solved or features to be improved. Many people wish to work on the subject, and they are all needed. It is of the highest importance that facilities should be provided for them, and surely one of the best methods for doing so is to provide some transmissions for them to work with. This will immediately remove a large difficulty from the path of such enthusiasts, that of great expenditure. Workers of this type have already proved their value when they performed their excellent work on short-wave communications over vast distances with exceedingly small power. That subject has advanced considerably since they did their remarkable work, and short-wave communication is now a daily occurrence. **Television needs these workers, and it is essential to provide broadcasting facilities for them to work with. In this subject there is plenty of opportunity for their energies and skill, thousands of times more so than in the case of the broadcasting of still pictures.**

"Hope of Better Things."

The time is ripe for some new application of broadcasting, and the possibilities introduced by the broadcasting of vision are making a strong appeal to the imagination of the people. It is essential that any form of vision broadcast shall not produce

(*Continued on page* 10.)

(*Concluded from page* 8.)

a feeling of disappointment, and the worst form that such can take is where it is not accompanied by hope of better things. **It is impossible to believe that the broadcasting of still pictures will produce anything but the bitterest form of disappointment,** whereas with the really live subject of television very many people will be highly satisfied with results right from the start, and all will retain their interest in watching or in taking part in the march of progress.

The subject is of such national importance that we are looking hopefully towards the Imperial Communications Committee to give due consideration to the problem, and to provide means for the broadcasting of television in the immediate future, and one method for doing so would be to form a new corporation to control the broadcasting of television with powers similar to those of the B.B.C.

TELEVISION IN AMERICA

Many Experimental Transmissions in Progress

By R. F. TILTMAN, A.M.I.R.E., A.Rad.A.

From time to time we have described television developments in America. In the following concisely worded article Mr. Tiltman gives us a résumé of what has been taking place recently on the other side of the Atlantic, and the impression he leaves us with is that, owing to the fact that broadcasting in the States is unrestricted, progress with television is going ahead with characteristic American rapidity.

TELEVISION has become a very prominent subject in America in the past months, for both professional and amateur experimenters have taken up the subject with great enthusiasm.

Several of the leading U.S. radio stations are broadcasting television regularly, many more stations have applied for licences and are planning radio vision tests, and there are thousands of keen experimenters constructing simple receivers. It is said that to-day there are over two thousand amateurs in New York alone equipped with experimental receiving apparatus. In addition, the syndicate which acquired the American rights of the British (Baird) system will very shortly commence broadcasting from a chain of radio stations in the U.S.A., Canada, and Mexico, and commercial receiving sets will be available.

Britain Still Ahead.

Although, of course, television was demonstrated in this country about fifteen months earlier than the first American demonstration, and Britain is still ahead in actual reception results, experimental work in America is now able to go forward at full speed, for broadcasting there is run by private enterprise and development is not hampered by a broadcast monopoly as appears to be the case in this country.

Television transmitting and receiving apparatus was one of the main attractions at the Radio World's Fair, which opened at Madison Square Garden, New York, on September 17th. It was at first rumoured that displays of television would be barred owing to the objections of certain radio manufacturers, but later plans were completed for a television transmitter to be installed, while a number of receiving sets were placed at various points in the exhibition to show images approximately five inches square to the spectators.

Numerous Applications for Television Wavelengths.

Some weeks ago the Federal Radio Commission was dealing with nearly a dozen applications for the allocation of wave-bands for experimental television work. The Radio Corporation made three applications, making a total of twenty bands, two applications were from the Jenkins laboratories, the Westinghouse Company asked for nine bands, and other applicants included W. J. Allen, H. E. Smith, and R. B. Parrish. It was said that more applications for licences were pending.

First Regular Broadcasts.

The first regular experimental television broadcasts in New York

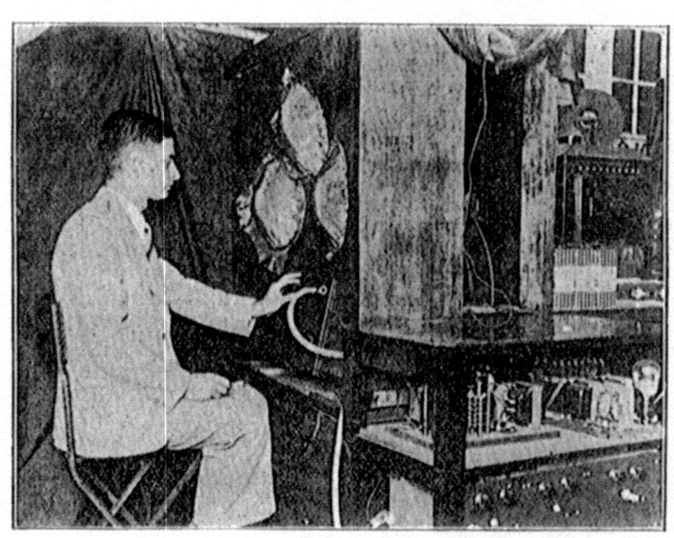

The television transmitter in use at WRNY. Readers will recognise the great resemblance to the Baird transmitter.

1*

started at station WRNY on August 13th last. This station is owned by the journal *Radio News*, and it was stated that thousands of letters had been received asking for the broadcasts. The tests were also sent out

A simple form of television receiver devised for the reception of images broadcast by WRNY. It is a skeleton form of Baird televisor, without any provision for synchronism except a hand-controlled motor rheostat. Images 1½ in. square are produced, according to reports.

by WRNY's associated short-wave station 2XAL, which operates on 30·91 metres. The first tests were received on apparatus installed in the home of H. Gernsback, president of the station, and, according to report, "The images were not perfect, but they were readily recognisable."

From August 27th WRNY incorporated television as a regular nightly feature, the impulses being sent out every hour, on the hour, while the station is working. Despite the opinion of some engineers that television signals in regular programme channels would be apt to cause interference with those operating on adjacent wavelengths, no "overlapping" was reported from these transmissions!

Station WGY, Schenectady, has been maintaining a regular series of television transmissions for several months. The programmes last from fifteen to thirty minutes and are intended for engineers and experimenters. In the middle of August this station claimed to have handled the image of Governor A. E. Smith delivering his speech of acceptance of the Democratic Presidential nomination. Engineers of the General Electric Company focused their transmitting apparatus on the Governor and broadcast the image from the Schenectady station. No detailed reports of the quality of reception have come to my notice.

On July 22nd last the American Continent was spanned by television, and people in Los Angeles reported viewing the image of a man transmitted from Schenectady, about 3,000 miles away. This long-distance reception was carried out by G. Lee and K. G. Ormiston. Special long-distance transmission and reception experiments are now being conducted by these experimenters.

A Play Broadcast.

On September 11th, according to the New York correspondent of the *Evening Standard*, a one-act play was broadcast and televised from Schenectady, and receiving televisors within a range of four miles tuned in both sight and sound. The experiment was said to be "a great success."

Station WOR recently broadcast images of puppets in motion with accompanying sound and music, and television tests have also been carried out by WLEX near Boston, and WCFL at Chicago. Television in sunlight was demonstrated by engineers at the Bell Laboratories, in New York, a short time after the British inventor, J. L. Baird, demonstrated television in diffused daylight in London.

Although eye-witnesses' reports which are to hand do not indicate that the actual reception results in U.S.A. are comparable with the remarkably clear images I have personally witnessed in the Baird Laboratories, London, America has taken up this new branch of science very seriously and very thoroughly, and night after night the ether is filled with "sights" as well as sounds.

Members of the British Association examining Mr. Baird's stereoscopic television transmitter at Glasgow.

TELEVISION AS "BOOSTER"

The Coming Revolution in Advertising

GIVE WIRELESS A CHANCE!

By SHAW DESMOND

In his usual inimitable style Mr. Shaw Desmond points out this month how useful a television broadcasting service could be to national advertisers. Mr. Desmond is very familiar with sound broadcasting conditions in America, and has observed how American advertisers have been quick to make the best use of this valuable new medium. In the following article he suggests how, similarly, television broadcasting might be made an even more valuable advertising medium, without, at the same time, impairing its entertainment value or causing offence to "lookers-in."

THE Upside-Down Man is known to each one of us. He is always standing on his head. He sees everything down side up. And when, by accident, he does stand the right way up and he sees things as they are, he says, triumphantly, "I told you so!"

It is from that man that television advertising is suffering. He is the man who says: "It can't be done. It won't be anything, if it does come." **He is the very identical gentleman who visits the Baird laboratories, sees things with his naked eye, admits he has seen them, and then goes out to tell the world he hasn't seen them at all!**

He is first cousin to those hoary old friends of mine who still parrot: "It can't be done. It won't be done. It isn't yet brass tacks." And then, that variation: "I believe a radical discovery is necessary before television is practicable."

Then there is our dear old friend, the B.B.C., with whom I will deal later, as with broadcasting. But, just for the sake of those "brass tacks," I want everyone concerned to sit up and take note of the following facts:

What are Brass Tacks?

First, that television advertising is so much brass tacks that to my knowledge at least two great newspapers have recently had it under consideration. Secondly, that chiefs of great stores are already weighing up its possibilities. Thirdly, that in the United States (always, unfortunately, first to seize upon new business ideas) radio-advertising has formed the natural bridge to that "television boosting" now being organised across the Atlantic. Lastly, for good weight, I may say that more than one politician is considering its potentialities for election advertising work. And, as super-weight, still holding to our sheet anchor of facts, Mr. Don E. Gilman, manager of the N.B.C. Pacific Coast Network, has recently visited the New York offices of the National Broadcasting Company for one single purpose—that of completing national plans for radio broadcasting, plans which will, I know, in due course lead to National Television Advertising.

The Rip Van Winkle Act.

Let a plain statement suffice, as Kipling would say. I will not burden the already overburdened brain of the B.B.C., or the brains of those who are doing their best to "knock" television publicity, by piling on the agony. But I could fill this page with similar facts.

Do these "candid friends" realise that stereoscopic television, with the figures, so to speak, "coming out of the screen," is an accomplished fact? Do they realise that daylight television is no longer just "a vision" or that the genius of Mr. Baird has made television in colours no longer a pipe-dream? Heaven knows I don't want to wake anybody out of *their* pipe-dreams; but I sometimes wonder, I who have not a halfpenny in television or anything to gain by the statement, whether some of our friends have not been doing the Rip Van Winkle act rather too effectively.

Ridiculous Objections.

Television *is* fact. Television in colours *is* fact. Daylight television *is* fact. Noctovision, or seeing in the dark at a distance, *is* fact.

I have discussed television publicity recently with various public men. Here are their objections taken in order of importance:

(1) That it is "still in the air" and that it'll be time enough to consider it when it comes down to solid earth; (2) that it is "only play acting anyhow"; and lastly, the objection made by those depending for their incomes largely upon the printed advertisement, that "it will kill advertisement by the printed word." (N.B.—This last is often adduced by intelligent men who have just told me that "it can't possibly compete with print.")

The reply to No. 1 is to be found in what has gone before. No. 2 is merely ludicrous and needs no answer. The reply to the last may be put in the words of the biggest

Pacific Coast publicity man: "On the contrary, successful radio (television, etc.) development will create more advertising for all other advertising media than it can possibly take for itself."

Before going farther, I want to get the B.B.C. off my chest (how many others have wished the same thing!).

Concerning the B.B.C.

I am wrongly supposed, if my mailbag does not lie, to be a sort of implacable critic of Savoy Hill. I am nothing of the sort. I have in print and otherwise repeatedly stated that the British Broadcasting Corporation have a most difficult task (under their present Charter an impossible task), that parts of that task they often carry out surprisingly well, and that the better-class items on their programmes are often unimpeachable.

But I don't believe in semi-State-run enterprises. I, being a believer in those "brass tacks," believe also that competition is essential to high standard so long as men are humans and not angels (of the angelic ancestry of the ladies and gentlemen behind the B.B.C. I have little doubt —they are so extra-terrestrial). And **I want wireless run by private enterprise upon ordinary commercial lines.**

If there be any other part of the moon for which I am crying, it is that the B.B.C. or its successors, as the commercial world generally, may recognise that television is going, and pretty quickly at that, to revolutionise the world of broadcasting as well as that of advertising, as I said in a recent TELEVISION article.

How could it be otherwise?

Pictures Oust Print.

My lords, ladies, and gentlemen: for the love of Mike, as my American friends say, how *could* it be otherwise? Pause, my brethren, pause! Consider, not the lilies, but the flowers that grow in the human garden—the flower of the human mind and its method of evolution.

Don't you see all about you how the picture is ousting print? Can't you guess why the cinemas, rotten though many of them be, are packed to the doors? Won't you shake the sleep out of your eyes and ask why is it that the greatest dailies are adding page after page of pictures to replace the printed word?

This may be good. It may be bad. But it is FACT.

People are getting used to thinking in pictures—not words. That is the outstanding fact of our day and generation.

Then what about television advertising now? Isn't it practical politics?

Is there any Rip Van Winkle of them all who can deny it?

The writer, Mr. SHAW DESMOND.

ON this page Mr. Shaw Desmond says: "I want wireless run by private enterprise upon ordinary commercial lines."

"... the picture is ousting print."

"... the greatest dailies are adding page after page of pictures to replace the printed word."

"The daily newspaper that first has the courage to take a lease on the ether ... will outdistance its competitors."

Some time ago I addressed the New York advertisers assembled in conference upon "Advertising and Art." I told them that the day of the picture advertisement, the "living" picture advertisement, was coming. I proved to them, by asking questions of my auditors, that the "dynamic" as opposed to the "static" advertisement was the stuff that would bring the bacon home. And they tacitly admitted that I was right, though I do not pose as an advertising specialist.

"Taking a Lease on the Ether."

The courageous advertisers who first "take a lease on the ether," as I have christened the new publicity, will have an enormous "pull" over the laggards. The daily newspaper that first has the courage to take this lease, when the law permits and the B.B.C., for instance, has undergone a spring cleaning, will outdistance its competitors, "leaving them standing."

In this, as in everything in life, it is the first step that counts.

If I were Mr. Selfridge or Sir Thomas Beecham (now looking for new advertising mediums for his opera scheme), or Lord Melchet, formerly Sir Alfred Mond, I would get to work right away. I would move heaven (that is, the Chancellor of the Exchequer) and earth (that is the politicians) to adopt my favourite scheme of metamorphosing the B.B.C. by turning it from a "peddler of licences" to a "transmitter of wireless."

Convert the B.B.C.

Let the big national advertisers of Britain compete for "time on the air" for the purpose of broadcasting their wares, first by ordinary wireless perhaps, later by that television publicity which is already in sight.

Let the big dailies run their own concert programmes, using, in the beginning, the ordinary broadcasting of to-day, and, later, especially now that television broadcasting is about to commence this autumn, the television screen. For, let there be no mistake about it, the television screen in the private home will soon be as common an object as the loud speaker itself.

Again, we are not dealing with fancies, but facts. Accomplished facts.

And the method?

First, the conversion of the B.B.C., as I have said, from a programme-maker to a programme-sender. Let the B.B.C. of that not distant day—be sure, a corporation of definitely different charter and concept—have charge of the transmitting stations.

And let the expert provide the programme — whether expert educator or entertainer. (And I would say that where education ceases to be entertaining it often ceases to be education !)

Secondly, the gradual reduction and final abolition of the licence fee paid by the public.

Wipe Out the Broadcasting Licence.

"Can't be done," you say? Well, it *is* done. America has done it. There they have one radio receiver to each five or six of her 110,000,000 population, and not a licence fee in the lot! There the revenue comes from advertising, etc. Let England do the same.

And don't forget one other thing, you people who say "it can't be done." In the Land of the Wooden Nutmeg, whatever its faults, there is no *Radio Times* or other programme monopoly publication. Any newspaper can publish the weekly programme of the wireless lists. And any newspaper does.

The third point in the method would be what I may call "the tuning picture," to match the present "tuning note" now broadcast to help listeners to adjust their receivers.

With television as an intrinsic part of wireless advertising a picture, the same as the curtain in a theatre, might be broadcast showing the names of the national advertisers who are to provide the different hours of entertainment during the evening.

Or an attractive little sketch might be presented, showing the actors using advertisers' goods. Properly done, such sketches could be made very interesting, and compel the attention of the television audience.

How to Save the Railways !

So much for the *method*, following my replies to the usual objections as to why "it can't be done." Now for a few major fields in which television advertising can be used with effect.

The British railways are losing a million a month we have just heard. One of the reasons, apart from freight, is that British people don't use their railways enough. Men and women in Tooting or Finsbury Park, in Bradford or Birmingham, Liverpool or Lower Mudcombe-on-Mud—these men and women will do anything possible to avoid travelling farther than fifty miles to their summer holiday resort.

2*

If you could throw upon the television screen, either in the cinema or the home, in the theatre or music hall, a living picture "actually happening" of a railway journey through delightful surroundings, with the train climbing between mountains or skirting great sea-swept bays, with the passenger seated in well-appointed carriages in which cheap, tasty, clean food was served—don't you think our Tooting and Liverpool friends might be induced to travel further afield to find freshness and fresh air?

But perhaps railway directordom won't think that brass tacks.

POINTS on this page: Mr. Desmond describes how the present broadcast licence fee can be wiped out, making listening-in free for all. The expenses in connection with broadcasting will be met, under his scheme, by national advertisers who want to "buy time on the ether." He also suggests how, under this arrangement, the Chancellor of the Exchequer can be assured of a greater income from broadcasting.

Yet it is these very directors who have been using widely the immensely less effective poster advertisement to get people to travel to "Breezy Bexhill" or "Sunny Skegness." **The "pull" of the television picture is, I should say, three or four to one as opposed to the "static" poster.**

Or you have a fashionable stores. You sell dresses—or you want to sell them.

To-day you buy a page in a society paper on which you print those often ghastly doll-like imitations of the female form divine, and underneath the price of the dress. To-morrow, the living *mannequin* will live and move and have her being before you upon the television screen.

Chic you say? What's the use of *chic* in a dress if you can't show it on the *living* figure?

Before this old earth has circled the sun half a dozen times more, and long æons before the Rip Van Winkles who, unfortunately, infest not only Government but other offices, have awakened to the fact that television is here, we shall be selling by television everything from soap to scent and from that golden powder now being used by the *Parisienne* as the *dernier cri* in complexion preparations, to puffs and pomatum. **For, again don't forget it, colour television (another of the Baird *tours de force*) is fact.** You will be able to *see* the mellow tinge of that soap, the fascinatingly dinted flask of that scent, and some day—who knows?—Baird or some other may find a way to send a scent over the wireless!

Possibilities for the Government.

In all this, it might not be a bad idea for Mr. Churchill, who is usually a trifle more wideawake than some of his dearly-beloved brother Rip Van Winkles, **to consider the possibilities of television for what I will call Government advertising. Also, let him get down to those brass tacks beloved of Chancellors and let him gently probe the possibilities of increased revenue from the same source.**

As regards the latter, all he has to do is to transform broadcasting as suggested above (and I defy any man who thinks to find it impracticable) and charge a fixed percentage upon all national wireless advertising, whether television or otherwise. It would be a thousand times easier, and more effective, to tax broadcasting at its source, than the present ludicrous and ineffective method of trying to tax millions of individual listeners. **Apply taxation to the bottle-neck, to those national advertisers who want to "buy time on the ether."**

Use the Television Screen for Emigration !

As regards the former, if he wants a solution to the overcrowding of England and the prevention of "forced emigration," now being put forward, let him look to the television screen.

If we could show in the overcrowded districts of this little island free life as it is lived *at the moment of living* in the Australian bush, on the Canadian farm, the South African karoo, or anywhere in Britain's wide-flung Empire, the Government

(*Continued on page* 20.)

(*Concluded from page 17.*)

would not have to use the whip to get men to emigrate. They would eagerly go themselves. *It is chiefly their fear of the unknown that hinders them to-day.*

"Live" versus "Dead" News.

But it is the showing of events *at the actual moment of occurrence* that is the supreme pull in advertising. The modern dummy film, made up of dummy men and women acting for the purpose, has a dummy, sometimes a dumbfounding, effect. Even the record of "natural" happenings, when they are past, loses its savour, as does what Fleet Street knows as "news." The value of news is in inverse ratio to its age.

The television advertising screen, recording things happening at the moment, will overcome all this. And the day is not too far distant when every section of our national human life, from a Westminster politician to a Westinghouse brake (so often having the same effect), from a department of emigration to a department of agriculture, and from a pound of sugar to a pound of soap, will find its place on the television advertising screen.

Did I hear somebody murmur again: "Can't be done?" If I did, it must have been Whitehall murmuring in its sleep. For no man who will take the trouble to re-read this statement of fact, no man who is a business man with imagination, will say it can't be done.

It is being done—and that's that!

THE ENTERTAINMENT VALUE OF TELEVISION *TO-DAY!*

Being an eye-witness's account of the public demonstrations—and the lessons they taught

By R. F. TILTMAN, A.M.I.R.E., A.Rad.A.

THE first public demonstrations of the Baird television system, which were held in premises adjacent to Olympia, London, during the time the National Radio Exhibition was open, naturally aroused immense interest.

Three commercial receiving televisors of the "super" type were installed in premises in Maclise Road, London, W., each being in a cubicle of blue curtains. Six people at a time were admitted to each cubicle, and after inspection of the receiving apparatus the light was dimmed and a short demonstration of combined sight and sound was given.

No more than eighteen persons could watch the reception at one time, and during the first day or so each little audience witnessed a fifteen minutes' programme, but afterwards, owing to the exceptionally heavy demands for tickets, the arrangements were speeded up and a demonstration was given practically every five minutes of the afternoon and evening. In this way about two thousand engineers, radio traders, and members of the public witnessed the demonstrations—although that does not represent one-twentieth part of the number that applied for admission tickets.

Eye and Ear in Step.

On the reception screens one saw the living head-and-shoulder image of the person being "televised," and the voice was heard from the moving-coil type loud speakers alongside.

The images were brightly lighted, appeared to be somewhere about 8 in. by 4 in. in size, and the synchronism of voice and movement was perfect. One saw the whites of the eyes and the eyelashes, the teeth could be counted, and every play of expression was clearly seen. The collar and tie of the men subjects could be seen, and in the case of several of the artistes dimples were noticeable.

WE reproduce below a facsimile reproduction of an account of the first transmission by television of an advertisement. The little experiment described here provides food for thought, and clearly indicates that we have in television an entirely new advertising medium. Out of this experiment, what will develop?

"DAILY MAIL" TELEVISED.

CONTENTS BILL SEEN THROUGH BRICK WALLS.

A *Daily Mail* contents bill was seen at a distance through several brick walls yesterday at the public demonstrations of the British television system which are being given during Radio Exhibition week at Olympia, Kensington, W.

Mr. R. F. Tiltman, a television expert, suggested that an attempt should be made to send moving lettering by television. He obtained a *Daily Mail* contents bill from a newsagent and had it sent to the transmitting studio a short distance down the road.

A few moments later the person being televised announced through the loud speakers, "We will now show you the contents bill of a London daily newspaper."

The image of the sitter faded from the reception screen, and gradually from the blur of orange light emerged in bold type the word "Daily" This moved across to the left and gave place to "Mail" The rustling of the bill could be heard from the loud speakers as the audience read from the slowly moving letters "24 Pages Again: County Prize Beauties Pictures"

The wording could be clearly read, and this hastily arranged test may be said to constitute the first advertisement to be sent by television in the world.

The three receivers were of the standard commercial type. A grill on the left concealed the loud speaker, and on the right was mounted a circular viewing screen about 12 inches across, on the lighted portion of which the *Daily Mail* lettering appeared.

I was reporting these demonstrations for a certain paper, and was therefore present every day.

Although a number of professional entertainers appeared on the screens at different times, a good deal of the demonstrations consisted of impromptu "divertissements" by members of the Baird laboratory staff.

In this respect the "palm" for the week must surely be awarded to Mr. A. F. Birch, a young technical assistant who spent hours before the transmitter. His features came through exceedingly well; he has developed an excellent "broadcast" manner, and he held the attention of the audience with choruses of popular songs and an easy flow of amusing chatter accompanied by facial gestures.

Television "Uncle."

It seems very evident that as the broadcast television service in this country develops we shall see and hear a lot more of this Mr. Birch, and when a children's hour is included in the programmes he should easily prove the most popular "Uncle."

A number of leading lights of the theatrical world attended these demonstrations and gave short "turns" before the transmitter, which for convenience was located in another shop a short distance down the road.

On the opening day, Saturday, the first of these was Miss Peggy O'Neil, the charming and popular actress who is so well known to all theatregoers. She was starring in "The Flying Squad" at the Lyceum Theatre, London, at the time.

Miss O'Neil watched one of the reception screens for about ten minutes while Mr. Birch was going through his amateur performance. She then agreed to be "televised" and was escorted by Mr. Baird to the transmitting studio. Immediately Miss O'Neil was seated before the transmitter her living image (instantly recognisable without the faintest trace of doubt) appeared on the screens and her voice was heard from the loud speakers.

A charming photograph of Miss Peggy O'Neil, one of the first professional actresses to appear before the televisor.

For about half an hour Miss O'Neil gave us a most charming entertainment, chatting and smiling at us, telling Irish stories, and, in response to telephoned requests, she sang "I'm a little bit fonder of you" and several other delightful songs. Finally, although it was not at all apparent, Miss O'Neil apologised for what she termed her "modest effort" on the grounds of having a severe cold.

The Television Face.

It was most fascinating to hear the familiar Irish-American brogue from the loud speaker and see on the screen the well-known features with captivating flashes of the eyes and dimpling of the face. Incidentally, among those who had the good fortune to witness this performance was Mr. Sydney A. Moseley, the well-known writer whose name is familiar to readers of this magazine.

I chatted for some time with Miss O'Neil afterwards, and she agreed that there was nothing to make anyone nervous in being seated before the electric "eye" of the transmitter. As one who had witnessed numerous television demonstrations previously I was able to assure Miss O'Neil with perfect truth that I had never seen a subject "come through" on the screen better.

On the Monday evening Mr. Harry Tate, of music-hall fame, attended the demonstrations with Mrs. Tate and their son. After inspecting the receivers and witnessing the reception for some time Mr. Tate senior and junior went round to the transmitter, while Mrs. Tate and I stayed before one of the screens.

Tate—Senior and Junior.

Very soon Mr. Tate's features appeared on the screen and the voices of him and his son could be heard in a short excerpt from one of the well-known music-hall sketches. It was found that the famous Tate moustache did not come through at all well on the television screen and it had to be removed.

Afterwards Mr. Tate junior sat before the transmitter for a quarter of an hour and kept an audience highly amused with a collection of funny stories and impersonations of Harry Champion. He came through on the screen very much better than his father, although that, of course, could have been put right in a normal television programme by the use of correct make-up.

During Monday afternoon I noticed that one of the most interested visitors to the demonstrations was Dr. J. A. Fleming, F.R.S., who spent over two hours inspecting the transmitter and receivers.

The Lyceum Theatre
Strand. London

Dear Mr Tiltman — —
Many thanks for your kindness in sending the cuttings from the Daily Mail and the other paper. My visit to the Television studio was most interesting and the whole thing simply fascinated me. Especially when watching the reception, or rather it was certainly no ordeal to sit before the Transmitter and the light in which I was televised caused no discomfort whatever. This Electrical sight is really too wonderful for words and it seems that it must develop into a really big entertainment force in time.

Sincerely Yours
Peggy O'Neil
Sept 28/28

We print above a facsimile reproduction of a letter which Miss Peggy O'Neil wrote to our contributor, in which she records her impressions of television, and what it felt like to be televised.

Miss Lilian Davies, the well-known actress, was the next stage celebrity to undergo a television test. She came and viewed the reception screen one afternoon and then went and sat before the transmitter and chatted to the audience. Miss Davies was another whose voice and features came through exceptionally well.

There was a good variety in the programmes on the Wednesday afternoon and evening. Harry Tate paid a second visit with his son, and with Mr. Marriot Edgar, the entertainer, kept things going with a swing, and one or two other professionals appeared on the screen. One of the real tit-bits, however, was when our friend Mr. Moseley was asked to go to the transmitter again. He went obediently to the studio and let us see his jovial countenance, and then as the pianist was picking out a few bars of various well-known airs Mr. Moseley gave us his *very* bass rendering of the tunes—he did not know the words, but that was a mere detail and nobody minded.

Miss Cicely Dandy, a singer with most expressive features, appeared many times in the programmes. I saw most of the artistes when seated before the transmitter, and none appeared to experience the slightest discomfort from the necessary lighting.

And now to come to the real purpose of this article. Is there any entertainment value in television as it is *to-day?* In view of the fact that at the time of writing it is expected that some form of broadcast television service will be started in this country in the immediate future, this question is one of prime importance.

Great Entertainment Value.

The majority of people who witnessed the public demonstrations of the Baird system for any length of time will be in no doubt on this point. Having personally been present for hours on end every day of the demonstration week I am not for one instant in doubt.

There is a real, definite entertainment value in television as it is to-day—allied with telephony, of course.

The dear old so-called "authorities" who have used up so much ink during past months in letters to the Press damning television will not like that statement at all. They have settled themselves so firmly in their arm-chairs and "proved" (to their *own* satisfaction) that television doesn't exist, it cannot be done, it needs a radical discovery before being a practical proposition, that—but why repeat all the old parrot-cries?

The fact remains, however, that while these ill-informed critics have *for months belittled a British television system which they had never seen demonstrated*, that system has been developed up to a point where it is well fitted to be introduced to the large body of interested amateurs.

Wonders of Electrical Sight.

I am not by any means alone in suggesting that television is entertaining, for I heard so many opinions expressed by members of the public as they left after witnessing the demonstrations, and they not only referred to the wonders of this electrical sight, but they had quite evidently *enjoyed* the brief sight-and-sound programme. I also had the opinions of other Press representatives present.

A few weeks back *Amateur Wireless*, one of our leading radio journals, gave a very fair report on two television tests specially arranged for them by Mr. Baird, and in the course of this they mention that the transmission "was in every sense enjoyable." One of these tests, by the way, was a wireless test with signals broadcast from the Baird London station in Long Acre.

Writing in the *Daily Express* some time back Mr. Moseley, after viewing a demonstration, referred to the experience of looking-in as "fascinating." He went on to refer to the test as the most stirring sight he had ever experienced, and mentioned the interest and awe aroused by this vision of distant objects.

There is not the slightest doubt that television *now*, in its early stage that is so pregnant with possibilities, has a real, enjoyable, entertaining, awe-inspiring interest.

[Photo by courtesy of G.E.C. (U.S.A.)

Izotta Jewel, former star of the American stage, and now the wife of Professor Hugh Miller of Union College, U.S.A., became first leading woman of a play presented by television. The radio audience, who had suitable receivers, saw and heard Miss Jewel in the presentation of "The Queen's Messenger," by J. Hartley Manners, which was broadcast by WGY of Schenectady recently. Miss Jewel is shown standing before the "camera," which consists of two smaller cases containing the photo-electric tubes and the larger and highest case containing the scanning disc and the light source. A microphone picked up the lines of the drama.

(Photo by courtesy of G.E.C (U S A.)

A PLAY BROADCAST BY TELEVISION IN AMERICA.

In the centre is the director, who controls each of the three television "cameras." By a twist of a knob he brings any of the three "cameras" into the circuit. At the left is Izotta Jewel before one "camera" (camera in this case refers to the unit containing the scanning disc and light source, and the two eyes or boxes containing the photo-electric tubes) and before the camera on the right is Maurice Randall. The two people in the right foreground manage the "props," which are placed in the view of the third "camera." In the left foreground, facing the director, is a television receiver, in which the director is able to check the image as it is broadcast.

Among the anti-television "die-hards" who will violently disagree with me must be included the editor of a certain wireless paper and some of his staff. That paper has for months been most virile in its campaign against British television, although it was ready to throw bouquets abroad at the slightest vague suggestion of any early research work by foreigners.

Uninformed Criticism.

But—note this carefully—I was given definitely to understand that up to the time of the public demonstrations in September *none of the editorial staff of this paper had witnessed a single demonstration of the system they were so gallantly crabbing*, and I certainly saw no sign of their presence during the demonstration week!

This state of affairs has become positively ludicrous to persons in touch with the real facts, but at the same time it is possible that there are some people who take such violent anti-television propaganda seriously—for at times it is dished up in a manner worthy of a far better cause.

Now to get back to a fact—television is an enjoyable entertainment in its present early stage. Bah! (or words to that effect) snorts the television critic, rapidly sorting over his parrot-cries. Very well then; let us analyse the matter a little closer.

Take, for example, one item on the long list of entertainments enjoyed at the Baird television demonstrations. As I said previously, Miss Peggy O'Neil went before the transmitter; this was the first time for her, and she was without any special make-up and had no schooling as to how to conduct herself in such ultramodern circumstances.

Perfect Synchronism.

At the receiving end I watched the screen and heard the voice in perfect synchronism for half an hour, and it was just real delightful entertainment, far more fascinating than could be any appeal to one sense only.

"That is no test at all," will insist the typical wilfully obstinate critic. "It was merely the novelty that had a fleeting appeal."

That theory of novelty only will not hold water in this particular case.

Viewing the television screen has got well past any fleeting "novelty" stage for me, for I have witnessed dozens of demonstrations in the past two years, and in the one week of public demonstrations I was watching the reception screen for hours on end each day—actually I have probably witnessed more television reception than anyone else in this country, excepting the regular staff of the Baird laboratories.

For point number two. It was no "novelty" for me to be entertained by Miss Peggy O'Neil, for I saw this star several times in her famous lead in "Paddy the Next Best Thing," and have seen her from time to time in her various London successes. It happens that I saw her in her latest rôle at the Lyceum Theatre only a very short time before witnessing her television performance.

Seeing for Themselves.

Lastly, it was no "novelty" to hear Miss O'Neil's voice via the loud speaker, for I have listened previously to her broadcasts from B.B.C. stations.

Altogether, this theory of a purely "novelty" appeal does not seem to fit in at all, and so we are left with the unquestionable fact that the appeal was on purely entertainment grounds.

Our illustration shows the Stand of "Television" at the recent Radio Exhibition at Olympia.

It was just the same in the case of Mr. Birch, the young member of Baird's staff whom I have watched for hours in all without any lessening of interest. I attended this week of public demonstrations from business reasons, but I found it mighty entertaining business throughout the time!

Now that television is so definitely here, and has been demonstrated fully and accepted by so many as a real entertainment force, let us leave the few remaining canting, carping, crabbing critics to waste their parrot-cries on thin air—for television has come out into the light of day; it has nothing to be ashamed of, and the public are seeing for themselves and are learning the true facts.

What should this television system be ashamed of when a reputable and impartial paper like *Amateur Wireless*, after a thorough investigation, used phrases like: "We must emphasise that the image is instinct with life . . . we must state with pleasure that it is miles ahead of the televised picture of three or four years ago. . . . Now we have the actual face . . . every movement is observed clearly . . . the image was a better one than we expected to find. . . ."

My Impressions of Television

By Dr. FRANK WARSCHAUER

Dr. Warschauer, who is a well-known scientific writer both in Germany and in this country, has had unrivalled opportunities of comparing the various Continental television systems with that of Baird in this country. In the following article he gives our readers a first-hand account of his experiences.

AFTER hearing of the great progress in television which had been made by Mr. Baird, I came specially to London to see his apparatus. Before leaving Germany I had already seen the apparatus of Karolus and Mihaly.

At the annual radio exhibition at Berlin this year three systems of television apparatus were shown, by the Telefunken Co., Dr. Karolus, and Denys von Mihaly. Details of the Karolus apparatus were given in the last issue of this journal, but no information as to the results obtainable was published. I have been fortunate in being able to attend both the Berlin and the London radio exhibitions, and also in being able to witness demonstrations of the three German systems mentioned above, and also of the system of Mr. Baird here in London.

Mihaly's Receiver.

On the screen of Mihaly's receiver, which was very small (about 4½ inches square) there was to be seen a shadow of a pair of moving scissors, and then single black letters, which spelt out the name "Mihaly," etc. These letters were moved at the transmitter.

In the case of the Telefunken demonstration the receiver and transmitter were about 40 yards apart. In this demonstration a transparent photograph (like a lantern slide) was placed before the transmitter. On first looking at the receiving screen one saw a flickering light, and on looking closer one could recognise the picture of a girl. **It was not a girl personally who was being transmitted, but a magic lantern slide which Dr. Karolus moved back and forth at the transmitter, and the movement could be seen at the receiver.** The image was, however, lacking in detail. The eyes, for example, could not be seen clearly, but merely as shadows. This receiving screen was about 80 centimetres (about 32 inches) square.

Karolus also had a small machine running which made use of the well-known Nipkow spiral-holed disc. To illuminate this image he used an arc lamp, the light of which is varied by the action of a Kerr cell, through which the beam passes. In the larger apparatus Karolus does not use a

[P. & A. Photo.

At the Radio World's Fair, held at Madison Square Garden in New York City, Miss Lita Korbe, "Queen of the Radio," was televised as shown in this picture, taken September 19th, at the show. This apparatus is made by the Carter Radio Co., of Chicago.

Nipkow disc, but a wheel around which are arranged little mirrors (see page 35, October issue of TELEVISION). The image seen on the smaller screen was much clearer than that shown on the larger screen. Again he showed the same lantern slides, and showed movement by moving the slides at the transmitter. **He did not give any demonstrations of actual television of objects or persons, but showed only the transmission of lantern slides.**

The German public, whilst appreciative of the results obtained, recognised that it was not yet television, but only the beginning. **I therefore did not see television until I came to London.**

What I saw made a tremendous impression, because it was indeed the first time that I had seen real television—all that I saw before was only the beginning. **I think the British people will be very interested to hear that Mr. Baird has been so successful,** because he has already placed on the market a commercial instrument.

Especially I was very interested to see how distinct were the persons transmitted, because the engineers and inventors in Germany say it is impossible to get the results which Mr. Baird is getting from the neon lamp. They think that the reproduction must be too dark to recognise the persons. But I saw very distinctly many particulars of the face, the eyes, facial movements, etc., and there is no doubt that the problem of television has been solved by Mr. Baird's system.

Baird System Pre-eminent.

I am very glad to have seen this apparatus, and I think that the future of the Baird system will be pre-eminent. For many years I have been writing about broadcast developments, because I believe broadcasting is one of the greatest inventions; but I think that the invention of television, and its practical development, so that every man can have television apparatus, is even greater, and still more important to the human race.

Broadcasting is international. When I sit in Berlin at my apparatus I can hear all stations, especially

(*Continued on page 44, col. 3.*)

(*Concluded from page 22.*)

London and Daventry. But there are also great difficulties, because most of the people in foreign countries do not understand the languages which are used in other countries. But in contrast to this television is indeed international. **There are no frontiers of language to separate the nations—you see with your eyes, and every man can understand sight.**

The invention of television will bring all nations together in a way that could never previously be imagined. When you see on the screen of your apparatus first the person or persons speaking in other countries, and later when you see the important events happening in all countries—then you will really be able to make acquaintance with all sorts of important events in all other countries. Therefore I think that television is an even more important invention than broadcasting, and I follow with the greatest interest all the stages of development. Mr. Baird's first public television broadcast will be one of the greatest moments in the history of this invention.

IMPRESSIONS AND OPINIONS OF A LAYMAN

By A. W. SANDERS

I'VE seen television; I've not only seen the living picture of a man on the screen of a televisor, but I've realised, with some effort, that this man was sitting about a mile away from me; that between him and myself there were many houses, trees, moving vehicles, etc., etc. This may sound funny, but it isn't; first because it is true, and second because I had to think of these things before I could grasp the meaning of television.

During the demonstration near Olympia I was interested, and naturally concentrated on what I saw on the screen; but unconsciously—I only know it now—I compared the televised image with a cinematographic picture, so that when I left the building I felt curiously dissatisfied and puzzled. In fact I felt quite angry with myself for not having that exhilarating feeling of having witnessed something absolutely marvellous. I looked at the others who had also been privileged to see the demonstration. I heard what they said and they all gave me the same impression. This unconscious comparison between early television and full-grown cinematograph cannot but do a lot of harm.

Having once realised all the objects that divided the televised person and myself, having impressed upon my unbelieving mind that my eye could not see him in the ordinary way, that I could not touch him, I began to wonder at it all, and I wanted to know all about it. So I bought a book on the subject and read about the selenium cell, the scanning of the picture, the optical lever, the synchronisation and phasing; and the more I read the more the comparative simplicity of the whole thing struck me.

My interest in television is not that of the experimenter, but solely that of the man in the street who sees in it something that will greatly enhance the joy of life, that will make all sorts of hitherto impossible things possible.

As a result two questions arise:—

1. How does this new invention affect me now?
2. What of the future?

It is clear that broadcasting of television from some station that I can pick up with my set is essential; but if that happened to-morrow I am quite sure that even in this stage of development it would give me distinct enjoyment; to see *and* hear Harry Lauder televised must be very much more fun than only to listen to him. Were it possible to get Chaliapin televised while singing his famous Song of the Flea (see him raise his eyebrows!), nobody who had the opportunity would miss it.

THE EDUCATIONAL VALUE OF TELEVISION.

Our illustration shows Mr. G. Holme, Editor of the "Studio," delivering a broadcast lecture on art pottery. By means of television he is able to illustrate his points in a thoroughly convincing manner.

These are not the only practical uses to which this invention can be put immediately. Dr. J. A. Fleming writes that television, even as it is at present, could give a lesson on lip-reading. A lesson on foreign languages would be of much more educational value televised and sound-broadcast (because the synchronisation is perfect) than broadcast simply as at present.

There are, no doubt, a great many more uses to which television can be put in its present stage, even when leaving the commercial application, such as advertising, wholly alone. When, however, the screen is made appreciably bigger the amusement and commercial application will be limitless.

Now that it is possible to "see in" the question of price becomes important. Compared with the prices asked for the first wireless sets, television sets are cheap; and the one which will be the most sought after will be Model B, priced £40, because it gives both speech and sight, has a moving-coil loud-speaker, and can be attached to two ordinary wireless sets capable of giving good loud-speaker results.

There is one point which, I have no doubt, has irritated prospective buyers of wireless sets as much as it has me. I mean that prices quoted for sets, till recently, never included valves, batteries, royalties, or other essential items, and these were not even priced. At the last exhibition this was corrected by quite a number of firms.

The televisors A and B require a voltage of 350, but the price of the required transformer is not quoted. I quite understand that this price varies considerably according to the supply current, while, I imagine, the full manufacturing problem of the televisors is not yet fixed. This question of all-in price therefore cannot yet be settled, it seems; but when once a bid is made for the greatest possible demand I do hope that the above remarks will be borne in mind.

How long it will be before I can enjoy both speech and vision by wireless I do not know. How long it will be before television will be perfected is equally unknown to me, but one thing I do know: THE SOONER TELEVISION IS BROADCAST THE BETTER.

NOW, THIS IS TELEVISION!

By Dr. ALFRED GRADENWITZ

In our last issue we published an article by Dr. Frank Warschauer, describing television demonstrations given at the Berlin Radio Exhibition by Mihaly and Dr. Karolus. Dr. Gradenwitz has also seen demonstrations of the Mihaly and Karolus apparatus, and in the following article he compares the results of these two systems with what he has seen demonstrated in London by Baird. Dr. Gradenwitz is very well known in this country, in many European countries and in the United States as a writer on scientific subjects. It is of interest to note also that the earliest trace we can find of the use of the word Television is in an article of his published in France as far back as 1904. Ever since that date Dr. Gradenwitz has followed television developments with a very keen interest. Considerable importance and interest, therefore, attaches to the following article.

WHEN, about a year ago, I came to London on a short visit, mainly with a view to studying at first hand the Baird system of television, I occasionally amused myself by asking people with whom I came in contact whether they had ever heard of television. In most cases I received negative replies, which gave me a welcome chance to deliver an impromptu speech on the subject.

I am now in London again on the same errand, and whenever I ask similar questions people seem to have become much wiser and know what I am talking about, and require no lengthy explanation to understand. In fact, television seems to have become a household word.

What is Television?

However, did I myself know what television actually was? Of course, I have always been familiar with what it was supposed to be, and what it was supposed to become, **but I never before met the real thing in such complete and perfect detail.** Many a time I have thought it was actually television that I was being shown. For instance, what Mr. Baird demonstrated last year well deserved that name, and in spite of unavoidable imperfections, it not only held promises for the future, but had already achieved something tangible and valuable.

At the recent Radio Exhibition in Berlin this year I was given the opportunity to examine very closely the apparatus demonstrated by Mihaly and by Dr. Karolus of the Telefunken Company—the great German wireless concern of which much has been heard. **Both Mihaly and Karolus merely transmitted lantern slides, and Mihaly sometimes sent the shadow of a pair of scissors; these transmissions were only shadowgraphs.** Mr. Baird has given me an equal opportunity of examining his system of television, and as a result of my examination I say that **now for the first time I have witnessed true television. I have seen the image of a living face with detail, both by wire and by wireless. Unquestionably the Baird system is immensely in advance of any system on the Continent.**

I had, of course, expected to find television greatly improved since I was here last year; I expected better definition and more detail, but I was not prepared to see the progress that has actually been made. The same persons with whom I had been conversing a few minutes earlier, one after the other appeared on the screen with an amazing truth to life, a remarkable clearness, and an astounding wealth of detail enabling, for instance, an outline of finger-nails, the smoke of their cigarettes, and even the lines on their faces to be seen distinctly.

The most wonderful effect of all, however, was an unexpected plastic effect endowing the figure on the

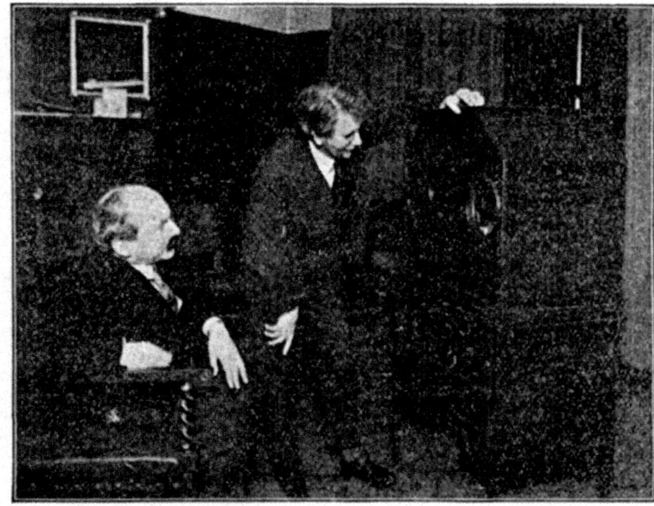

Mr. Baird demonstrating television by wireless to Dr. Gradenwitz (seated).

screen with an almost uncannily life-like appearance. **The image was almost stereoscopic,** although not deliberately designed to be so in the case of the apparatus I was examining. Last, but not least, I was greatly impressed by the perfect combination of sight and sound which was bound to enhance the illusion.

I think the simultaneous reproduction of voice and vision mutually assist each other. While one looks at the image the loud speaker seems more perfect. When listening to the loud speaker the visual effects of television are even better appreciated.

Nor should the fact be overlooked that only a commercial apparatus was being demonstrated this time. **Everything had definitely outgrown the experimental stage and was in a form which could safely be put in the hands of the wireless amateur or the man in the street.** In Germany every encouragement is being given to research workers in television. In fact, one reason for my visit to London is because **the German Post Office is at present negotiating for the purchase of the German rights in the Mihaly system** with a view to incorporating television broadcasting with the existing sound broadcasting service.

Dr. Gradenwitz seated before the televisor in the new Baird studios. Standing behind Dr. Gradenwitz is Lord Angus Kennedy, Vice-President of the Television Society.

"I am amazed."

In view of this fact I am amazed at the attitude of the British Broadcasting Corporation. I have learned with interest that steps are being taken to transmit English sound and sight programmes from the Continent. This, of course, can be done very readily from existing stations without in any way interfering with the present wavelength arrangements, but **it seems to me extraordinary that a British invention should be unable to obtain facilities for its development in the country of its birth.** However, I feel sure that this state of affairs will only be transitory, because the British authorities are certain very soon to appreciate the value of television.

There is no language in sight. Television broadcasting, unlike speech broadcasting, is truly international because a face is just as thoroughly recognisable to a German whether it be an English face, a French face or a Dutch face. And that, I think, is one of the greatest powers that television possesses, being capable as it is of bridging frontiers and bringing the nations of the world closer together than ever before. **Those who in the near future will be wont to see men and women of all nations appearing on the television screen will not be inclined to nourish unfriendly feelings against foreigners ;** and there will be a sounder basis than ever before in the past for mutual understanding and goodwill.

Incidentally, I may point out that here is one reason why the "Talkies," as you call them over here, will never supersede the silent film. Silent films are international, whereas the "Talkies" are limited to the country in whose language they speak.

But I am not here to criticise the latest film invention. I am here to investigate the Baird system of television, the importance of which does not appear to be recognised, and I am truly *amazed* at what I have seen. There is an immense future for this invention. At the same time television will prove an invaluable adjunct to wireless broadcasting by adding visual perception to sound received by the ear. An educational lecture delivered by a person visible on the screen and accompanying his or her words with appropriate gestures and effects will impress far more, and produce a far more lasting effect than broadcasting which appeals to the ear alone. Such a combination will, in fact, be an almost perfect substitute for the actual presence of the lecturer or orator. **A singer, actor, or comedian, seen as well as heard, will impress a radio audience much more profoundly than the same person perceived only by the sense of hearing. This is particularly so in the case of broadcast humour.**

Mankind has ever been intent on freeing itself of the fetters of space and time. In fact, much of the progress of human civilisation is the direct or indirect outcome of those endeavours. I believe television to be the crowning step of the efforts of humanity to annihilate space and time.

The Future of Television

By Dr. C. TIERNEY, D.Sc., F.R.M.S.

Vice-President and Chairman of the Executive of the Television Society.

WE have recently read in a leading London newspaper, the *Morning Post*, a report of an interview between its representative and an official of the B.B.C, which purports to challenge Mr. J. L. Baird to come forward with any new development which will enable them to reconsider the question of affording facilities for broadcasting television. Without entering into any discussion as to the merits or demerits of this mode of negotiation, or whether the desired end is not more likely to be achieved by co-operation rather than by pseudo challenges, a few observations upon the results of some recent tests will be of interest.

While theorists are still debating whether the disc and spot-light method of exploring is capable of scanning anything more than a very small object, and whilst they are theorising on the speed of the disc in terms of millions of revolutions per second which make one giddy to read, what are the facts? J. L. Baird has repeatedly demonstrated to scientific and other competent observers, as well as to the public, the adequacy of his own method for the transmission and reception of televised images.

The image of the head and shoulders of the subject is received with complete satisfaction to all, and more recently he has transmitted *a whole stage scene* showing two athletes giving an exhibition boxing-bout to demonstrate the practical application of his system to larger scenes. The latter demonstration, which the writer, amongst others, was privileged to witness, was carried out from a stage some 15 feet by 10 feet, temporarily constructed for the test.

The scene, received in another room of the same building, clearly depicted the small, but recognisable, images of the combatants and their every movement, which at times were particularly rapid, as blow upon blow was exchanged, and one or other would speedily dodge or retreat in order to escape an impending disfigurement. The reception only needed the loud-speaker attachment to render audible the exchange of blows, and perhaps the remarks, to complete the realism.

Dr. C. Tierney, D.Sc., F.R.M.S., Vice-President and Chairman of the Executive of the Television Society.

A further difficult and exacting test was carried out with equally satisfactory results. **A cyclist, riding a bicycle round a ring, illuminated by the same method, was transmitted to the same receiver, which accurately showed every movement, both of the machine and the rider in motion and without any question as to identity or direction of movement, which abundantly justified our expectation.**

I have referred to these two experimental tests in order to show the practical application of the Baird system to extended scenes, and if further evidence were necessary to emphasise the possible development and potentiality of this system I may perhaps be permitted to refer to the subject of projected television, i.e, the projection of the image on to a large screen.

In company with a number of distinguished visitors to Mr. Baird's laboratories, I subsequently witnessed the received image of a well-known person **projected on to a screen some four feet in diameter, which could be seen and recognised by a large audience.** The result, though as yet not fully developed, was astonishing Not only was every movement of the head, the eyes, the lips, etc, reproduced with fidelity, but also those subtle expressions of pleasure or annoyance, of joy or grief, truthfully portrayed.

These few facts alone are sufficient

"WHO'S ROOM IS THIS ?"
A scene from the play "Box and Cox," which was broadcast, both orally and visually, from the Baird Company's experimental transmitter at Long Acre. The entire scene, as photographed, was televised.

to show that **there is in these developments a potentiality as yet unappreciated in this country.** Foreign governments and powerful organisations from abroad are concerned to acquire rights and privileges in these which our own authorities are so reluctant to secure, and which, in the opinion of those experts most competent to judge, are more than sufficiently advanced to justify trial through any of the British broadcasting stations, all of which, for good or ill, the B.B.C. is granted the monopoly.

It is alleged, but without adequate reason, that television broadcasts would cause interference with the already overcrowded broadcast wave-band, which allegation has been repeatedly disputed. It is not denied that the broadcast wave-band is congested, and none know better than the B.B.C. engineers and listeners alike the disappointment and difficulty of maintaining twenty-one stations in this country free of interference. But the obvious remedy which is at last being adopted is fewer and higher-power stations as contemplated in the Regional Scheme.

What, then, can be the reason for refusing the public demand for a broadcast television service ?

One hesitates to think there is any design in this refusal, but the fact remains that progress is obstructed, and neither the experimental worker nor the public, who are the final judges, are afforded the service which is demanded. It is seldom that any good comes from such obstruction. On the contrary, progress is more readily achieved by co-operation and goodwill.

Our present broadcast system was not perfected in its first year of service, nor indeed, while fully appreciating its excellence, is it perfect yet ; but if, in the matter of television, the British public is expected to be satisfied with the transmission of still pictures when the living image is as readily available, then **those responsible for the delay are failing in their duty to their employers and to the public in whose hands the ultimate remedy lies.**

Many listeners must have heard more than once the announcers regret that they could not see some particular artist giving a performance before the microphone. I recently heard a well-known actor introduced to listeners as follows :—"Here is Mr. ——, who has called at the studio on his way to the theatre. He is dressed up in all his war-paint ; **I wish you could all see him,** he is going to sing ——." Now there must be a large number of listeners who know that that is the very subject which could have been broadcast by television with satisfaction had the facilities been available for the purpose.

To challenge Mr. Baird to produce anything new may appear very

REHEARSAL OF "BOX AND COX."
Right to left: Mr. Gordon Sherry (producer), Mr. Lawrence Baskomb (Box), Miss Vivienne Chatterton (Mrs. Bouncer), and Mr. Stanley Vilven (Cox)—*and* the Television Cat.

heroic, but it sounds rather like the smarting boy who whistles to keep his courage up. In any case, it is not "challenges" but co-operation and a fair trial that the public would welcome.

The present attitude of the B.B.C. is by common consent a fundamental mistake, and it is to be hoped that their mode of remedying this is not so insincere as it at first appears, and that the British public may yet have a British system of television which, in spite of ill-formed opinions to the contrary, is more advanced and more promising than any other.

None but a very young or very stupid person would either claim or expect perfection from these early developments. There are, unquestionably, difficulties which demand combined effort and a fuller knowledge for their solution, and when it is remembered that it has taken over thirty years' concentration of the world's best brains, and millions of public money to develop our present wireless services, and these are not perfect yet, it is the more amazing that television has reached so remarkable a stage of development in so short a space of time.

It is safe to say that television is many, many years ahead of where wireless was when that first started. We have no hesitation in stating that had this country turned down wireless in those early days because of the crudity and imperfections of the coherer and decoherer, Britain would not hold the position it controls to-day in wireless telegraphy and telephony; and when a prominent member of the B.B.C. staff, who wisely, or otherwise, posing as an ***unbiased technician***, states in reference to the Baird system of television that "quantities beat it," whatever that might mean, he is speaking with insufficient knowledge and without authority.

We have instances enough in this country of the folly of waiting for perfection before we condescend to consider important inventions which have gone abroad for development, resulting in huge industries. The cinematograph is a conspicuous evidence of this, and if through garrulous ineptitude television is similarly compelled to go abroad for its development and practical application, then the loss will be to British workers and a scandal to British enterprise and British genius.

How the "War" ended

By A STUDENT OF PROGRESS

We print the following important article by "A Student of Progress" without necessarily endorsing his views. The personality of the writer warrants the publication and consideration of his opinions—particularly since they help to an understanding of the attitude of the B.B.C. The article, as the writer himself suggests, should be read in conjunction with that written by Sydney A. Moseley.

THE fact that I occupy a position of detachment from the controversies which have raged round television in the past two years may have influenced the Editor in inviting me to contribute to his columns. Anyway, my sole interest is to foster the real progress of the application of science to the needs and happiness of humanity.

In the days when it was customary to throw ridicule upon the broadcasting of voice and music, I was glad to find myself on the side of the pioneers of what has now become, in a few short years, a tremendous beneficent force in the community. That the instrument of broadcasting should have been so wisely conceived and conducted in this country is a matter for gratification among all British people. I hold no brief for the B.B.C., but when I notice the intemperance of some of its critics I am moved to suggest a readjustment of perspective. When all is said and done, the B.B.C. IS carrying out its policy of providing in nearly 90 per cent. of the homes of Britain the best available entertainment, thought, culture, and general enlightenment. Moreover, the B.B.C. is miles ahead of any other similar organisation in the world both in the efficiency of its entertainment and in the social value of its work. Having said this one is entitled to criticise as much as one likes!

B.B.C. Procedure.

I do not agree with all that the B.B.C has done and failed to do in connection with scientific development. But I understand and appreciate the motives that lie behind its necessarily conservative attitude. Many new inventions and ideas are constantly being brought to Savoy Hill, where they are examined with care and consideration. If they are regarded as being of potential value in a general service sense then they are taken up, developed, and ultimately incorporated.

This general procedure, unfortunately, was not applied early enough in the case of Baird television. Unhappy concatenations of circumstances conspired against the invention. Some of the most zealous and coherent of its earliest advocates happened to have been bitter opponents of the B.B.C., and with their advocacy of television linked a demand for the dissolution of the B.B.C. On the one side irrational claims were made; on the other side was incubated an attitude of intolerance and resentment. The normal channels to coöperation were closed and secured. This was four years ago. The merits of Baird television became obscured in the repeated skirmishes of personalities.

Uninspired and uncontrolled by the Baird companies, the "City" took a fancy to television. There was speculation with all the consequent suspicions. Meanwhile the inventors and research engineers were steadily "plugging away" in the Baird laboratories. But still other considerations stood in the way of coöperation with the B.B.C. Effort on both sides seemed to be much more concerned with increasing the gulf than in building a bridge.

"Spirit of Distrust."

To get the discussion placed on a

THE LAST WORD IN TELEVISION STUDIO DESIGN.
A new television broadcasting studio which has just been completed in the Baird Laboratories is completely lined with copper for the purpose of screening off interference.

proper technical basis was the task undertaken by well-disposed people last autumn. But it was not possible to exorcise the spirit of distrust at once. The inspection by the B.B.C. in October was a real advance, significant of the first practical contact in four years. The verdict was not liked by the Baird people, who failed to understand that, in view of all that had gone before, it was the only possible one. The B.B.C. said, in effect, that they were distinctly impressed with Baird television, but that it was still in the laboratory stage. If, however, it was improved they would be glad to come and examine it again. Implied in this verdict was the offer of technical assistance, if sought.

Two important mistakes were made on both sides. The Baird engineers declined to allow Captain Eckersley to see the transmitter, thereby losing a golden opportunity of interesting him personally. The Baird interests regarded the verdict as a declaration of war, and acted accordingly, paying no attention to the real possibility of constructive coöperation contained in the B.B.C. verdict, and declining to recognise the *bona fides* of the B.B.C. engineers.

The B.B.C., on the other hand, might have been more cordial, and gone farther towards coöperation; and secondly, should not have published its verdict in advance of communication to the Baird Company. So, on the whole, there was not left very much but the wreckage of good intentions.

War was resumed; as sterile as most wars. And then after a period of this unprofitable exchange of acerbities, some *thinking* was resumed. Common sense slowly asserted itself. Forces of reconciliation and understanding were set in motion. Executive chiefs were brought together in friendly conference for the first time. Various proposals were considered in an entirely new spirit. At the time this is being written considerable progress is being made. It has been possible, for instance, for both the President of the Board of Trade and the Postmaster-General to call attention to the new spirit in the House of Commons. Pledges of secrecy must be observed as to the details and nature of concrete proposals. Nevertheless, the outstanding fact is that the almost traditional war between the Baird companies and the B.B.C. has been brought to an end—it is hoped for good.

What of the future? I would venture to predict with some confidence that even if the Baird system is still regarded as in need of development to fit in with a broadcasting service in conjunction with the broadcasting of sound this will be undertaken with greatly strengthened technical resources, and with the goodwill of the B.B.C. readily available. And one final word.

One of the Baird Company's engineers checking the wavelength of the wireless television transmitter on the roof of the Baird Laboratories.

Many hard things have been said about Captain Eckersley in this connection. I want to impress readers of this journal and friends of television generally that I am sure Captain Eckersley has done nothing that he did not regard as his absolute duty from the beginning. It would be well if some of his traducers would stop to reflect how much easier it would have been for him to have allowed his technical doubts to be overcome in order to share in the limelight and glamour of hastening the advent of a revolutionary invention.

When the full history of these transactions and events emerges from the facile pen of Sydney Moseley—himself a principal protagonist for the cause of recognition—I believe it will be discovered that, taking the long view, Captain Eckersley's honest doubts and opposition have really served the progress of the application of science to the needs and happiness of humanity.

Those whose duty it is to think in terms of service to millions must put the brake on the natural enthusiasm of those whose sole concern is inventions—the more wonderful and compelling the more calculated to invert perspective.

MIHÁLY'S TELE-CINEMA

By DR. ALFRED GRADENWITZ

In our issues for October, November, and December last we published the latest available details concerning television developments on the continent. In the following article our contributor gives us the latest information concerning the work of Mihály, the Hungarian inventor.

DÉNES v. MIHÁLY, a young Hungarain electrical engineer of undoubted technical skill and scientific genius, claims as far back as July 7th, 1919, to have transmitted pictures to a distance in such a way as to make them immediately visible. Reproductions at that early date, of course, lacked definition of outline, and, moreover, entailed the use of rather complicated transmitting and receiving apparatus. Nor were the first accounts of this achievement, which, of course, was still far removed from actual television, received without considerable scepticism, which did not give way before, in 1922, Mihály's book on "Electrical Television and the Telehor" was published in the German language.

The appearance of this book stimulated other inventors in the same field to renewed attempts in the same direction, the more so as the advent of radio broadcasting about that time opened up new uses for a successful television scheme. At the same time, however, economic conditions in Hungary compelled Mihály, at least momentarily, to give up his work and turn to other pursuits, until in the summer of last year, having resumed his early experiments, he brought out a new and simplified system able to transmit and receive lantern-slides and shadowgraphs.

Early Apparatus.

Mihály's original "telehor" was based on the use of highly sensitive oscillographs—i.e., electrical instruments comprising minute oscillating mirrors, for scanning the original picture at the transmitting end and rebuilding a reproduction of it at the receiving end. The rotating disc, as used by Baird, he regarded as a crude makeshift able at most to achieve some rudimentary television, but utterly incapable of even moderately delicate results.

However, realising the drawbacks of his oscillograph, which did not seem likely in the near future to give even partly satisfactory results, he resorted last year to the once-despised rotating disc which, he said, though obviously inferior to his own telehor principle, might at least prove useful in demonstrating the practicability of his system. As, however, his work was progressing towards a satisfactory reproduction of lantern-slides, silhouettes, etc., he came to appreciate the possibilities of the disc, and now seems to have made up his mind definitely to cling to it, the more so as the production of cheap apparatus was, from the outset, his foremost endeavour.

Berlin Radio Exhibition.

This is how, in September last, at the Berlin Radio Exhibition, he was able with a remarkably inexpensive type of receiving apparatus to show results in no way inferior—nay, even slightly superior—to those which his German rival, Professor Karolus, of Leipzig University, working in conjunction with the Telefunken people, had been achieving with a much more elaborate and costly outfit. Though only including the reproduction of lantern-slides, demonstrations at that exhibition were with mild exaggeration termed television, and gathered round Mihály's stand crowds of interested visitors.

Personally, I must confess that my interest in these results suffered a severe set-back when, in November last, I was given a few demonstrations of Baird television in London. In another article* I have given a

* December 1928 issue. See also Dr. Frank Warschauer's impressions of the Karolus, Mihály, an Baird systems, published in our November 1928 issue.

Mihály's transmitting apparatus.

2*

candid statement of the well-nigh overwhelming impression received during these first demonstrations, and I have tried to give readers an idea of the enormous difference existing between the transmission of lantern-slides and actual television —i.e., the immediately visible reproduction of people and their doings.

Mihály (behind) photographed with his tele-cinema transmitter.

London Demonstrations.

By a fortunate chance I was also present when, about the middle of December, Dr. Bredow, Secretary of State and High Commissioner of Broadcasting, the highest German official in the wireless field, together with two of his technical advisers, came to London and was given some demonstrations of Baird television. This semi-official visit resulted in the Baird Company receiving an invitation to come to Berlin and demonstrate television from broadcast transmitters there, the three experts having been very favourably impressed with what they had been allowed to see.

Renewed Activity.

I had heard nothing from Mihály and the progress of his work till, a few weeks ago, I was called up on the telephone and asked by the Hungarian inventor to come to his new laboratory and attend a demonstration. Being always anxious to see experiments in the television field, I gladly accepted, and, forthwith, betook myself to the laboratory, where I met Mihály himself and one of his assistants, Mr. Faragó. After a few introductory words the inventor left the room, and shortly afterwards a television image of a human face appeared on the screen of a receiver. The face was visible with only moderate flicker and a fair amount of detail, moving to and fro, smoking a cigarette and behaving to some extent like television images seen elsewhere, though the sharpness and definition of outline was as yet distinctly inferior to what had been shown me in London.

While thus examining the picture I was asked by the inventor's assistant if I recognised the man, to which I replied that, though he seemed familiar, I was not able to make out just who he was. "Why, this is Mihály himself!" Mr. Faragó exclaimed, and I had to apologise for my lack of perspicacity, drawing attention to certain shadows in the picture which seemed to stand in the way of recognition.

Cheap Apparatus.

When Mihály, a few moments afterwards, returned to the room I had to congratulate him upon his recent progress and his demonstration, which, seeing the crude apparatus intentionally used by the inventor, was quite satisfactory. During the discussion then following Mihály strongly advocated the use of cheap apparatus, claiming that, with larger and better receivers, in connection with an expensive and elaborate transmitter, he would be able to do as well as any other inventor in the same field. But, so he said, limitations of the available wave-band would prevent any but the crudest images containing a minimum number of elements (say 900), from being transmitted. Nor, he said, would any other inventor for this same reason be in a position to show anything better. He then went on to discuss the chances of the tele-cinema, which he had been engaged in developing, and when I asked him to give me a demonstration of this he replied that I could have one in the course of a day or two.

A few days elapsed and nothing was heard from Mihály, when suddenly the German daily press began publishing a large number of articles, partly illustrated, about the Hungarian inventor's new television cinema. Representatives of the press had obviously, in the meantime, been invited to witness demonstrations, and now commented rather favourably upon Mihály's recent achievement.

Tele-Cinema.

Though the word "television" had possibly not been mentioned in

Mihály's small tele-cinema and television receiver.

connection with the demonstration just given me, I had used it myself, and the inventor, who had obviously seen that I took it for granted that it was television, had done nothing to contradict my belief. Again, a few days after the press demonstrations, when I asked Mihály to give me a private demonstration of the tele-cinema, I was told that, as his engineers were away, I would have to wait a short time. Still, seeing that practically the whole of newspaper comment turned about the tele-cinema, actual television being only referred to in passing, I could not help thinking that, after all, the demonstration was not, as I first had reason to think, actual television, but tele-cinema, in which case I should, of course, have to state that even the broadcasting of cinema films on the Mihály system is still at an initial stage. Others who have had a better chance to investigate the Mihály system told me that he had to show at the present moment mainly the tele-cinema, that as a matter of fact he had done actual television, but that this was yet of a very crude description, though the light used at the transmitting end in scanning the person to be televised—viz., powerful Jupiter lamps—was rather unpleasant both by its blinding effects and the enormous heat developed.

The Difference.

The main difference between actual television and the tele-cinema, of course, is that while in connection with the former, the various parts of the persons televised have to be scanned by *reflected* light, the case in connection with the tele-cinema is the same as with transmission of lantern-slides—viz., that only *transmitted* light is used. This, of course, entails an enormous simplification of the television transmitter.

Another difference is that the cinema film already comprises a decomposition of the original movement into successive stages, each separate picture corresponding to a different stage. In fact, all that is required is to scan the whole length of the film once at a certain minimum rate—i.e., ten individual images per second.

The Picture Element Question.

Mihály, like other inventors in the same field, has in the course of his work come to the conclusion that the effect of television images cannot be gauged by mere reference to a still picture consisting of the same number of elements. In fact, television images made up of even a very small number of elements will produce much better effects than a still picture similarly composed. This is why Mihály's 900-element pictures show much more detail than one would be inclined to suppose from so coarse a texture.

Mihály's large receiver.

In order, now, to decompose a given individual film picture into 900 elements—i.e., thirty rows each of thirty dots, a rotating disc is provided which near its circumference, comprises thirty small openings spirally arranged at distances apart which each correspond to the distance between two consecutive rows.

If this disc be set rotating at a rate of ten revolutions per second the beams from a constant source of light behind the disc will once during one-tenth of a second pass over the whole of one individual film picture, scanning its various shadings.

Greater Detail.

According to a recent improvement, some sort of diaphragm (stop) is provided which in succession covers up the upper and lower halves of the original picture, thus enabling the half actually bared to be scanned—with a correspondingly greater wealth of detail. The mutual distance of successive disc holes then is only one-half of what is otherwise required. While the size of each perforation in turn is reduced by half, the disc will rotate at a speed twice as high. This is how a greater fineness, or the subdivision of the picture into the double number of rows, is obtained, each of which in turn comprises twice as many picture dots.

Mihály, like other inventors in the same field, has, of course, long given up the use of selenium cells as light-sensitive devices, using photo-electric cells instead, which are not only more sensitive, but entirely free from any lag or inertia, responding instantaneously to any variation of luminous intensity by a corresponding variation of electric current. The special photo-electric cell used by Mihály has an alkali metal cathode facing a grid-shaped platinum anode, both of which are enclosed in a tube filled with some rare gas such as argon. The mode of action of this cell, according to tests by Hertz and Hallwachs, is due to the cathode on the impact of light, giving out a flow of electrons. The intensity of the electron stream is directly proportional to the intensity of the light falling on the sensitive surface. The current intensity in the dark is practically nil.

How the Picture is Scanned.

Now, the light beams allowed to pass through the holes of the rotating disc will, in accordance with the above, scan in succession the various points of the original picture —in the present case the cinema film picture. Passing through the film, they are reduced in intensity the more as the point of the picture actually struck is darker; bright spots will be traversed without any appreciable reduction. The sequence of current impulses corresponding to the various points of the original picture can, of course, like microphone currents, be transmitted either across a line of conductors or by wireless (by modulation of the waves sent out from a broadcast transmitting station), in order at the receiving end to be received by an ordinary radio receiver.

The Super Frequency Lamp referred to in this article. This lamp is used in the receiver instead of the more familiar (to our readers) neon lamp.

What there remains to be done at the receiving post to reconvert these variable current intensities—by a reversal of the process above described—into variable light intensities, and to reconstitute out of these a semblance of the original picture.

A New Lamp.

To this effect Mihály uses a peculiar source of light—a tungsten lamp which with its incandescent electrodes works without any lag, lighting the more strongly as the current intensity actually feeding it is higher. When traversed by a slight current impulse it will give out no appreciable light, while in the case of a strong current it will shine with a bright, intense light.

Now, as the current impulses arriving at the receiving end correspond to the alternately bright and dark elements of the original film picture, this lamp—termed super frequency lamp—is bound to burn brightly whenever there is some bright spot of the original picture being scanned at the transmitting station, and *vice versa.* Its beams, however, are allowed to pass through the series of holes of a rotating perforated disc similar to that of the transmitting station, and thus to reconstitute the original film picture.

Persistence of Vision.

Each individual cinema picture—i.e. each phase of the moving scene represented on the film, is scanned within a maximum period of time of one-tenth of a second, and, during the same interval of time, recomposed at the receiving end into a reproduction of this picture. Our eye, of course, has sufficient inertia to perceive these successive light spots of variable intensity simultaneously—beside and above one another—and, accordingly, to receive the impression of a continuous picture.

The tungsten lamp has recently been improved so as to give out a brightness two to four times higher than the one demonstrated at the Berlin Radio Exhibition. By means of a lens about 25 cm. diameter it gives fairly clear images that can be viewed from two to five metres distance. Ten pictures per second is, of course, a minimum, and it is thought that by speeding up the apparatus to, say, sixteen pictures as in accordance with the usual cinema practice, the flicker still remaining could be disposed of.

At the time of writing it is being rumoured that Mihály is already engaged in installing his transmitters at the various Berlin wireless transmitting stations. His first demonstrations, which I understand are to be limited to tele-cinema shows, are therefore likely to coincide with the Berlin demonstrations of the Baird Company, and to afford to Germans an unique opportunity of comparing the respective merits of the two systems.

A Simple Error.

(The leader of the little German band raised his baton and counted: "*Ein, zwei, drei.*")

British Fan, excitedly: "Crikey! I've got America."

'Arriet: "Got America, on that one-valver? Garn!"

British Fan: "Not 'arf I 'aven't! The blooming announcer just said DRY—and that's America."

—*Radio News.*

An audience "looking-in" at the large receiver, behind the lens of which a photograph would appear to have been placed for the purpose of the taking of this picture.

The Postmaster-General's Decision

[Below we reprint from " The Times " of March 28th the full text of the Postmaster-General's letter to the Baird Television Development Company, embodying the decisions which he arrived at after witnessing the official test of television which we reported in our last issue.]

SIR,—The Postmaster - General has considered the results of the recent television demonstration, in conjunction with the British Broadcasting Corporation and his technical advisers, and he has reached the following conclusions, which accord generally with the opinions of those who witnessed the demonstration. The demonstration showed that the Baird system was capable on that occasion of producing with sufficient clearness to be recognised the features and movements of persons posed for the purpose at the transmitting point. It is not at present practicable to reproduce simultaneously more than perhaps two or three individuals or to exhibit any scene or performance which cannot be staged within a space of a few feet in very close proximity to the transmitting apparatus.

In the Postmaster - General's opinion the system represents a noteworthy scientific achievement; but he does not consider that at the present stage of development television could be included in the broadcasting programmes within the broadcasting hours. He bases this view not so much upon the quality of the reproduction which further experiments may be expected to improve as upon the present limited scope of the objects which can be reproduced.

The Postmaster-General is, however, anxious that facilities should be afforded, so far as is practicable without impairing the broadcasting service, for continued and progressive experiments with the Baird apparatus, and he would assent to a station of the British Broadcasting Corporation being utilised for this purpose outside broadcasting hours. He understands that the Corporation would agree in principle to this course, provided satisfactory terms were negotiated between the Corporation and the Baird Company.

It will probably be essential that any experimental demonstrations of television should be accompanied by the broadcasting of speech, and in consequence two wavelengths and two transmitters would be required. It will not be possible to provide a second transmitter in a suitable locality which will avoid interference with important wireless services in Central London until the completion of the new station of the British Broadcasting Corporation at Brookman's Park, which is expected to be ready in July. In the meantime, it is suggested that the company should open negotiations with the Corporation as to the financial and other arrangements which may be necessary, and it would probably be advantageous to them to enter upon discussions of the technical aspect with the Corporation's Chief Engineer.

In order to find room for a television service in broadcasting hours, it will probably be necessary to utilise for the reproduction of vision wavelengths outside the bands now being used for speech broadcasting. These bands, as you are doubtless aware, are already highly congested, and it is important, therefore, that the company should press on with experiments on a much lower band, which will be notified to you in due course.

In conclusion, it is necessary to emphasise that in granting facilities for experimental demonstrations in which the public can if they so desire take part, neither the Postmaster-General nor the British Broadcasting Corporation accept any responsibility for the quality of the transmission or for the results obtained. The object of the demonstration is to afford the Baird Company a wider opportunity than they at present possess for developing the possibilities of their system of television and for extending the scope and improving the quality of the reproductions. While the company will not be precluded from selling apparatus to anyone who desires to purchase it, the purchaser must understand that he buys at his own risk at a time when the system has not reached a sufficiently advanced stage to warrant its occupying a place in the broadcasting programmes.

I am, Sir,
Your obedient Servant,
G. E. P. MURRAY.

General Post Office,
March 27th, 1929.

TELEVISION Broadcast by the B.B.C.

The Inaugural Programme Transmitted on September 30th

ON Monday, September 30th, 1929, the inaugural broadcasts of television were sent out by the British Broadcasting Corporation from their London Station 2LO. The event was made the subject of a special programme, the details of which we give below :—

11.4 a.m.—

Mr. S. A. MOSELEY, in his capacity as announcer, said :—

"Ladies and Gentlemen,—You are about to witness the first official test of television in this country from the studio of the Baird Television Development Company and transmitted from 2LO, the London station of the British Broadcasting Corporation.

Sir Ambrose Fleming and Mr. J. L. Baird before the television transmitter on Sept. 30th.

"On this inaugural occasion we are very fortunate in having with us Sir Ambrose Fleming.

"I must explain that as the facilities for broadcasting both *speech and vision simultaneously are not yet available*, we shall transmit first of all speech, and afterwards those of you who have televisors will have an opportunity of seeing the speakers. Listeners not yet in possession of televisors should leave their sets tuned in in the ordinary way.

"Before asking Sir Ambrose to address you, I have a message from Sir William Graham, the President of the Board of Trade, which I shall read to you."

MESSAGE FROM THE RT. HON. WILLIAM GRAHAM, P.C., M.A., LL.D., M.P., PRESIDENT OF THE BOARD OF TRADE, MEMBER OF THE BRITISH CABINET.

"It was with great pleasure I received the invitation to speak and be seen on this occasion of the first public experimental broadcast of television and I deeply regret that circumstances prevent me from being present.

"I look to this new applied science to encourage and provide a new industry, not only for Britain and the British Empire but for the whole world.

"This new industry will provide employment for large numbers of our people, and will prove the prestige of British creative energy.

"In this first public broadcast, we have a beginning which will be historic in the evolution not only of a science but of an art which will encourage closer relations between communities at home and abroad and provide a new avenue for educational development."

11.8 a.m.—ANNOUNCER : "Sir Ambrose Fleming will now be televised for two minutes."

11.10 a.m.—ANNOUNCER : "It is now my pleasure to call upon Sir Ambrose Fleming to address you. It is hardly necessary for me to say that it was due to the untiring work of Sir Ambrose, in the early days of wireless, that radio broadcasting, as we understand it to-day, has become possible. Now Sir Ambrose Fleming will say a few words."

SIR AMBROSE FLEMING : "As President of the Television Society, I have great pleasure in saying a few words of congratulation both to the Television Company and to the British Broadcasting Corporation on the inauguration of a new departure in the art of television. It will, I am sure, contribute to the pleasure of countless persons and assist in the creation of a new industry which owes so much to the genius of Mr. Baird."

(*Continued on page* 471)

Television Broadcast by the B.B.C.

(*Concluded from page* 450).

11.12 a.m.—ANNOUNCER: "Professor Andrade will now be televised for two minutes."

11.14 a.m.—PROFESSOR E. N. DA C. ANDRADE: "I am glad to be able to take part in this interesting experiment, which may be compared with the occasion on which the records of the early phonograph were publicly tried. The voices that then issued from the horn were not of the clarity which we now expect, and the faces that you will see to-day, by Mr. Baird's ingenious aid, are pioneer faces, which will, no doubt, be surpassed in beauty and sharpness of outline as the technique of television is developed. One face, however, is as good as another for the purpose of to-day's demonstration, and I offer mine for public experiment in this first television broadcast."

ANNOUNCER: "Now we come to the second half of the proceedings which I fancy will be in a rather lighter vein."

11.16 a.m.—SYDNEY HOWARD. Televised for two mins.

11.18 a.m.—SYDNEY HOWARD. Gave a comedy monologue.

11.20 a.m.—MISS LULU STANLEY. Televised for two minutes.

11.22 a.m.—MISS LULU STANLEY. Sang "He's Tall and Dark and Handsome" and "Grandma's Proverbs."

11.24 a.m.—MISS C. KING. Televised for two minutes.

11.26 a.m.—MISS C. KING. Sang "Mighty Like a Rose."

11.28 a.m.—MAJOR A. G. CHURCH, M.P.: "This is the first official occasion on which the features of living persons have been transmitted through the British Broadcasting Corporation's station by arrangement with the Baird Television Development Company.

"It is a great occasion, and one on which it is a great privilege to be present. As the General Secretary of the Association of Scientific Workers I welcome this triumphant application of science to life. Mr. Baird's is a comparatively new invention. He does not assert that the images which you, who are in possession of televisor receivers see, are the final product of his brain, or those of other inventors interested in the development of this new means of communication. You yourselves will appreciate the nature of his triumph. You will not be satisfied until you can have transmitted to you visible images of activities in which so many of us are interested—for example, a classic horse race, the boat race, a football match, or a great public ceremony.

"We have heard recently a great deal about conversations across the Atlantic. The development of television will enable people not only to converse over great distances, but also to see each other at the same time. The influence this may have upon the relations between the heads of States throughout the world is incalculable. It must be beneficial.

"I desire to congratulate Mr. Baird warmly on the success which has attended his efforts. I should also like to express the hope that those who wish to see this new industry developed, and television must be regarded as a new industry, will, so soon as possible, instal a receiving apparatus in their homes."

BAIRD "TELEVISORS."

In order to save time and unnecessary inconvenience, will those seeking information with regard to the Baird "Televisor" Receivers make application direct to the Baird Television Development Co., Ltd., at 133, Long Acre, London, W.C.2, (Telephone: Temple Bar 5401) and not to the offices of this magazine.

Simultaneous Sight and Sound Broadcast. *Television Makes a further Advance*

SIGHT TRANSMISSION 261 m.
SPEECH TRANSMISSION 356 m. } FROM BROOKMANS PARK.

ONLY those who have taken a keen interest in, or have been intimately connected with the development of television, can realise the anxieties that assail the members of the staff when an important demonstration is about to be given.

Every weak link, imaginary or real, is known, and due care taken to safeguard it; still the possibilities pile up in the mind to an alarming mountain; tempers, even the most equable, become strained to breaking point, and even the best of friends may quarrel—for a moment.

Knowing these things, yet not suffering from them, TELEVISION'S representative went with considerable interest to the Baird Company's experimental station at Hendon on Monday, March 31st, to see and hear the first complete broadcast of television, *i.e.*, sight *and sound*, from the dual stations of the B.B.C. at Brookmans Park.

To say that nothing happened would, perhaps, be misleading; but it is also correct, for nothing in the way of trouble or disappointment did happen. The transmission was a complete success.

Since the opening of the two B.B.C. stations at Brookmans Park one has read so much in the wireless papers of the difficulties of broadcast fans in getting one station without the other that the question naturally arose: "As the sight and sound signals are to go out from these stations, is it not likely that a little of the sight signal may stray into the loud-speaker, and a little of the sound signal find its way on to the screen of the receiving 'televisor'?"

It has also often been stated by some critics that television wants such a wide band that it would make transmissions on a large number of adjacent wave-lengths impossible, even if only a limited image was being sent.

These doubts made the demonstration additionally interesting and important, for neither image nor sound interfered in the least with one another.

At the request of our representative, first, loud-speaker and associated circuits and then the "televisor" and its associated circuits were cut out. The fact of either being on or off had no effect whatsoever on the other.

Promptly at 11 a.m. the blank screen of the "televisor" showed signs of a signal coming along, and it was synchronised just as Mr. S. A. Moseley, who was announced to open the programme, began to say:—

Miss Gracie Fields and Miss Annie Croft before the Televisor.

"Ladies and Gentlemen,—This inaugurates an epoch in television transmission. For the first time in history we are putting over television simultaneously with sound. I want to explain, as simply as possible, what is happening. I am seated in the Baird studio in Long Acre before a microphone and the television transmitter. My voice is being carried along one line, and the vision along another line to the B.B.C. control room at Savoy Hill. There both voice and vision are connected to the Brookmans Park Broadcasting Stations, where, in turn, they are radiated through two separate wave-lengths.

"Those of you who are looking-in should be able to see me as I make this announcement."

At Hendon there were a good many who knew Mr. Moseley. Judging by their comments, they had no difficulty in seeing and recognising him—in fact, they found his appearance on the "televisor" screen a "speaking" likeness. But to go back to his remarks:—

"The interest in these proceedings has been very great," he continued. "Several eminent people have some specially to the studio to take part in what they regard as an historic occasion. I now propose to make way for Sir Ambrose Fleming. Sir Ambrose's wonderful invention of the thermionic valve made wireless in the home possible. His interest in television is so well-known that an occasion such as this would be incomplete without his presence.

"After Sir Ambrose has spoken I shall ask the Right Hon. Lord Ampthill, Chairman of the Baird International Television Company, to say a few words. His pioneer work in bringing television before the public will occupy a proud position in the history of this remarkable new science.

"Nor am I forgetting Mr. John Baird, the man who made practical television possible, but whose innate modesty makes it uncertain whether he will appear before you publicly or not. At any rate, I hope I may persuade him to come and reveal himself to you before the end of this transmission.

"In order to balance the experiment, on the entertainment side we have procured the services of the popular stars, Miss Annie Croft and Miss Gracie Fields.

"I have now great pleasure in introducing Sir Ambrose Fleming. . . ."

Sir Ambrose Fleming's Speech.

After a brief pause, the head and shoulders of Sir Ambrose appeared, and he said :—

"We are assembled here to-day, in the Baird Laboratories in Long Acre, London, to conduct a very interesting demonstration of a simultaneous broadcast of speech and visible images of the speaker, with two frequencies, as arranged by the British Broadcasting Corporation.

"We have now had, for some time, the broadcast of speech and music, but there is no manner of doubt that appeal to the eye is, in general, more interesting than appeal to the ear alone, and appeal to both eye and ear at once is much more powerful than appeal to the eye alone.

"All advertisers know that a visible image or picture holds attention far better than a mere printed sentence or even happy verbal slogan.

"The celebrated picture by Sir John Millais, called 'Bubbles,' with its curly-headed little boy blowing bubbles, was vastly more successful in capturing popular interest than any mere recitation of the saponaceous virtues of an article in frequent domestic use.

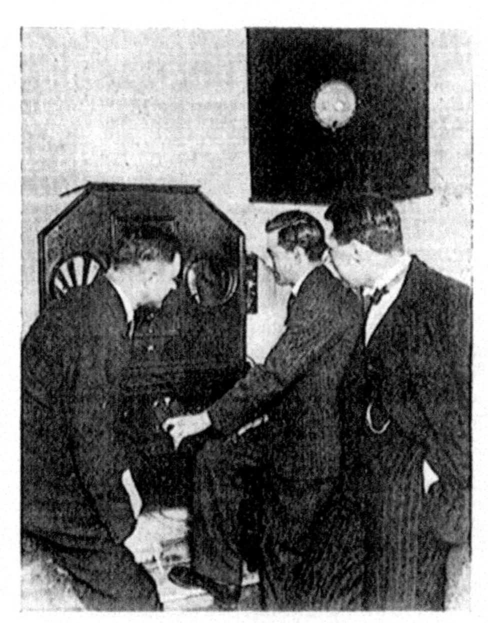

Just before 11 a.m. at the Baird Laboratory at Hendon—waiting for the signal to receive the first simultaneous broadcast of sound and vision.

Lord Ampthill, Mr. Baird and Miss Lulu Stanley (one of the artistes) looking-in at Long Acre, on March 31st.

"To be able to see the speaker adds something to the effect of his speech, not only because we all do a certain amount of lip-reading, but because much meaning can be better conveyed by the looks than by words alone.

"To be able, then, to see as well as hear will add a new interest to broadcasting.

"Most persons gain more knowledge by the eye than by the ear. Anything moving or giving evidence of being alive at once exercises an attraction, and we all know how quickly a crowd gathers round a shop window when there is anything to see in it which is in motion.

"It is, then, one of the latest achievements of applied science to be able to give this simultaneous broadcast of audible speech or song and visible reproduction of images of the faces of living speakers or singers.

"I beg to congratulate heartily all those whose ingenious scientific labours have made possible this remarkable feat."

Sir Ambrose was followed by the Right Hon. Lord Ampthill, who, to judge by his appearance in the televisor, has a remarkably good "television face." His speech, too, was clear and deliberate, and came over well. He said :—

(*Continued on page 122.*)

Simultaneous Sight and Sound Broadcast.

(*Continued from page* 119.)

"This is a very memorable occasion for all those who are interested in television, or, indeed, in the progress of science, as we have reached a goal at which we have been striving for a long time and are actually inaugurating the double-wave transmission.

"I desire to offer my hearty congratulations to my friend, Mr. Baird, and to all those who have assisted him in the development of his wonderful invention.

"As Chairman of the Baird International Television Company, I welcome this opportunity of thanking the B.B.C. for all that they have done to assist us. As you may know, this transmission has been made possible by the close co-operation of the B.B.C. engineers with the engineers of the Baird Companies.

"I greatly regret that my colleague, Sir Edward Manville, cannot be present this morning, as he was concerned with the development of television long before I had the privilege of being associated with the Baird Companies. Sir Edward Manville has a better right than I have to speak about the progress and prospects of television, and I hope that on some future occasion you will both see and hear him with the aid of your televisors.

"Meanwhile, I must not delay you any longer, as there are others who are as proud as I am to be present on this occasion whom you ought to see and hear."

After Lord Ampthill came the variety section of the programme, opened by Miss Annie Croft, who sang a medley of popular songs from musical comedies.

An Entertaining Programme.

Miss Gracie Fields was the next performer; she gave "Nowt about 'Owt'" and "Three Green Bonnets," and left no doubt in anyone's mind that it was Gracie Fields and not a clever impersonator.

Mr. R. C. Sheriff, author of "Journey's End," was to have concluded the programme, but was unable to attend. His place was taken, at a moment's notice, by Miss Lulu Stanley, and so ended the first complete television broadcast.

Throughout the reception was uneventful. The synchronising scarcely varied from start to finish, and did not require adjustment. There was no interference with the picture through fading, local "howlers," or any unwanted atmospherics or other electrical disturbances.

The programme was received on two "televisors" at Hendon, while one loud-speaker served for both. The room in which they were placed was not specially prepared, and might equally well have been any room in any house. The blinds were drawn to cut out excess daylight, and that was all.

Any amateur could have done what was done at Hendon, and, in fact, many did, as reports elsewhere in this issue prove.

The demonstration was but another proof added to the many already given, that Baird television can do what it claims to do, and do it successfully, without any fuss or special attention.

W. C. F.

The First to be Broadcast

ADVANCE

Exclusive to

No. 1.

The first Television Play will be transmitted at 3.30 p.m. on 14th July, from the Baird Studio at 133, Long Acre. The play is called "The Man with the Flower in his Mouth," by Luigi Pirandello.

The Man	- - -	VAL GIELGUD.
A Customer	-	LIONEL MILLARD.
The Woman	- - -	(*not yet cast*).

Adapted and produced by Lance Sieveking (B.B.C.) and Sydney A. Moseley (Baird Television).

Scenery by C. R. W. NEVINSON.

EVERY new thing has its beginnings, things on which at a later date it is interesting to look back, which seem, in retrospect, amusingly primitive.

There are two kinds of people without imagination. The first kind cannot appreciate the fact that the new thing is primitive and elementary, but thinks that it is born in a state of perfection. The second kind of unimaginative person is unable to perceive the inevitability of the ever-developing presence of the new thing. He dismisses it as a clever toy for fun in the laboratory. But those with imagination and vision (that is to say, the readers of TELEVISION) not only realise that the new thing is in a primitive

No. 3.

B.B.C. Play by Television

DETAILS

Television Readers

No. 2.

No. 4.

condition, but see it in their mind's eye as it will be when later developments have brought it to its fullest function. These people are interested and amused to be kept *au fait* with the early stages as they occur—just as the unimaginative world is eager to know about the early days of things which have years ago indisputably come to stay.

We show you here several photographs. Nos. 1, 2 and 3 demonstrate the use of the fading board (as Mr. Sieveking calls it) which was made two days ago. In the first, Mr. Gielgud, the productions director of the B.B.C., is seen seated in front of the television transmitter, with the sound microphone beyond him on the left. He is just coming to the end of a speech in "The Man with the Flower in his Mouth," which is the cue for the appearance of another character in the play. In the second picture Mr. Sieveking is seen fading Mr. Gielgud out, who in the third picture is seen *hastily but silently* making way for whatever is to succeed him in the visible side of the play. In No. 4 you see Mr. Sieveking holding his scenery frame in front of the transmitter. It has, as the photograph shows, a sliding leg used for adjusting the height of the frame according to how much it is desired to show. The scenes or objects are going to be painted by Mr. C. R. W. Nevinson, the famous artist, on thick pieces of millboard, and slid in and out at the side of the frame farthest from the handles. Mr. Nevinson, in exploring this new field of art, will be to a certain extent bound by the limitations imposed by the present stage of the transmission, which will only permit of designs of the boldest and simplest nature. *It is hardly necessary to add that the scribble contained in the frame when the photograph was taken was not by Mr. Nevinson, but hastily dabbed in by Mr. Sieveking as an indication.*

In No. 5 the same frame is seen reversed and used for titles and captions. [The caption has nothing to do with the play on July 14th.]

As regards the fading board itself, this may possibly be regarded in a year's time as a clumsy and antiquated method of fading things in and out during television programmes, but it is essentially the most practical method with which to begin. To have a small metal shutter or contracting iris (once much used on film cameras) was deemed unsuitable to begin with, for two reasons. The first is that it

No. 5.

would be a difficult and unsatisfactory job to manufacture a contracting metal iris of sufficient dimensions to mask the photo-electric cells, and, if made, would probably be rather hard to use, though this will be experimented with. Also, the movement of the shutter close to the cells would set up reactions very difficult to control, so that the next object to be televised after the shutter opened would be subject to a number of distortions and shadows before it became clear. To avoid this, it was thought better to have a fading board operated within the same focal distance as the various objects to be seen both before and after its interception.

The second reason is that from the non-technical point of view (namely, the artistic side) the producer will have the fading apparatus under his own control, and be able to slide the actors, the scenery frame or anything else, to and fro in such a way as to ensure the smooth continuity of the play.

It is probable that several devices will be made, not only improvements on this, but to suit fresh necessities as they arise.

A Question of Colour

Research is also being carried out on the question of what colours transmit most successfully. The actual moving picture transmitted contains all shades from absolute white to absolute black, but in many cases various shades of green, blue and yellow create a far more intense reaction in the cells than actual black, grey or white.

Mr. Sydney Moseley, Mr. Sieveking and the productions director are therefore experimenting with all the varieties of grease paint make-up which are possible. It might be found, when they conclude their inquiries, that the make-up required for the cinema is not so suitable for the television transmitter, though, at the present moment, it seems likely to be so, with certain accentuations, such as the strengthening of all the raised lines of the face (nose, chin, and temples) with strokes of blue or dark green.

It should not be forgotten that the television play will be to the familiar broadcast play roughly what the talkie is to the silent film. That is to say, that the ordinary B.B.C. listener, though he will be able to get the audible side of it on 261.3 metres, will obviously find it less complete than the plays to which he is accustomed; and the same with those who only have television receiving sets. They will, in the same way, see something which cannot stand by itself without its complement of sound. The method of production, therefore, which Mr. Sieveking is investigating will demand a good deal of ingenuity, and the script of the play will be a cross between that of an ordinary radio play and the scenario of a talking film, with certain additions. A more elaborate technique than the mere televising of the face of each successive speaker will obviously come into use.

Looking Forward

As the play proceeds the audible side will contain announcements, dialogue, music, and sound effects, under the direction of Mr. C. Denis Freeman, while the visible side will be made up of printed captions, scenery, the heads and shoulders of the speakers, close-ups of their hands, and objects which are related to the play. For example, the spectator will see sometimes the speaker's face, sometimes the face of the man he is addressing, sometimes his hands, or the scene which the speaker is contemplating, and so on, while between all these things the fading board comes down and goes up.

It is possible to look forward to a day when the fading technique of the multiple studio play, familiar to B.B.C. listeners, will be in use elaborated by means of multiple television transmitters, fading into and across each other. Perhaps, too, we shall live to be audiences of a television which transmits the actual colours, and then . . .

THE FOURTEENTH

By *Lance Sieveking*

Lance Sieveking, D.S.C., joint producer of the first Television Play.

The first Television Play was transmitted on 14th July from the Baird Studio at 133, Long Acre. "The Man with the Flower in his Mouth," by Luigi Pirandello.

The Man - - - EARLE GREY.
A Customer - LIONEL MILLARD.
The Woman - GLADYS YOUNG.

Scenery by C. R. W. NEVINSON.

Adapted and produced by Lance Sieveking (B.B.C.) and Sydney A. Moseley (Baird Television).

WHEN I think how one used to go along a dusty road on a motor cycle in 1912 and '13, to a field, and there stand with a group of oddly-assorted, rather unplaceable people, I am forcibly made to compare it in my mind with what happened on the afternoon of the 14th of July, 1930. The field, all those years ago, was just rough grass, and on one side of it a small and rather insecure-looking tarpaulin had been rigged up. We enthusiasts talked among ourselves, and flicked cigarette ash about. Every now and then someone held up a handkerchief to judge the strength and direction of the breeze. Soon a flap was drawn up from the side of the tarpaulin, and out came three or four men, laboriously wheeling something that looked like a gigantic daddy-longlegs, with translucent wings spread out. We gathered round, examining the thin fabric of which it was made, and commenting eagerly upon the cunning way in which the piano wire was twisted about the bottoms and tops of the struts. We nodded sagely over the bicycle wheels underneath, and gingerly felt the edge of the propellor.

"Contact!" said the pilot, grasping the little joy-stick, and thrusting his elevator backwards and forwards. We noted how he tested also the ailerons. He adjusted his golf cap on back to front and, fixing his cigarette firmly to his upper lip, prepared for the jerk.

The engine started. The men at the wing tips let go, and away it went, bumpty-bumpty-bump, across the field. We held our breath. It rose, *it undoubtedly rose.* Now it was down again. Again it rose. Up,

No. 1.

of JULY, 1930

up, ten feet, fifteen feet. It sank abruptly. It was approaching a tree. We held our breath. It sank beyond the tree. We turned to each other. Someone began to run, and then we were all running.....

The 14th of July is celebrated in France in connection with a revolution. The 14th of July, 1930, had its revolution too, and I wonder how many people who stood and looked at that flickering picture, and heard those voices which now boomed, now scraped, now rattled, realised just the import, the significance, of the thing they were witnessing. There was just a group of them, all sorts of people. Some sat in rooms, and were shown the first play produced by television by means of the little commercial sets which Baird Television Ltd., have put on the market. Others, a few friends of mine, and some more who were interested, just in the same way as those men were interested in the early flying machine, came up in a great open lift, on to the windy roof of the Baird building. Here had been erected a long tunnel made of tarpaulin.

No. 3.

No. 2.

We scrambled inside, and stood about a little awkwardly. The late comers flashed into our darkness from the blinding sunlight outside. The wind blew, the tarpaulin rattled, shafts of sunlight shot across our vision.

And then, on a cue given by telephone to Savoy Hill, the first television play began. From first to last the audience never stirred or made a sound. I think there was something in all their minds, which gave them the ability to see beyond that which their physical eyes and ears were receiving, something which does come upon groups of people sometimes, and which is called prophetic. At the end of the long tunnel, where it narrowed, the big screen leaped into life. It had only been tried a few times before, and we none of us quite knew what to expect. But it held from first to last that oddly-assorted audience, standing or crouching as best they could, and certainly it was not only the work of Pirandello, nor the acting—though it was very good acting—of the cast, nor the production which had unified all the little bits of the play—which held them. No, it was something more......

The problem to be faced in setting out to produce a play within the mechanical limitations at present imposed was a problem which needed all the patience and ingenuity we could bring to it, but here we had the sound judgment of Mr. Gielgud, the productions director, who chose what proved to be a most suitable play. The fading board, which was described in the July number of TELEVISION, was scrapped, as it

No. 4.

was impossible to use it, since, whether raised, lowered, or done anything else with, it merely put out the rhythm of the synchronisers; also, no matter what design in black and white was painted upon it, the photo-electric cells, in some way rather like a nervous horse, shied at it, and sent the picture skidding wildly in all directions. So a new one was made, which slid backwards and forwards along a groove in a firm trestle. It was thus enabled to enter the picture along a horizontal line, and to remain firm when it had completely arrived there. By this means it followed the example of the electric impulses, which also pass across the picture horizontally. The chess-board design painted on it in black and white was found to be the best relationship of black and white for the purpose, disturbing the photo-electric cells hardly at all.

You will see it in one of the pictures; on the left you will observe the handle with which it is pushed in and withdrawn. The bottom row of squares are worn away by continual friction in the groove.

The other four pictures are reproductions of the four scenes, or "sets," which were specially painted for "The Man with the Flower in his Mouth," by Mr. C. R. W. Nevinson, the famous artist. He was asked to do it because he is one of the few living artists of any importance who is really interested in the developments of the modern world. There are quite a number of artists whose technique and manner is, in the true sense, modern, but for the most part they fear the actual objects which go to make up modern life, and concern themselves entirely with things which belong to all ages—the human figure, and the natural landscape. Mr. Nevinson, however, is keenly interested in introducing into his paintings designs which are significant of the modern world—aeroplanes, motor cars, trams, wireless masts, battle-ships, skyscrapers, and so on. It was not surprising, therefore, that he consented to attempt to make scenery for the new medium of television, without cavilling at the limitations of simplicity which it imposed.

The four scenes he painted specially for the first television play were:—

1. Conductor's score and café tables.
2. The dark street outside the café.
3. The table at which the Man is sitting; with glass, etc.
4. Close-up of tumbler into which he stares.

A great deal more might be written about the make-up of the actors' faces, their limitation of gesture, and voice; and all the effects—music, train, traffic, and so on. But I will conclude by recording the feats of understanding and efficiency performed both in front of the transmitter and also in the little darkened room beyond its scope. Mr. Earle Grey, Mr. Lionel Millard and Miss Gladys Young gave a fine performance. Miss Mary Eversley, as announcer and stage manager, executed feats of such difficult dexterity, with the help of her assistants, that it was nothing short of astonishing when, at the end of the play, not even the minutest mistake had been made. Mr. Freeman conducted the music and effects with his usual sure touch.

At the end, Mr. Baird, Mr. Gielgud, Mr. Moseley and I looked at each other in silence.

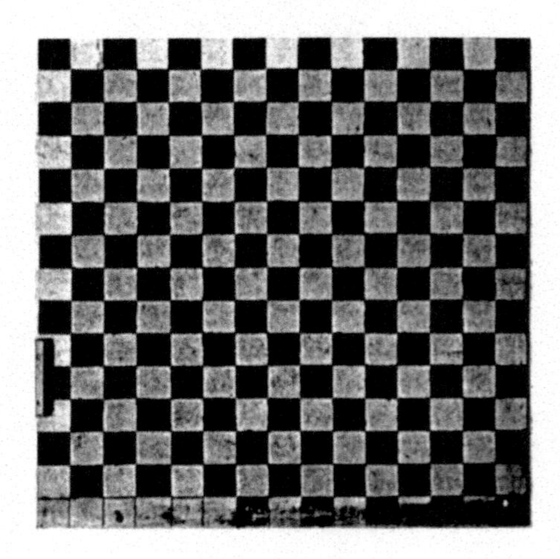

The sliding fading board which was moved to and fro by the handle on the left.

Television in the Cinema

PUBLIC DEMONSTRATION

Read what *The Times* special Wireless Correspondent has to say on this wonderful advance.

(Reprinted from " The Times," dated 16th July, 1930.)

"ARRANGEMENTS have been made for a demonstration of television in the cinema to form part of the programme of the Coliseum in the week beginning July 28th and onwards, as was announced in *The Times* yesterday. The demonstration will consist of the recent developments of Baird's system of television, whereby an image can be seen by a large number of persons simultaneously. One of the chief disadvantages of the home television sets which are now on the market is the comparatively small size of the reproduced image, which makes it impossible for more than two or three observers to be accommodated simultaneously. This has led Mr. J. L. Baird, the inventor of the Baird 'Televisor,' to concentrate recently on the problem of increasing the size and brilliance of the reproduction so that any number of onlookers can see the televised images at great distances. The relative importance of this problem of 'television in the theatre' compared with the more familiar problem of 'television in the home' has already been emphasised in these columns, mention being made of the fact that the large performance factor of safety necessary in home 'Televisors' could be dispensed with in the case of cinema or theatre television because in the latter case any complicated apparatus could be worked by experts.

Conveying a Message

The new type of reproduction, which is based on a device patented seven years ago by Mr. Baird, emphasises in a striking way one of the essential elements of all methods of picture transmission in that the object to be televised is to be regarded as made up of a large number of elements of equal size. The television 'eye' can be said to scan each element in turn, running over the whole of the object about twelve times a second. Corresponding to each element in the object there is an appropriate element in the image, the brightness of which is controlled by an electrical impulse passing from transmitter to receiver. For each element scanned an electrical impulse must be transmitted, conveying the 'message' whether the element is light or dark.

In Mr. Baird's new apparatus the receiving screen is broken up into as many as 2,100 elements, each of which consists of a cubicle in which is situated a small metal filament lamp such as is used in pocket electric torches. The front of the cells is covered with a sheet of ground glass. Each of these little lamps is connected to a separate bar of a gigantic commutator, which switches on only one lamp at a time, and, as the contact of the commutator revolves, each of the little lamps is switched on in succession. The contact switches on and off the whole of the 2,100 lamps in one-twelfth of a second.

Operation

In operation the incoming television signal is first of all amplified, and this powerful current is then fed to the revolving commutator, which switches it to every lamp in turn. The current is strong at a bright part of the picture and weak at a dim part, so that the little lamps are bright or dark accordingly, and the picture is built up of a mosaic of bright and dark lamps. This device differs from any other television device previously shown, in that the lamps are not instantaneous in their action; they remain alight for quite a considerable time, and it is on this fact that, to a great extent, the success of the new device depends, great brilliancy and reduced flickering being easily attained.

The screen demonstrated recently on the roof of the Baird Laboratories in Long Acre has been specially designed to receive the standard Baird transmissions now being sent out through the B.B.C. station at Brookman's Park, and, as these transmissions are limited by broadcasting regulations to a certain amount of detail, the screen has only a limited number of lamps. There is nothing, however, to prevent a screen of any desired magnitude from being built for the use of the cinema and transmissions of greater detail supplied by means of land lines. When standing quite close to the screen the coarseness of the scale is so marked that the image is quite unrecognisable, but at a distance of about 150 feet the picture seen compares very favourably with that obtained with the normal 'Televisor.'"

How the First Television Play was Received

By the *Managing Editor*

SINCE writing my usual article on another page I have received many messages, letters and telegrams from all over the country concerning the reception of the first television play. The following is a varied selection :

ALEC A. KEEN, 48, Broad Street, Chesham, Bucks.

"After looking at and listening to the play 'The Man with a Flower in His Mouth,' transmitted by Messrs. Baird and the B.B.C. this afternoon, I feel I must congratulate them on such a wonderful success.

I saw the play from beginning to end, but I am afraid that we—that is, friends who were with me and myself—paid very little attention to the speech, being more interested in the acting which was excellent, the hand movements being particularly good.

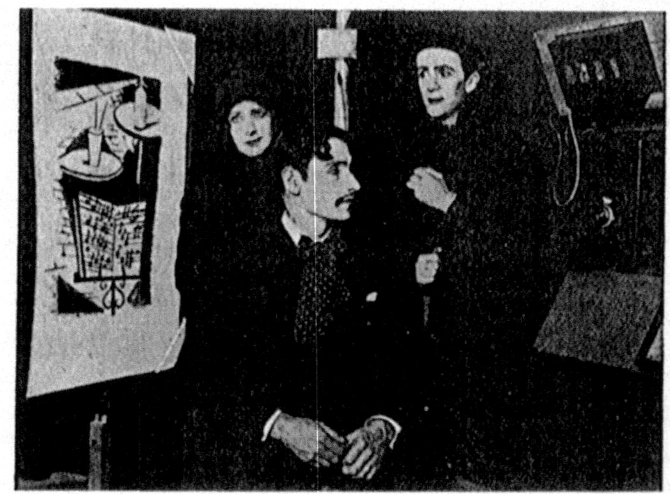

An illustration marking a milestone in Baird Television. "The Man" (seated), Earle Grey, "The Customer," Lionel Millard, and "A Woman," Gladys Young, are shown in their make-up Notice also the scenic effects on the stand, at the back, and the four photo-electric cells.

I had difficulty in distinguishing the scenery, but I think this was due more to the trouble I experienced to-day in 'holding' the picture, synchronism being very erratic, than anything else. My motor is mains driven and the voltage fluctuates badly in the daytime owing to factory motors being fed from the same mains. On the midnight transmissions synchronism is everything that can be desired.

Why the chequered or 'draught-board' fading screen or board ? This, in my opinion, rather interrupted the 'atmosphere' and smooth continuity of the play.

The 'make-up' on the faces of the players made their features more prominent, and was an improvement on the ordinary transmissions. I noticed that the picture or image became much clearer towards the end of the play.

I should imagine that Mr. Sieveking—whom I recognised by pictures of him in TELEVISION—had not 'made-up' his face, as I noticed a distinct difference between the 'contrast' of his features and that of the players. Why did Mr. Sieveking show us his profile only ?

On the whole a great success and I hope the forerunner of many more such plays."

W. A. A. PAGE, Unthank Rd., Norwich.

"With reference to the play broadcast yesterday afternoon, we were provided with a very interesting entertainment and we hope you will continue the idea.

May we be allowed to make a few suggestions in order to assist you. The scenery provided came out very well, but for some reason Mr. Van Gielgud's head was situated in the middle of the synchronising strip and only his mouth and eyes were visible. It appears that you use a mechanical device for obtaining this synchronising strip and its position is not always to be relied on. Cannot you use a red or black strip on your background so that its position is apparent, and place the sitter accordingly? This synchronising strip at times has a definite time lag and spreads into the top part of the picture; for this reason it appears to us to be mechanically operated, as opposed to line effects due in the picture itself which show no trace of time lag, and are, of course, electrically obtained.

Another point which, perhaps, deserves attention is the microphone used and the noise in the studio. The microphone gives the impression of one employed about 1923, but this may be due to vibration or the surrounding noise, which is not unlike a spark transmitter at times. It cannot surely be the disc motor, as these can be made to run absolutely silently. Neither should it be the brush gear, as no doubt this is totally enclosed. In fact it is puzzling to know what it can be.

I hope you will not be offended by these criticisms and suggestions which are given without any idea of the apparatus used or the surroundings. I merely give some idea of the improvement we should like at this end.

I should very much like, when at some future date I am in London, to see your studio and plant as I am quite ignorant about television."

H. R. JEAKINGS, Mill Street, Bedford.

"Congratulations on the first play by television. It was received here in Bedford splendidly. During the whole of the transmission every movement of the artists could be easily followed. Both the close-ups and extended view of the images were very good, also scenery and various other subjects.

We were pleased to notice that during the close-ups and the extended view the image never once went out of focus, as is so often the case with ordinary transmissions.

We experienced some trouble with synchronising when the changes of scenery, &c., were taking place, and we should be interested to know if any other 'telegazer' experienced the same trouble.

We hope we are going to see more transmissions on the same lines as the above, as one gets more scope for experimenting than with the eternal soprano and tenor, who, but for the movements of their mouths, are as still as dummies.

We suggest that a competition along the lines of the B.B.C. in the early days of broadcasting would be interesting and instructive. For instance, we would suggest that a number of ordinary everyday articles should be televised and that we, at the other end, should be asked to describe them in our reports.

Since writing the above we have received the morning transmission (Tuesday, 15th), and we must say that it was the best and clearest pictures we have yet received, the two young ladies being exceptionally good—the one playing the piccolo and the other a concertina and 'uke.' Not only were the faces good, but the strings on the 'uke,' the keys on the piccolo and the folds on the concertina could be quite easily made out. We are very pleased with the way the transmissions are improving and we feel sure that before long it will be a feature in every home.

Wishing you all every success."

[In view of our competition, the suggestions of Mr. Jeakings are remarkable. Is this a case of telepathy?—ED.]

H. H. LASSMAN, 427 & 429, Barking Road, East Ham.

"I am very pleased to inform you that the first television play was a huge success. The play was well thought out and directed in a wonderful manner. I should, however, have liked the lady to have screamed to make it more exciting.

We had a large crowd here to see the play and all were delighted with the clearness of the transmission."

Have you turned to the Competition page?

Do not miss this unique opportunity of winning a

BAIRD "TELEVISOR"

BAIRD SCREEN
THE COLISEUM

Demonstrations form part of a Public in the World's History

As was reported in the August issue of TELEVISION, an experimental television screen had been in operation on the roof of the Baird Laboratories at Long Acre for some weeks, but demonstrations of this apparatus were limited to a few prominent people and members of the Press.

At somewhat short notice, however, the engineers were given instructions to install the apparatus at the London Coliseum, so that demonstrations of Baird television could form part of the normal programme to start on Monday, July 28th, three performances daily being the schedule. Previously all the apparatus had been contained in a temporary laboratory which held, in addition to the screen and commutator, a fairly extensive array of amplifiers, batteries, meters, switchgear, etc. It was fairly obvious, therefore, that for the test in view the complete outfit would have to be portable as a single unit, and in consequence a large caravan trailer was decided upon as the best medium to press into service,

The transportable nature of the apparatus can be gathered from this view of the "van" and screen as it was installed on the Coliseum stage. Notice the large loud-speaker.

Transportable Nature

Then followed days spent in the design and manufacture of the trailer, and the assembly and wiring up of what can be most appropriately called the world's largest "portable Televisor." By dint of hard work and close co-operation between the staff engaged, the apparatus was at last completed and tested out on the evening of Saturday, July 26th, and on the following day it was removed to the Coliseum and run on to the stage. Since the Coliseum has a stage which revolves in order to effect scene changing, it was not possible to connect permanently the various signal and power lines to the van. In consequence an ingenious arrangement of long flexible cables terminating in plugs was devised.

TELEVISION TRIUMPH

Theatre Programme for the First Time

Monday afternoon of July 28th came and found several engineers (not to mention other people) suffering from a severe attack of stage fright, but determined to do their best in this new departure for Baird television. Happily for all concerned, matters went like clockwork.

The announcer, Mr. Radcliffe Holmes, appeared before the curtain and made a short introductory announcement as to the purpose and value of Baird screen television, pointing out that it was the first time in the world's history that a paying theatre audience had been privileged to witness television on such a large scale. The curtains parted and on the screen the audience saw Mr. Sydney A. Moseley, holding a small telephone receiver to his ear, in order to keep in touch with Mr. Holmes on the stage. Mr. Moseley gave a brief but general explanation of what the audience were seeing and hearing. On this particular afternoon's performance Bombardier Billy Wells was also seen, and he gave his views on the Scott-Stribling fight, due to take place that evening at Wimbledon. Then followed songs by Miss Pearl Greene, Miss Lulu Stanley, and Mr. Frederick Yule. Naturally the applause which followed this first demonstration was most encouraging to the staff who had spent so much time in erecting the apparatus.

The Press, in the reports which appeared up and down the country and abroad, were unanimous in stating that Baird screen television was the forerunner of epoch-making developments.

Working Details

Readers will undoubtedly be interested to learn a little of the inside workings of this spectacular event. Photographs are included in this issue showing the special studio and control room which was employed, and it is interesting to note that both these rooms will eventually supersede

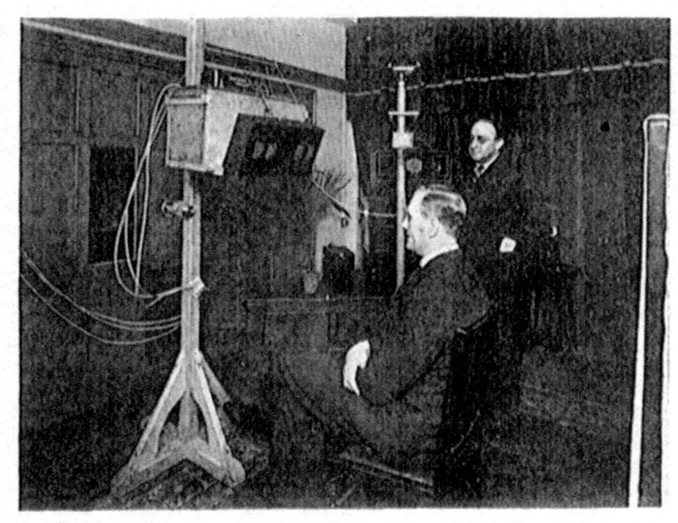

A view of the studio used specially for the Coliseum demonstration, the control room being situated behind the partition aperture on the left. The microphone, photo-electric cells and an amplifier box are conspicuous, while seated we have Bombardier Billy Wells, and standing Mr. Sydney A. Moseley

the existing ones now employed by the Baird Company for their daily television broadcasting.

Telephonic communication with the Coliseum was effected through two distinct telephone lines. One of these passed from the control room at Long Acre to a control board at the Coliseum, from which the engineer in charge had full view of the stage. The second line was from the studio to the stage, it being possible for the announcer on the stage to ask questions of the particular person who was being televised at the transmitting end.

The procedure was much the same as for ordinary television transmissions, of which readers of TELEVISION are now *au fait*. It should be noted, however, that for the first time in England a special method was employed for tilting the projection light, that is, raising or lowering the beam as required according to the different heights of the artists. This was effected by a special tilting head, and not by mirrors, as in America by the Baird Television Corporation.

The outside of the Coliseum by night, with the brilliantly illuminated sign advertising Baird television.

At about 3.15 the engineer at the back of the theatre asked Long Acre to put the signals on the lines. These were listened to and adjusted, and the O.K. or otherwise given to the control room. Then, while the act preceding the television demonstration was finishing, occurred a wait which to the engineers seemed of enormous length and comparable to the approach of "zero hour." The cue was then given from the Coliseum wings to the Long Acre studio, and the quality of the picture was reported upon. After Mr. Holmes had made his entry and introduced the subject, the curtains were swung back and the audience saw the television screen in the centre of the stage, the illuminated screen standing out in a black background.

Operating the Controls

One of the accompanying photographs shows quite clearly what the apparatus looked like on the stage, two loud-speakers of the public address type being seen beneath the screen itself. The portability of the apparatus can be easily gathered from the nature of the van's exterior, while, in addition, interior views are given to make the story more complete.

The strength of the picture and the speech volume was controlled in the Coliseum itself, while in addition there was a line between the engineer watching the picture at the back of the auditorium and an assistant in the van itself. This assistant could see whether the picture was properly synchronised by means of a "peep-hole" in the large black screen seen in an illustration. He watched a reflection in a mirror, and if the picture showed any tendency to run out of synchronism he was able to make rapidly any readjustment necessary to "hold" it in place.

It is hardly necessary to add that every audience throughout the fortnight's run was keenly interested in the experiment, and delighted to see British television so far to the front. The turn lasted for about a quarter of an hour, and invariably concluded with "question time." This proved a very popular part of the demonstration, for members of the audience were invited to put questions to the announcer and, after these had been telephoned through to the Long Acre studio, they were answered from the screen. This was sufficient to convince anyone who displayed any scepticism. The requests which were passed on generally took the form of "Put out your tongue," "Ruffle your hair," "Shut your eyes," "Put your hand in front of your face," "Undo your tie." When the person being televised was in evening dress, and the tie a made-up one, it caused roars of laughter to watch the efforts of the man on the screen in his endeavour to undo the knot at the back.

During the fortnight a number of prominent people appeared on the screen, and amongst these may be mentioned the Rt. Hon. the Lord Mayor of London, Mr. A. V. Alexander, P.C., M.P., the Rt. Hon. George Lansbury, P.C., M.P., Miss Ishbel MacDonald, Mr. Frederick Montague, Colonel L'Estrange Malone, M.P., Mr. H. W. Austin, Young Stribling, Lord Marley, Sir Oswald Mosley, Miss Ruby M. Ayres, Miss Irene Vanbrugh, Bombardier Billy Wells, Sir Nigel Playfair, Mr. Herbert Morrison, Miss Ellen Wilkinson, Mr. Robert Young, M.P., Lt.-Commander Kenworthy, Sir Francis Goodenough, Mrs. Wentworth James, etc.

Apart from the apparatus itself, which, of course, was pre-eminent, the success of the undertaking was contributed to in no small measure by the splendid co-operation which existed between the engineers and programme department of the Baird Company and the theatre staff at the Coliseum. Then again the artists who were appearing at the Coliseum sometimes went to Long Acre to be televised to the audience.

Miss Beryl Beresford was the first artist to do this, while on two occasions Miss Jonstone appeared on the screen while her partner, Miss Yorke, was on the stage, the two artists singing a duet together although four hundred yards apart.

Baird screen television has taken the lead, and the invention will open up a new field of entertainment and utilitarian value and developments are now in progress to increase the size of the screen.

The Big Screen in Germany

By Sydney A. Moseley

READERS who followed the epoch-making demonstrations of the big screen at the Coliseum, London, will no doubt be equally thrilled to hear that similar demonstrations were given recently in the Scala, Berlin's largest music-hall.

A good many wiseacres shook their heads when it was suggested conveying the whole paraphernalia from London to Berlin. They pointed out that, while possible under the best conditions at home, the difficulties that would be encountered in attempting an experiment of such a nature in a foreign country, where all the conditions were new, and after the apparatus had had to withstand the joltings and joggings of the lengthy journey, would be well-nigh insuperable.

Miss Maud Hansen, the first German girl to appear in Berlin with the Baird screen, is here seen at the studio in Friedrichstrasse.

The pessimists, however, were, as usual, confounded, and the transmissions which took place twice daily for a fortnight were a pronounced success.

At the same time certain inevitable snags manifested themselves at the onset. Local electricians, who were confident that they could supply all the needs of the Baird engineers, were rather misinformed as to the true state of affairs.

For instance, at the Scala Theatre those responsible stated that the electrical supplies were 200 volts A.C. and 200 volts D.C. When the Baird engineers arrived in Berlin and were settling down to begin the experiment, lo and behold it was found that the supply was 125 volts A.C.

Did they despair ?

They did not. With the aid of the Baird associates in the Fernseh A.G.—you know, the sister German company in which the Baird Company is a partner—these difficulties were soon overcome and in particular my friend, Dr. Goerz, who is managing director of the Fernseh A.G , proved a tower of strength.

Let me give you some idea of what was needed in order to show the German people that big screen television was something more than mere talk.

There was, first of all, the studio to fix up. One had to find two or three empty offices and completely equip them in a short space of time with benches, tables, heavy curtains, floor matting and a thousand and one gadgets which the amateur especially will appreciate.

Within a week three offices were transformed into a studio, transmitting room and engineers' offices. In addition, of course, there was a reception room.

All that was at *one* end.

At the other end, the Scala, equal difficulties were encountered. Here again some of the data supplied to the Baird engineers proved to be incorrect. For instance, take one " detail." The measurement of the stage door was given as such that it would have enabled the caravan containing the screen to enter on the stage without mishap.

When the caravan at length arrived at the stage door it was found that the door was too small !

Again, did the pioneers despair ? Again the answer is, they did not ! The stage manager of the Scala and his men set nobly to work and knocked down many inches of the wall of two or three feet in thickness ! All through the piece the whole staff of the German theatre helped enthusiastically. General-Director Marx proved a first-rate sportsman in the whole business. He readily appreciated the fact that in presenting television on the big screen for the first time in Germany he was making history for the Scala—by the way, one of the finest theatres I have ever seen.

The greater question of electrical power was met by Dr. Goerz supplying immediately a special generator which was brought post-haste from the Zeiss-Ikon works at Zehlendorf. Then, mark you (and this was a new experiment so far as foreign land lines were concerned), the distance between the studio in the Friedrichstrasse and the screen on the stage of the Scala Theatre was no less than seven kilometres. Wiring to the extent of three additional kilometres was necessary, so that between the studio and the screen there was no less than ten kilometres of wire. In London the distance between the Baird Studio and the Coliseum was a matter of about a quarter of a mile.

Then, again, a new amplifier was being tried out and one would not be able to tell until the last moment whether it would "work" or not!

Well, this gives an impression of the magnitude of the experiment, and of the risks taken by the Baird Company to put Britain on the map in Germany so far as practical television was concerned.

After all, in order to achieve big things one must take some risks.

The arrangements were that the demonstrations should start on the Monday, and actually all the difficulties were overcome by that time, although some disappointment was experienced in not being able to give a special press demonstration the previous day.

But when the screen was seen for the first time in public in Germany general amazement was expressed, not only by the lay public, but by the body of interested scientists, both in Berlin and other parts of Germany.

No fewer than 26 performances were given, and from beginning to end there was not a single hitch. Indeed, it was the unanimous opinion of the Baird engineers that the picture was even better than that shown at the Coliseum, London. This, I think, is a remarkable achievement considering the language difficulties and the local technical problems.

There was, of course, another trouble to overcome, and that was in regard to artistes. But here Captain Pogson, of International Productions, Ltd., came in very handy with his perfect knowledge of German, and the artistes whom he engaged proved most satisfactory.

There was Max Steidl, the well-known comedian, who sang one of Germany's popular songs and could make extraordinary grimaces; Mr. Wolff Scheele an accompanist, whose face came over very well; and last, but not least, an excellent young German soubrette, Maud Hansen, who sang not only with charm but with an intelligent understanding of the needs of the "Televisor." Fraulein Hansen, in fact, scored an immense success, and was immediately offered engagements on the strength of these television performances.

Despite the rush in which proceedings were arranged, many of Germany's notable politicians and artistes came to the studio in the Friedrichstrasse to be televised. Among them were Dr. Bredow, the "Sir John Reith" of German broadcasting who appeared twice, Max Steidl, Senta Soneland (German "Marie Lloyd"), Dr. Herold, editor of the *Munich Medical Weekly Review*, physician to ex-King Amanullah, Evelyn

Dr. Paul Goerz, Chairman of the Fernseh A.G., who rendered such valuable assistance while the large screen was being shown in Berlin.

Holt (famous screen and film star), Hans Erwin Hey (famous opera singer), Felix Josky (popular author and poet), Kurt Vespermann (Germany's "Charles Hawtrey"), Manni Zeiner (revue star), Henri Lorenzen, N. A. Pogson, Hans Jungermann (German "Sir Charles Wyndham"), Marianne T. Winkelstern (the German "Pavlova"), Max Mensing (from the Berlin Grand Opera House), Gertrude Hesterberg (the German "Lily Elsie," leading in *The Merry Widow*), Ines Monlosa (stage and screen star), Leo Monosson (from the Neues Theater am Zoo, in his famous song *The Valse in the Sleeper*), Manny Ziener (revue star), Ludwig Manfred Lounnel (famous comedian), Paul Heidemann (of the State Opera, Berlin), Dr. von Bredow (Secretary of State), Erika Aderholt (famous actress), Minister of State Dominicus (Minister of Health and Air), Wilhelm Bendow (operatic singer), Captain Hermann Kohl ocean flyer ("Bremen." Kohl and von Huhnefeld), Elisabeth Pinajeff (Germany's "Greta Garbo").

Altogether one more notable chapter in the fascinating history of Baird television.

SEEING *the* DERBY *at a* DISTANCE

A Promise Fulfilled to the Confusion of the Sceptics

THE mention of the Derby inevitably calls forth exciting visions and invokes the possibility of making a small fortune on the winner, but this year the great race had an importance quite apart from its betting medium. It was the occasion of one of the greatest experiments yet made in television—the first experiment of its kind in history. This scheme was built up in the brain of Mr. John L. Baird, the inventor of the new science, and was carried out by the Baird Company in conjunction with the B,B.C. On the afternoon previous to Derby Day, preliminary tests were carried out and a description of what took place is of paramount interest.

The Baird caravan—which is now almost symbolical of the Baird Company's external activities—was moved to the racecourse at Epsom and positioned against the rails almost opposite to the grand stand and winning post.

Post Office lines were laid under the course to the stands and from there they travelled direct to the Baird television control room in Long Acre. The signals were passed from here through to the B.B.C. and so to Brookman's Park, from which station they were broadcast by the National transmitter on a wavelength of 261 metres.

The preliminary test excited the greatest interest in the minds of those who witnessed it. Both the transmitting and receiving end of the experiment were fully revealed to members of the press who were invited to participate in this last-minute rehearsal. In the morning they were taken to Epsom to inspect the caravan and learn a little of the technical intricacies from the engineers on the job, then a return was made to Long Acre, where a formidable array of "Televisors" gave a land line picture of scenes from the course within the natural range of the transmitting apparatus. The horses, jostling crowds, in fact all the panoply of the famous racecourse was plainly seen on the "screens" of the instruments in question and, further to excite the senses, were the multifarious noises which go so far towards building up a picture of Epsom in full swing.

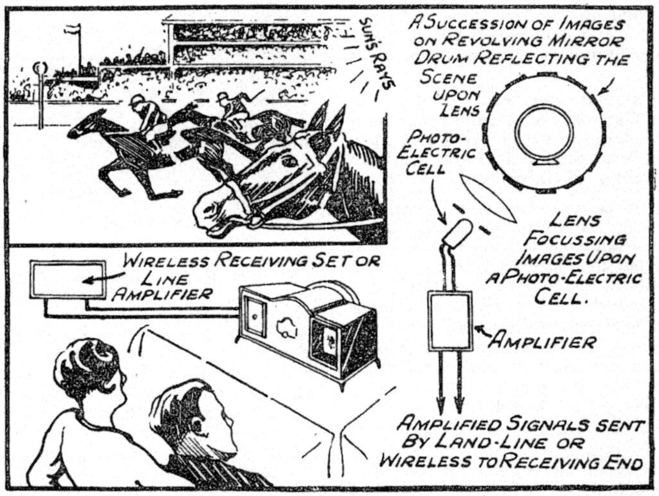

In order to secure a change of vista, an ingeniously contrived mirror had to be brought into play. The mirror was fitted on the side of the caravan furtherest from the course and, when set at various angles, reflected different pictures of the adjacent activities. It was this reflected image that was scanned by the revolving mirror drum, carrying thirty mirrors around its periphery. This latter piece of apparatus was used in place of the customary disc with its 30-hole spiral, each of the mirrors referred to being set at a slightly different angle from the preceding one. As the drum revolved each mirror sent a strip of the scene through the lens on to the photo-electric cell, the individual

Portraying in a simple manner how the Baird Co. brought the Derby scenes into the comfort of the home.

mirror inclinations relative to each other being such that the picture was split up into thirty strips. The whole process was repeated twelve and a half times per second, and the electrical result of the image was sent along a line in the ordinary manner, after being amplified.

The apparatus at the receiving end needed no emendment and was thus in perfect readiness when the great hour arrived. At 2.45 p.m. on Derby Day the first scenes came through and were built up in the " Televisors " in use at the Baird offices. These varied, sometimes showing clearly and at other times appearing somewhat indistinct. Occasionally interference from the telegraph and telephone lines wiped out the picture, but, in spite of this unavoidable fact the parade of the horses and jockeys was seen by all present, while, now and then, the close-up of a man or woman passing across the line of vision showed on the screen.

A short pause and then we heard the announcer telling us that the horses were rounding Tattenham Corner. We held our breath and, before we realised it, the first three horses flashed by the winning post to the frenzied roar of the crowd. The rest of the field followed in close pursuit.

And so we were able to see what was happening at fifteen miles distance. The experiment proved beyond doubt to the most sceptical of die-hards that television had come out of the studio into the sphere of topical events. The restrictions of four walls and artificial light had been put aside, and Mr. Baird had fulfilled that promise which had once called down such a storm of disbelieving contempt on his head.

A few of the extracts from the press comments on this wonderful experiment are printed below :—

The Times, 4th June.

" Yesterday afternoon the Baird Television Company, in co-operation with the B.B.C., broadcast a television transmission of scenes from the Derby, including the parade of horses before the start and the scene at the winning-post during the race. This broadcast is important in that it is the first attempt which has been made, in this or any other country, to secure a television transmission of a topical event held in the open air, where artificial lighting is impossible."

The *Daily Telegraph*, 4th June.

" Fifteen miles from the course, in the company's studio at Long Acre, all the Derby scenes were easily discernible—the parade of the horses, the enormous crowd and the dramatic flash past at the winning-post.

* * *

Looking from behind the caravan we see the grandstand on the opposite side of the course. Note the swivelling mirror which enables the scenes to be changed and the movements of the horses followed.

* * *

"After the transmission Mr. Baird said that he was quite satisfied with the experiment.

" This marks the entry of television into the outdoor field,' he said, ' and should be the prelude to televising outdoor topical events.' "

The *Daily Mirror*, 4th June.

" I did not go to Epsom for the Derby yesterday. Instead, I formed one of a score at an epoch-making gathering in the West End.

"Nevertheless, I saw Cameronian, closely followed by Orpen and Sandwich, win the great summer racing classic !

"This I was able to do by means of the two latest and greatest marvels of science—wireless and television . . . . . . At last the starter got the field away, a fact that was recorded on the ' Televisor ' by the crowd beginning to jostle and push one another for a good view.

"They were now at Tattenham Corner and racing down the hill for home !

"Then—two horses dashed across the centre of the picture we were looking at, closely followed by another. Cameronian and Orpen followed by Sandwich ! "

TELEVISION

For the transmission of "The Man with the Flower in his Mouth," in July 1930, special scenes were drawn.

IN a visit recently to the Baird studios I was surprised to realise how much development had really occurred since my previous call. The first of these was the advent of the extended scene. Instead of merely head and shoulder pictures a full-length view of those who appeared in the new Studio was put over. The type of transmission which was regarded as impossible had not only become feasible, but was part and parcel of the everyday programme.

There were, for instance, plays, sketches, dancing, illustrations of tennis, physical culture, and so on. In fact, it was possible to show that every talk could be illustrated if it lent itself to it. The usual talk on how to play tennis or how to play golf, or the latest dance steps, which were given from the B.B.C. studio, interesting as they are in themselves, always seem to me to lack the finishing touch.

In the actual illustration it was possible not only to *hear* how we were to take our daily exercise, but we were able to see Captain Muller go through arm bending and extending and Indian club exercises. One saw the movements distinctly, and the same thing happened when a lesson in tennis was put over by Mr. Last.

He was not only able to say "This is how the racket should be held," but to show how it could be done. One observed the correct position of finger and thumb. The same thing applies to the feet with dancing. In my opinion dancing transmissions by ear alone had little value, but if one sees, as I myself saw, Miss Coleridge Taylor dancing the pretty steps on a little platform, then the whole thing takes on a totally different aspect. One is captivated, one is able to concentrate and marvel at television's advance.

The plays and sketches which at one time aroused tremendous interest because of their rarity are now almost everyday occurrences. It is necessary only to go to the publisher of little plays containing two or three characters, arrange for their transmission, and put them over. Readers will remember the tremendous interest that accompanied the transmission by television of *The Man with the Flower in his Mouth.* It doesn't seem to be very long ago, and yet this play can now be transmitted on totally different lines. Instead of merely seeing the head and close-up view of the hands of a character, we can observe the extended scene of the café where the action takes place. Other little plays, *The Wrong Door* and *Great Expectations,* have been put over

Another scene used to illustrate the sketch where only head and shoulder images were observed.

AS AN ENTERTAINMENT

more than once, and the Marionette Players, whom we have encouraged in their splendid work, not only presented *The Man with the Flower in his Mouth*, but their own fascinating show.

How crude this original production appears beside the production of such little plays as *The Double Cross, Another Pair of Spectacles, Dining Out,* and *A Touch of Truth.* We now see two or three people on the stage at once, and the detail is sufficient to include actions which are necessary to the development of the plot of the sketch.

Those of you who have not looked-in at these productions will be interested to learn that excellent reports have been received not only from London and the provinces, but from countries abroad.

Referring once more to the dancing transmissions, the Baird Company transmitted a series of national dances in appropriate costumes. Ballet dances in which the dancer appeared in ballet attire, cabaret and Scotch dancing, as well as tap, buck, and wing dancing.

There was also a demonstration of ju-jitsu, and it does not require much imagination to see how useful this would be if "Televisors" were installed in police-stations as well as in every home. Ju-jitsu is a form of exercise which has been associated

Illustrating the big stride forward which has been made now that extended scenes are available for transmission by television.

rather wrongly, I think, with methods of defence. As a matter of fact, it is one of the best all-round exercises that could possibly be indulged in, and the fact that its home is Japan is sufficient recommendation for it from a physical culture point of view.

What lookers-in tell me is so interesting normally is to see artists playing the piano. The movement of the hands on the keys are clearly discernible and add considerably to the interest of the artist's playing. Watch the tense interest of a huge audience at the Queen's Hall when a pianist is playing. There is no need to call for concentration then. The difficulty of the average B.B.C. transmission is that there is no sight to accompany the sound. That explains to a large degree the unfair criticism with which the B.B.C. has to contend from time to time.

I see I have omitted to refer to the very good transmission of a cricket coach. The movements of the bat and wrist were seen clearly and, I imagine, must have helped ambitious young cricketers who were able to look-in.

There was also a talk on "Naval History" in which the models of ships were actually shown in order to accentuate the interest of the talk.

Once upon a time when duetists were "giving a turn" they were observed one at a time. Now two people can be shown singing together, and are able to put over comedy acts much more easily. The same with dancing. Two figures are shown with distinctness.

In other words, the development of television from the entertainment point of view goes on apace while developments in other directions are on the way.

In "The Wrong Door" a step forward was made, for in addition to one artiste alone two artistes were seen together.

B.B.C. and Television

Official Statement of Policy

We print below the first official statement of the British Broadcasting Corporation's policy on television. This exclusive article has exceptional interest at the present time, as television transmissions are to be a regular feature of the programmes from Broadcasting House. Editorial comment appears on page 126.

THE attitude of the B.B.C. towards television has been the subject of a good deal of misunderstanding. The B.B.C. has to bear in mind always the priority of its normal service to listeners, the derangement or mutilation of which

Sir John Reith, Director-General of the B.B.C.

in the interests of a small minority would be unjustifiable. The listening constituency is certainly not less than 20,000,000, and probably a good deal bigger. It has been said truthfully that broadcasting has become part of the essential machinery of our civilisation. Accordingly, all the greater vigilance is necessary to maintain an uninterrupted service.

Some of the earlier advocates of television sharply criticised the B.B.C. for not including television transmissions in normal programme hours, even at the risk of withdrawing wavelengths from the ordinary service. The attitude of the B.B.C. was that this course would be wrong; that before television could be included in the programmes, even experimentally, it must be developed considerably beyond the stage of five years ago. When, however, some progress was noted in the Baird laboratories, it was decided to arrange with the Baird Company for a limited series of transmissions outside programme hours. These began about two and a half years ago. They were supplemented from time to time by feature transmissions, such as that of the Derby last year and this year.

Recently sufficient further progress was recorded to make possible an extension of facilities. Beginning at an early date, there will be a new series of half-hour transmissions on four nights a week at eleven o'clock. The present experimental series will close down for about a month, and it is expected that the regular transmissions will begin between July 15th and 20th. The programmes, hitherto provided by the Baird Company, will be taken over and adapted by the B.B.C. The Baird Company, for its part, will concentrate on research, and on perfecting the apparatus of reception. To give reasonable security to purchasers of Baird sets, the B.B.C. has agreed not to discontinue transmissions by this process sooner than March 1934.

Until experience has been gained with the new series of transmissions, it would be rash to offer definite conclusions; but it is not the opinion of those concerned that television, if and when developed to a service stage, will revolutionise programmes. For one thing, there would not be the automatic televising of every broadcast programme; even if the scarcity of wavelengths did not preclude this, it would not be appropriate, for the reason that only some programmes lend themselves to vision as well as sound. It would be undoubtedly thrilling to see the finish of the Grand National as well as to hear it, but it does not follow that it would be advantageous always to see a band performing or an announcer reading the news. As television progresses, it will be incorporated here and there, illustrating only those programmes which would be improved thereby.

There are, of course, technical difficulties in broadcasting television. One of these is the need of more elbow-room for the operating channel. Another, of course, is that two channels are required, one for vision and one for sound.

In a letter to the publishers of TELEVISION, *Sir John Reith writes:*

"The association of an independent publishing house with television coincides with the development of B.B.C. experimental work in the same connection. While it is impossible as yet to determine when television will become of general practical value, there is no doubt that it has already reached a stage at which it merits serious consideration. The B.B.C. wishes your journal the success which your enterprise deserves."

The B.B.C. "First Night"

A Successful Programme

On August 22, the first television programme was transmitted from Broadcasting House, London. By special permission of the B.B.C., a representative of TELEVISION *was present, and he gives his impressions of the performance as seen in the studio. This is followed by reports from our special correspondents who were "looking-in" to the programme in various parts of the country.*

As I walked into Broadcasting House on the night of August 22, I was more than ever struck with the atmosphere of "mechanised art" which the B.B.C. has created in its new headquarters. The machinery on the roof stared gauntly at the sky and the pink geraniums seemed to blush in their boxes outside the lighted windows. The impression was increased on entering Studio BB in the sub-basement, where the artists were assembling to give the first television programme from Broadcasting House.

To the accompaniment of the piano a girl dressed in white, with white cheeks and *black* lips, was rehearsing an American jazz song, while the engineers were making last minute tests with the television transmitter which cast its eerie flicker across a large screen. Moving from one group to the other were the stage directors in faultless dinner jackets, jostled by engineers in shirt-sleeves and headphones, and carefully avoiding the wires which connected up the photo-electric cells. Through a long window in the wall of the studio the transmitter was pointing, and as Mr. Baird said to me while waiting to open the broadcast, it looked "just like a machine gun." Five feet away from the window was a large white screen, forming a background to the television "stage."

Like a Dream.

Watched from the corner of the studio, it was all like a dream in which the actors were half human and half machine, the jazz and the flickering lights forming one of those mad backgrounds that haunt a restless sleeper.

Then came the "Quiet, please" of an announcer, who stepped in front of the microphone at 11 p.m. Stating that this was the first of the new broadcasts to be given four nights weekly, he explained that, although television was still feeling its way, it had such potentialities that the B.B.C. was lending as much of its time and resources as was consistent with the normal demands of the programmes. "As it is the Baird process which is being used," the announcer added, "we are glad to welcome its inventor in the person of Mr. John L. Baird, who will now be televised."

Holding a sprig of white heather the inventor then stepped into the beam, and after thanking the B.B.C. for inviting him to speak on this first night, he expressed the hope that the new transmissions would lead to developments of broadcasting, increasing its utility and adding to the enjoyment of the great listening public.

New Technique.

Then followed a half-hour's entertainment in which some music-hall turns were performed in front of the vision transmitter. Those who broadcast for the first time are conscious of the difficulty of speaking to an instrument, without the response from the audience which means so much to a platform speaker. Sitting in the studio a few yards from the television artists, I got a vivid impression of what it feels like to *act* under these conditions. If it is hard to be natural in front of a microphone, how much more so to act to a glass window, with an orange light flickering across one's face the whole time. Apart from choosing songs and sketches which are suitable for the small television "stage," the artist must acquire a new approach to the technique of acting.

The programme was opened by Miss Betty Bolton, who gave some French and American songs. A change of hat or scarf was handed to her between each item, and it seemed strange to think that, though every sound in the studio would reach the wireless audience, they could see only the small area covered by the vision transmitter. Instead of the properties being concealed in the wings, they were arranged on a table nearby, ready for the artist to step out of the beam of light for a few seconds to change into another guise.

On the thick soft carpet of the studio the engineers quietly moved or adjusted the photo-electric cells so as to pick-up the scene to the best advantage. Meanwhile the beam of light continued to scan the "stage," and I wondered how much of the flicker was being seen by the wireless audience, for at a few yards range in the studio it was naturally very marked. I imagine

Controlling the television signals, which are now being transmitted four nights weekly from Broadcasting House.

that Mr. Fred Douglas, who next sang some nigger songs, " came through " well, as his black face contrasted sharply with the white background. In place of the traditional thick red lips, his were left white and there were large white rings round the eyes. During one of his songs Mr. Douglas raised his hands in expressing emotion, and I wondered if the flash of his gold ring was noticed by any of the wireless audience.

When Miss Louie Freear took her place in front of the transmitter her long stage experience was used to advantage. She sang " I want to be a lady " and other numbers from " The Chinese Honeymoon " in which she made her name at the Strand Theatre in 1901. Every experienced actor knows the need for exaggerating every gesture and not cutting words short, and Miss Freear made the most of this in her cockney " asides," which were not lost on the audience. Miss Betty Astell followed with some love songs.

After these head and shoulder monologues the small screen was removed to make way for a dance, and the flickering light now illuminated an area about 10 feet square. Dressed in black and white, the predominant colours for a successful transmission, Miss Bolton did a Spanish tango, the transmitter being moved on its swivel base to follow her about the stage, just as a spot light is used in the theatre. Rather heavy shadows were thrown on the wall behind, which at times made the movements a little confused, and I asked myself whether they would be noticed when the scene reached the vision receiver as a small picture. As a final act each of the four artists danced across the stage.

I could not help wishing that the programme had been rather less American and that Miss Freear's cockney songs had been accompanied by something else that was typically English. This criticism in no way minimises the skill or enthusiasm of the actors, who put every ounce of effort into this " first night," which was doubtless even more exacting than many "first nights" on the stage.

As I sat in the studio it seemed uncanny to think that the same actors were being watched by hundreds of " lookers-in " throughout the country, and I felt very curious to hear what they made of this interesting evening.

London

Belsize Park

I was one of a party of press representatives who had been invited to Mr. Baird's house in Belsize Park to look in to the first television programme from Broadcasting House. At 11 p.m. we heard the B.B.C. announcer state that he was about to introduce the inventor in person, and immediately all eyes were centred on the television screen, an air of excitement pervading the company as we were about to watch this inaugural ceremony by television. The words " television screen " are used advisedly, for we were privileged to watch the proceedings on the new model Baird " Televisor " receiver, which projected black and white images on a translucent screen about 9 inches high by 4 inches wide.

We were not disappointed, for the image of Mr. Baird appeared and we heard him open the programme in one apt sentence, in which he expressed the hope that the development of television under the auspices of the B.B.C. would add to the enjoyment of the great listening public. The image came through clearly and extremely well defined, the only adverse comment one could make being that his natural

Operating the television transmitter, which scans the scene in the studio (shown opposite).

reticence made him confine his appearance to a few seconds.

Then we saw in turn the artistes contributing to this first B.B.C. programme. Betty Bolton sang one or two songs, and her head and shoulders as a semi-extended image were extremely clear, the lip movement and mannerisms peculiar to the vocal numbers rendered being distinct. Following this came Fred Douglas, made up as a "coon." The image in this case was not so good, due primarily to the absence of feature contrasts owing to the all black colouring, and it seemed rather inappropriate for the first transmission to include a subject of this nature. Louie Freear, the commedienne, who came next, was broadcasting for the first time. She sang "Twiddley Bits" from "The Chinese Honeymoon" and here we had images of a very high standard, while Betty Astell in the last semi-extended scene came through with flying colours.

An extended view with Betty Bolton and Betty Astell dancing in turn concluded the programme, and contrary to general expectation the results in this case were of a high standard. Naturally, facial detail was absent but the interpretation of the dance tune as expressed by the rhythm of the body movement could be followed and appreciated.

Speaking generally, the whole transmission was an undoubted success and augurs well for the future. The programme material was exploited to the fullest effect, and there were no breaks in the transmission while changing from one artiste to the other. No fading from the London National station broadcasting the vision signals was apparent, although the accompanying sound from the Midland Regional faded a few times during the course of the transmission. Undoubtedly the B.B.C. deserve a measure of congratulation for their first effort, and future development will be watched closely by all those who have the cause of television at heart.

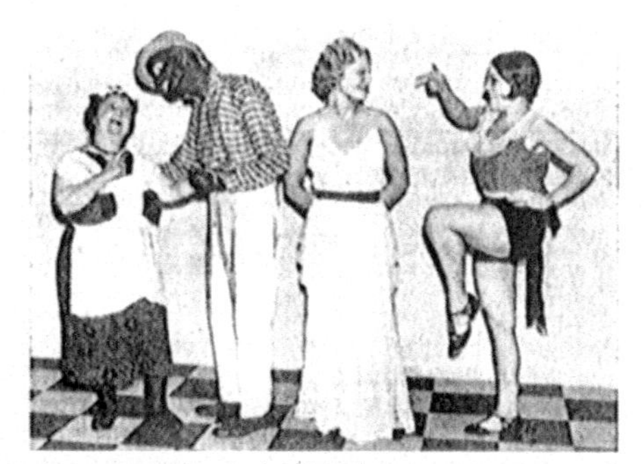

Artistes performing in the first B.B.C. programme: (left to right) Miss Louie Freear, Mr. Fred Douglas, Miss Betty Astell and Miss Betty Bolton.

Performing in the studio in front of the transmitter, which is seen through the window in the wall. On each side of the actor are the photo-electric cells.

Hammersmith.

After looking-in to the first television broadcast from the B.B.C., I am most enthusiastic about the results obtained. It was quite the best reception of television I have ever had, the semi-extended pictures being incredibly clear. Definition and detail were extraordinarily good. This applied also to the extended pictures, although in some cases there were slight shadows. The B.B.C. are to be greatly congratulated on the success of their first transmission. The old type of Baird Televisor was used for reception and it was run for ten minutes before the broadcast began, in order to get the apparatus "warmed up," which is always a wise precaution.

Leeds

At the start of this new series of transmissions on August 22, it was hoped by those of us in the North that the experiment of transmitting in the evening instead of in the daytime would make it easier for them to be seen at greater distances than in the past. Unfortunately, however, the conditions existing during this first broadcast were exceptionally bad. To begin with, the recent hot weather has ended in a condition of severe atmospheric storms and the interference on London National, which is a comparatively weak and distant station, was very severe. Had Midland Regional been selected for vision and London National for sound, a much better opportunity would have been available. Apart from the atmospheric difficulty, quite a powerful receiver is required to receive London National successfully even after dark at this time of the year in this area, and such a set must also be possessed of considerable selectivity. A receiver possessing these characteristics is unfortunately not ideal for the reception of television transmissions.

When these difficulties have been taken into consideration, it will not be anticipated that very good results were obtained, but although it must be admitted at the outset that the results were poor, nevertheless they created quite considerable enthusiasm.

The type of programme chosen was a good one. I found that the " close ups " which formed the major part of the programme were of great assistance in enabling the vision receiver to be correctly adjusted. It was possible to distinguish the faces and attitudes of the performers, but they were caricatured by the distortion introduced.

A curious incident occurred during the short period of the transmission which effectively blotted out the programme for some minutes. The " televisor " was being operated with the cover removed, and a moth chose this moment to commit suicide amongst the brush gear of the motor, which immediately caused the motor to lose speed and synchronisation was lost for the time being.

I am looking forward to the winter months, when conditions will have improved and there will be a better chance of good results.

Newcastle-on-Tyne

The first reception of the B.B.C. television transmission was, on the whole, fairly well received at Newcastle. The " coon " vocalist was seen very distinctly. His black make-up. thick lips, white eyes and what appeared to be the black and white stripes of his costume, were most outstanding, also his large hat." The lady next appearing was easily recognised as an old performer, her facial expressions being very plainly seen, also her comedy hat with small feathers.

The artiste who appeared next, full figure, executing first of all a Spanish dance with tap dancing, was very clear, and her second number, in which she was wearing " shorts," also came through very well. I find that full figure transmissions are received very well and provide a pleasant change from the head and shoulders picture. I am therefore hoping that full figure transmissions will fill a fair share of the future programmes from Broadcasting House. This transmission was seen 300 miles from the London transmitter under adverse conditions, including considerable morse interference, and considerable fading and very bad atmospheric disturbances

I " looked in " again on August 26, to the third B.B.C. programme, which was also very successful. The image of Rupert Harvey himself was well received, followed by his drawings which were also clearly seen at times. The next item, a man doing a step dance, came through very well, particularly when he introduced his dog and caressed it, then the dog stood on its hind legs reaching up to lick the face of its master. Following this came a lady, full figure, doing various acrobatics, dancing, etc.

A feature of the transmission was the remarkable distinctness of the black and white checks on the stage or floor of the studio.

Crieff, Perthshire

Our apparatus was undergoing adjustments and tests at commencement of the new series of broadcasts, and the first night of actual reception took place on Wednesday, 24th, and proved very successful indeed. The image was maintained during the entire transmission and showed excellent detail, with a complete absence of fading, which is one of the greatest difficulties we have to contend with, as we are situated about 400 miles from the London transmitter.

A slight difficulty was experienced in holding the image, as the automatic synchronising gear did not appear to be quite so effective as during the following broadcast, which took place Friday, 26th. Reception on this latter date was rather disappointing as during the entire transmission a heterodyne signal was noticeable and periods of fading were very frequent. Some peculiar effects were noticed, as the image was subject to a complete fade out while the synchronising signal remained strong.

While signal strength was maintained the image was quite good, and showed considerable detail; and we are looking forward to an interesting season.

Future Programmes

Further reviews of the B.B.C. broadcasts will be published in the next issue of TELEVISION. In order to make the criticisms as comprehensive as possible, the editor will welcome the co-operation of readers, who are invited to submit their comments on the programmes. A number of these letters will be published, and each month the correspondent who sends the best criticism will receive a year's subscription to TELEVISION free of charge. Where a prizewinner is already a subscriber, his existing subscription will be extended for a further year.

Television programmes are broadcast every week on Monday, Tuesday, Wednesday, and Friday from 11 to 11.30 p.m. Vision is transmitted from Brookman's Park on 261.3 metres, and the accompanying sound is transmitted from Daventry on 398.9 metres.

Last Month's Programmes

By "Spectator"

THE chief novelty in the past month has been an exhibition of palmistry by Miss Nell St. John Montague, who is distingushed for having provided the first programme to receive general recognition from correspondents. Counting heads is not always a fair test and, as comparisons are sometimes disastrous, the B.B.C. prefers not to publish figures of its post; but it is only fair to this lady to make it known that her performance evoked a volume of mail at Portland Place that would have been gratifying to the producer of a popular revue.

Superstition

Superstition is always with us. Who would not foretell the future if they could—and it is not necessary to be a serious student of chiromancy to wish to learn how to read a hand. If part of Miss Montague's success must be attributed to these factors, a great deal of credit must be given to the method and manner in which she presented her subject. Using a charcoal pencil, she sketched on white paper the outline of an ordinary hand and then, explaining each line as she progressed, she drew in bold definition lines of life, fate, head and travel. Next she took an unusual palm and showed the markings of suicide, murder, fame, and danger from air, fire and water. Each hand could be clearly read in the receiver in the projection room, and the interest was so keen in the studio that when Miss Montague finished, everyone was gazing at his palm. Seizing Eustace Robb by the left wrist, the seer then gave an impromptu reading of the producer's palm, and we applauded when we learned that he was to overcome difficulties for five years!

Clairvoyance is a delicate topic, and while this demonstration was entertaining and informative, I confess to surprise that it was permitted. The producer feels that the response vindicates its inclusion, and I am glad to find in the incident fresh evidence that he is to be given a free hand. It would be a mistake to fetter these new television productions with conventions that have grown up around the ordinary programme. In any case, an art that has survived for five thousand years and still thrives over the greater part of the East, is worthy of at least a cursory study, and Miss Montague is an authority on the subject. After the transmission she told me that she was reared in an atmosphere of superstition, in one of the Central provinces of India, of which her father was then governor, and that her mother came of Scottish highland stock which had a reputation for claivoyance extending over hundreds of years. Sceptics claim that the size of the "mountain of Venus" is determined by the muscular development of the ball of the thumb. I am putting this to the test by means of a course of thumb twiddling exercises, but to date can record no improvement. Miss Montague will be seen and heard again.

Writers often liken television to the cinema in its early development, and as both forms of entertainment appeal to the eye, a resemblance does exist; but the introduction of the talkies seems in one respect to provide a closer parallel. Vision was already established when sound was added to the cinema. In the case of television, sound is already established and it is vision that is being added. Am I mistaken, or do I now hear the same people who said that sound was not needed in the cinema, moaning that broadcasting is enough without vision?

More Receivers Wanted

It is a misfortune that a crystal set cannot bring television to the home. Lookers would increase by hundreds overnight as listeners did in 1923, if the humble, cheap and efficient cats whisker could throw an image upon a screen. The apparatus for receiving television programmes is somewhat formidable and many listeners must be without vision because they are unwilling to purchase and operate an additional wireless receiver. It is a shame that so much enterprise should be frustrated by the inability of listeners to become lookers, and as I have sat in the studio watching auditions, rehearsals and transmissions, I have pondered this problem of reception, and now in a diffident, artless way throw a suggestion to the experts.

Receivers in wireless exchanges carry broadcast programmes into sixty thousand homes. In return for one or two shillings per week a subscriber to a relay service is provided with a loudspeaker which he plugs into a socket whenever he wants to listen, and in many areas alternative programmes are on tap. Programmes are carried by telephone line from the receiver at the Central Exchange to the listener's homes. Is there any reason why television should not be relayed in this way? Listeners attached to these

" Vladimir flung Vanda out of the picture once or twice."

wireless exchanges would probably become lookers if facilities were offered. I am told that relaying of television is practicable but not simple by this means. It might be worth investigation. People will hire when they will not buy.

Irving Berlin, who called on Henry Hall at Broadcasting House and was most impressed with all he saw, said they had nothing like it in the States. Hall naturally took Berlin to see Studio BB and gave a demonstration. Standing before the screen he sang one of Irving's songs and he was moved by the chorus to try a step dance. In the flickering light he kicked over the microphone, which cracked. Asked when he was going to televise, Henry, who is not accustomed to stage performance, said he disliked the make-up which is rather forbidding; and for those who prefer not to use it a blue lamp is placed in the projector. The blue light is less vivid than the white arc and as a result the image is not so well-defined. For this reason, it is rarely used.

Ventriloquists

I have always been partial to ventriloquists and even without vision they always raise a laugh. Coram is one of the best and his " Disorderly Room " sketch is really funny. Music-hall patrons would not have recognised Jerry, his dummy, if they had seen his television debut. His face was livid and his mouth was black. A good effect was contrived when Ralph Coram and his wife faded out of the picture on a tandem cycle at the end of " Daisy Bell."

Guelda Waller and Vera McConochie were chosen to televise from a photograph taken in their old-world costume in which they sang nursery rhymes and songs of another era very prettily. The see-saw on which they sat for " Margery-Daw " was produced at short notice by a resourceful official who now looks after these things. He said a little job like that was easy, ask him for something difficult next time! I thought the puppets were a little too small to be really effective. Their faces are only about four inches in diameter, which accounts for the relative lack of detail. It is surprising that so much was seen. The props—especially the magic tower—were seen well, and the gollywog behaved just as a full-sized gollywog should. The speech was a little thick at times.

Meduria sang " A Peanut Vendor " with great verve and danced the Rumba on the Spanish-Mexican night, but Val Rosing suffered from having a bad cold. His make-up was good; though fair, he looked a Mexican, and his voice was well-suited to his songs. Val is going on the concert platform where his father, Vladimir Rosing, had a distinguished career. We must now look forward to seeing his successor as dance band vocalist, Les Allen, who also plays the saxophone.

Camaraderie in the Studio

Fred Douglas was seen without a black face in a programme with Nat Lewis, fat and hearty, and Betty Burke, a blithe dancer. These three separate acts were blended into one in rehearsal. Fred and Nat appeared together in one scene, and Betty joined Fred in a cockney dance for his number, " Lambeth Walk." This camaraderie is good to watch. All the time artists in the television studio are helping each other out. They are always ready to change costume and play in character with another act at the suggestion of the producer. No sign here of jealousy or striving for effect at the expense of other turns. Eustace Robb must be complimented upon the friendly informal atmosphere that prevails.

Vladimir flung Vanda out of the picture once or twice in their acrobatic dance, but it was not his fault that he was too quick for the man handling the projector. Vanda looked sweet and managed her crinoline to the manner born, but her acrobatic dance was the biggest thrill.

Mary, the performing ape, ambled about as apes do. Did you detect the deception? We are seeing a lot of animal life, and I hear that a greyhound may be here soon, in which case an expert will explain the winning points.

The " Party " was a slight disappointment. At times the fun seemed to be a little forced. Namara came in costume from the theatre and as her dress was suitable she just had to sing " Carmen." Having extravagant gestures she televises well. Betty Pollock is a competent mimic; her make-up as Gracie Fields was best. Namara wore a mantilla for her bolero and her Cuban costume was graceful in the rumba.

The Newells

The projection light plays queer tricks. Elsa Newell's red dress looked pale in the receiver, blue appears as black and the filmy black tulle —or was it chiffon?—used by Betty Burke for her dance was so transparent that it was barely visible. The dancer was seen to great advantage. Bill and Elsa Newell have some saucy patter. They form one of the best light song and cross-talk double acts on the radio to-day. They gave an effortless, finished performance that looked as well as it sounded. Bill said that he wrote most of the patter himself, so there should be good original material here for future shows, and both he and Elsa know how to make the best of their clothes.

Dell and Dene in the same programme were a surprising pair. Nan Dell has walked for a mile and a half on her toes. I believe it now that I have seen her hop round that studio, and Jerry Dene, besides this remarkable toe dancing, can get a tune out of almost anything from a wine glass to a saw. Melodious stuff, too, don't you think?

Jass, of Jass and Jessie, kicks like a real pony. Ever since my first pantomime, I have loved a human horse.

The introduction of a second figure in Max Templeton's shadow play complicates the picture and I prefer to concentrate on a single object until images improve.

Short gaps have occasionally appeared in the programmes for which no apology has been made. I am no stickler for etiquette in presentation detail, but listeners have become accustomed to expect an apology or mention of any short break that may occur in regular programmes, and I think that lookers should be treated with the same courtesy.

A Stage Effect

During the month the producer has taken a stage effect to finish several turns. Instead of stepping back from close-up to extended position and fading out, artists, after backing to the extended screen have then walked forward to a close-up and following a curtsey or bow have slid out of the picture. This " curtain " is more impressive and serves to implant the artists' features in the memory.

Acoustical balance is not yet perfect, but I can see no way out of this problem with the existing facilities. The accident when Henry Hall kicked over the microphone by catching his foot in the cable demonstrates the difficulty. The microphone used is of the Western Electric condenser pattern and is housed in metal torpedo-like casing, from which it derives its nickname of " the bomb." The Western Electric instrument is an omnidirectional, good, all-purpose microphone that has been chosen for its robust qualities, but no condenser apparatus can resist a knock-out blow. It is awkward to move the microphone while a performance is in progress, and so it is necessary to compromise by placing the instrument before projection in such a position in the studio that it will give adequate reproduction of sound throughout the scene, whether the performers are in close-up, intermediate, or extended positions, dancing or stationary.

Placing the Microphone

The task of placing the " mike " is further complicated by the need for a good acoustical balance on the piano or other instrument that accompanies the performance. By means of a movable stand and cable, mobility is secured, as with the familiar standard lamp and flex in domestic use. But this arrangement is far from ideal, and I wonder whether the use of two or more microphones, suspended from the ceiling, would not improve matters? Engineers employed

Miss Montague, the palmist, provided the chief novelty in last month's programmes from Broadcasting House.

on outside broadcasting meet the same problems in theatres from which excerpts are occasionally relayed. The technical difficulty of getting good acoustical quality is one of the reasons why these relays are now heard so seldom. The experience of outside broadcast engineers in theatres might be valuable to the television people. Perhaps they could be consulted.

It is surprising that so much was seen of the puppets, considering their size.

November Programmes

Scotsmen may like to note that the highland fling and other Scottish items will be given on St. Andrew's night, November 30. Gasson, the human bird, will also be seen this month while Nina Devitt and Jack Browning, Louie Freear and Lauri Devine are all appearing again. Penelope Spencer is to dance to Lord Berners' "Funeral March for a Rich Aunt" and the composer has been asked to come to the studio and accompany on the piano. A play, "Posters," is being specially written by Val Gielgud and Holt Marvel. The characters are three men and one girl, and the authors tell me that their play will employ every device of the medium discovered to date. Late November or early December should see its first performance. As previously, the B.B.C. programmes will be broadcast this month every Monday, Tuesday, Wednesday and Friday from 11 to 11.30 p.m. Vision is transmitted from Brookman's Park on 261.3 metres, and the accompanying sound is transmitted from Daventry on 398.9 metres.

By the time these notes appear in print, a mannequin parade will have have been seen by television for the first time. Reville is the enterprising firm of modistes, and I hope to deal with this programme in the next issue.

Last Month's Programmes

By "Spectator."

SIX months have passed since the first television programme was broadcast from Broadcasting House, and in this time many reputations have been made and lost in the basement studio. A new criterion is applied to artists wishing to televise and in the case of regular radio acts performance must be re-assessed in terms of sight and sound.

February has been a period of consolidation in programme building and throughout the month auditions have been given at the rate of three or four a day. Many established broadcasters have faced the projector for the first time with a critical audience of two, the producer and his assistant, and much latent talent has been discovered. Often results have been disappointing and sometimes surprising. Cases have been seen of artists who, through long concentration upon an aural effect, have allowed their ocular appeal to atrophy. Such applicants must study afresh and come again; the television transmitter demands more than a subtle voice and an insinuating wit, and new material has frequently shown more promise.

Through the process of audition several fresh acts have reached the programme during the month. Passing the newcomers in review, the broader and lighter elements predominate and I predict success with lookers for Mabel and John Marks, a genuine brother and sister act, appearing together in a studio for the first time. For several years Mabel Marks has been a favourite of listeners and regularly every three months or so she has given her songs at the piano. Though the double act is known on the music-halls, she has not before brought her brother to the microphone. It is a joy to watch this partnership in action, whether singing or dancing; their combination is perfect, their method is snappy and there is humour in their feet. Clearly they enjoy their work and I do, too.

Professor Makaroff's "Lady," who smokes a cigarette, wears pince-nez glasses "when reading,"—and plays yo-yo.

Another of the month's pleasures was to watch Professor Grigori Makaroff and his "Lady," a Royal fox terrier and a fine example of canine intelligence, with wide round brown eyes and the kind of look that drives a fellow to join the anti-vivisectionists. As lookers know, "Lady" can wash her face and smoke a cigarette and wears her pince-nez when reading. Believe it or not, she can also play Yo-Yo.

The professor is a distinguished figure of the old school. Trained as a singer, he has taught this art and dramatic elocution for many years and in other days has sung at La Scala, Milan. At the tender age of two weeks "Lady" joined the professor and for eight years she has been his hobby and his pet. At first he had no idea of training her, but later he was struck by her unusual comprehension, found she understood his language, and so taught her tricks. The professor was persuaded to take "Lady" to a cabaret, where she was seen by an artist and recommended to Eustace Robb, television producer. In a way, we were lucky to see this act, because the professor, on the point of leaving Broadcasting House after a morning rehearsal, was under the impression that his job was done.

Nat Travers made his first appearance in the same programme. This coster-comedian was vastly intrigued by the similarity in make-up for all acts, and claimed that he looked just the same as Fredrika Tree, an Anglo - Viennese singer. On the stage he would have had a crimson face and a black eye and Fredrika would have looked a peach, while in the studio they both had white faces, blue lips, noses and eye-brows. Nat was not convinced that it was natural until I led him to the receiver, where Fredrika was seen to be looking her normal self and rather like La Seidl. Yet another recruit, "Richie," made his bow the same evening. This coloured dancer, with the stamina of his race, was ready to go on dancing until asked to stop. He knows his stuff and can give a song, too.

Nina Mae McKinney, a coloured star, with a personality that shone through the receiver.

Dancing has reached a high standard during the month. Good news spreads quickly in the profession and fresh talent is continually coming forward for trial, but there are disappointments. "I hate films and can't stand lights," said an applicant on reaching the studio. When the producer and his assistant were down with influenza, the engineers laid down their switches and, seizing make-up box and props., prepared the artists for the show.

As soon as I saw a visual announcement I knew we were in for a good time, for the black type on white cardboard has become a sure sign of a gala programme. It was used to introduce Nina Mae McKinney, a coloured star of the film, "Hallelujah." This artist has a pronounced personality that shone through the receiver. It was lucky that the programme opened on time, because she had to get to the Leicester Square Theatre by 11.15 for her last performance of the day. She topped the bill at this theatre and opinion was divided on her performance. I think that she came over.

W. W. Reville-Terry paid a return visit to the studio to show advance Spring fashions. A gown designed for the Courts was worn by Marchesa Malacrida and some striking evening dresses, mainly in black and white, were displayed by mannequins and other graceful creatures. The exhibition of ballroom dancing by Laurie Devine, in a new gown from the Reville-Terry house, was particularly arresting. I was sorry to have to say 'Goodbye' to Cyril Smith, a polished and flexible pianist whose work as accompanist has been praised by every artist. Already known as a notable soloist, he has left to concentrate upon concert engagements.

The accompanist is now screened from the main part of the studio by means of a black gauze curtain, which has been hung up following the discovery of slight intermittent interference from the piano lamp. As the pianist turned over, the flick of his sheet music oscillated the light, and caused a yellow flash in the picture.

Adeline Alexandra was discovered at the Prince of Wales Theatre, where they have an eye for youth and beauty such as hers. I liked her naive, light-hearted air. "Women haven't any mercy on a man" is a song that has long amused me, and shot from the mouth of George Lane the words lost nothing of their point. George is a ripe comedian who has toured in Huntley Wright parts. Charles Gasson makes a fine bird. He chirps, hops, sings and flaps his wings, in fact does everything but lay an egg, yet ten minutes is a shade too long for this act. Concentration for this length of time upon an unfamiliar and often quickly moving object places a strain upon a looker. Vision and a volatile expression lay emphasis upon the sentiment of "Please, Mr. Hemingway," sung by Yolo de Fraine, a versatile artist with deep and satisfying voice. Words that might shock, if spoken, amuse in a song.

High cheekbones tend to add several years to an artist's face as seen in the receiver and Robb takes special care in making up such cases. False eyelashes soften the expression and have become popular with nearly all feminine players for close-up work. These synthetic hairs, about one inch in length, are stuck on to nature's eye-lashes. Once fixed, an artist must move with care, as her face must be made up afresh if the lashes tumble off. Lucyenne Herval, a lovely rounded blond, was one of the first acts to make use of artificial lashes. In this artist a pleasant voice is allied to a sensuous form, and may we see her again? Her rumba was unusual.

I watched Enrico Garcia in rehearsal and during transmission. Garcia has sung with the Russian Ballet and he has that kind of voice. Obviously cramped by the limitations of the picture, I feared in the morning that this singer might not achieve the visual effect that his talents deserved, but he proved to be amenable to production, as intelligent artists are, and in the evening gave a polished and wholly acceptable performance. An instance of the value of expert direction.

Violet Victoria appeared in charwoman's bonnet and a nervous twitch. Here is a comedienne who deserves to be better known. Her character-acting is not only good but confident and her comedy is not forced. " Nerves " is a clever number that offers a temptation to overdo the fright, but Violet got it just right, building up her effect in a crescendo of agitation as the song progressed. A good robust comedienne is a grateful item in any bill.

" Fanfare " deserved a longer run if it contained other numbers of the quality of " I wonder why Poor Nellie never writes," which Violet chose for her third appearance this evening. The title tells the story and the lyric merely amplifies in the richest bathos the contrast between the condition of the unhappy mother fretting for her daughter, and that of Nellie who has left the squalor of her home to seek fame and fortune among the bright lights of the city. The theme is as old as the stage, yet the anguish of the mother, soliloquising on the temptations that beset her daughter is always funny. This song is a fine example of its kind and Violet Victoria squeezed the last tear from it. By the way, I find " Don't sell any more drink to my father," and songs of this sentiment equally irresistible.

Jass and Jessie

Jass and Jessie provided an unconventional interlude in the only extended shot that we had in this evening's programme. Jass, a human being, but you've guessed it already, always pleases as a performing pony. Jessie should crack her whip more often, there's always a chance that she will hit something in the studio sooner or later, and the more often she cracks it, the greater the risk.

Robb has been experimenting with black lines drawn upon the white screen used for extended pictures. First wavy horizontal lines were shown, and later, when lookers complained of sea-sickness, the lines took the shape of steps. The object of the design was to emphasise the lateral movement of figures in the pictures. As the beam swings across the background when a player makes an exit to the side, some detail in the screen would certainly clarify the picture, but the producer is not satisfied with his steps and the waves distressed lookers, so he has to think of some other device. Horizontal designs having failed, he now favours a vertical outline and we may expected to see a few trees soon. Just a little copse, I think; too much black paint would

Jass and Jessie provided an unconventional interlude. Jass always pleases as a performing pony.

obscure the picture. A pretty little property shrub in a wooden box embellished the scene for " Virtue Triumphant "—a melodrama that gave scope for burlesque to Charles Wade, Theo and Gibson. Simple scenic effects, such as this, are welcome as a relief.

I place cartoonists among the most successful acts and Harry Michaelson's method of drawing faces around the initial letter of his subject is effectual. His drawing of Ramsay MacDonald was an excellent likeness and his picture of Maurice Chevalier was good caricature.

An audition for the Children's Hour was arranged for James Liversedge after his African adventures before the television transmitter. There seemed to be a lot of lions about, must have been in a game preserve, but I wished that he had relieved his jet black nigger costume with more daubs of white.

Isabel Chisman hails from Bath and brought a Regency air with her to the studio from the town beloved of Beau Brummel. This evening we enjoyed a quartet that was engaged to add tone colour to her authentic eighteenth century steps. Her dance, composed for the Duchess of Portsmouth, " to be danced in a stately manner," and her single dance for a young lady, conjured visions of more graceful times. How could deportment be otherwise in such costume? Monica Waddington is now an experienced television artist and looked very sweet in her Regency bonnet when singing her old-world songs. Olive Groves in tunes of charming sentimentality and Gustave Ferrari in good voice completed an unusual programme that was memorable for its restful dignity.

On the eve of his sixty-first birthday, George

Mozart turned up to revive old memories. Like other giants of his day, George Mozart needs to be seen to be appreciated, and his facial work might well be taken as a model by several light comedians half his age. His gags were old, but they wear well and his imitations were entirely convincing. E. Kelland Espinosa brought Strube's Little Man to life for us the same evening. I had always wanted to meet him, had not you?

Newcomers, Dinks and Trixie, are frankly vulgar, but this does not spoil their act for me. With her coy voice and baby face Trixie gets a laugh, and Dinks' vamping is quite funny when he is dressed as a dame in the pantomime style. It required no prescience to foresee that he would split his skirt before he left the picture.

A good curtain is essential to any entertainment and sometimes lookers ask why the walk-up finale is not discarded more often. The difficulty is to find a method as good. The walk-up is neat and gives each performer an equal opportunity, and the producer is right to stick to it until he finds a better finish.

Elizabeth Pollock is a competent mimic and in the way now popular with impressionists assembled her victims at an imaginary party. It was a distinguished gathering of theatrical stars that included Marie Tempest, Lillian Braithwaite and Edith Evans, Cleo Nordi wore a striking costume for her dance to Beethoven's "Polichinelle." Did lookers see the little bells that hung from the cloth?

Once in six months has the programme started late as the result of a technical fault, and only for a matter of moments has a show been interrupted from this cause. For this happy record thanks are offered to D.C. Birkinshaw, the engineer in charge of television development at Broadcasting House, his enthusiastic staff and the single Baird instrument that has been in use for several hours each day since August last.

B.B.C. Television Policy

Rumours and Facts

AUGUST is known in Fleet Street as the " silly season " and the past month has not failed in its reputation. In a letter to *The Times* Mr. Samuel Courtauld launched an exaggerated attack on advertising, which was soon reduced by more expert pens to its proper proportions. In the realm of radio, it was left to the *Daily Mail* to work up a " stunt," and the opportunity was provided by the B.B.C.'s action in taking a census of the television public. During the week of August 14, the programme included a request by the television announcer to those " actually looking in to send a postcard marked *Z* to Broadcasting House immediately." It was added that " this information is of considerable importance."

The Postcard Census

With that enterprise for news which has always characterised the *Daily Mail*, it published the next morning a prominent paragraph dealing with this B.B.C. " Postcard Census Mystery." The announcement from the television studio, it was explained, " may well prove to be the thin edge of the wedge in the move to banish television from the B.B.C. programme." Mr. Collie Knox, the radio correspondent of this newspaper, went on to say that " The blow to the industry which is at last gaining a foothold, would be incalculable."

If this last sentence was intended to get the television manufacturers up in arms, it could not have been better worded, and its immediate effect was the summoning of a meeting the same afternoon at Radiolympia to protest at what was interpreted as an attack on television by the B.B.C. Representatives of various firms interested in television sets and components passed a resolution recording their " deep resentment at the possibility of the B.B.C. abandoning the regular television transmissions." The resolution added that the " taking of a postcard census by the B.B.C. at this season of the year is of very doubtful value, if not definitely misleading."

The subsequent report showed how exactly the whole " stunt " had worked out to plan. The heading " B.B.C. Change of Mind " was followed by the sub-title " Result of *Daily Mail* Criticism," and readers were informed that the action of the newspaper in being the first to call attention to the mystery broadcast to television lookers had effected an " entire change of policy by the B.B.C." ! Explaining that the feeling aroused by the disclosure had been " intense," Mr. Collie Knox stated that the television transmissions " will continue just as before," adding that in the autumn experiments would be made at Broadcasting House with the Baird and H.M.V. 120-line transmission systems.

This last information has, of course, been common knowledge for many weeks, so the net result of the *Daily Mail's* excursion into television politics was the remarkable fact that " transmission will continue just as before "—a statement which, in the absence of anything to the contrary from the B.B.C. meant precisely nothing that was not already known either to the television trade or the public !

The episode did, however, draw a statement from the chairman of the Radio Manufacturers Association, Mr. W. W. Burnham, who, referring to the close liaison of this body with the B.B.C., said that " the discontinuance of the transmissions is not under consideration. On the contrary, both the B.B.C. and the R.M.A. recognise the value of the present transmissions as an initial aid to the continual development of the art, the potentialities of which are fully appreciated."

Readers of TELEVISION will recall that the official statement of B.B.C. policy, first published in this journal in June, 1932, announced that " to give reasonable security to purchasers of Baird sets, the B.B.C. has agreed not to discontinue transmissions by this process sooner than March, 1934." The programmes, which have now been running for just a year on the 30 line system, have provided valuable experience both in the technique of transmission and in the staging of programmes.

In the past year great strides have been made in research on 120 line transmissions by ultra-short waves, and it remains to be seen whether the 30 line broadcasts will be continued beyond the next six months. In the statement issued by Mr. W. W. Burnham, it is significant that reference to any particular system for future use is avoided, but it may be taken for granted that the B.B.C. will be guided by the public interest in its policy, and that the possibilities of television in one form or another will be still further developed by the broadcasting authorities.

News from Abroad

From our Own Correspondents

The United States

The formation in the United States of the National Television Association is an indication of the faith which the leaders of the industry have in the future of the science. Many of the mistakes made in the early days of broadcasting were largely due to the absence of a " guiding hand " and television is likely to avoid many of the same pitfalls. Among the officers of the new association is Mr. John V. L. Hogan who has done much to advance the knowledge of television in the States.

Removing Restrictions

With its nucleus of some 50 radio engineers and representatives of television equipment manufacturers, the new association is apparently well fitted to cope with the problems that will arise when television " hits its stride." It is reported that one of the first activities of the group will be to agitate for the removal of the ruling that ties television to experimental use only. The Federal Radio Commission is firm in its stand that television shall not fall into the evils that beset the path of sound broadcasting, and therefore has so far refused to issue television transmitting licences for purposes other than experimental. It thus remains to prove that television is sufficiently advanced to warrant " going into business for itself."

If the new Association is successful in its attempt to secure licensing regulations for commercial television stations, it is fondly hoped by all those who look to television as a vast potential force for education and entertainment that the same Association will function to keep the inevitable advertising within reasonable bounds from a standpoint both of quality and quantity.

Commenting upon the future possibilities of advertising by television, Mr. Orrin E. Dunlap Jr., writes as follows in a recent issue of the *New York Times*: " With so many listeners complaining after 13 years of broadcasting that there is far too much ' speiling ' on the air, predictions have been made that advertising by television, if sanctioned, will be restricted to definite channels. Then if anyone chooses to ' look in ' on the market place or the automobile mart, he knows such information, both aural and visual, is available on a particular wave. But should every wave vibrate with pictures of products for sale or commercial demonstrations, the step is not likely to encourage the public to purchase home television receivers for 100 dollars or more."

That the educational advantages of a perfected television system are not being overlooked is made quite evident by the recent action of the American Association of University Women in adopting a resolution that, since radio and television as media for the advancement of education and culture are destined to become increasingly valuable, the State, division, and local branches are urged "to be alert to conserve in every feasible manner these agents for the purposes of education and culture, and to protect them and the public from undesirable development and exploitation."

The Don Lee Broadcasting System in Los Angeles reports that 3,600,000 feet of cinema film have been televised by their transmitters W6XS and W6XAO during the last two and a half years. The films have presented a varied list of subjects. Close-ups and full-length shots of " stars," outdoor views, a seven-reel feature, and regular news reels (including those of the Long Beach earthquake) were transmitted. The latter scenes were rushed through the transmitters in record time.

Accurate identification of the films has been reported by " lookers " as far away as Santa Paula, fifty-five miles from the Don Lee Building. Other reports include useful data on signal strength of great help in television research.

Germany

The tenth annual Radio Exhibition held in Berlin last month, marked the first step towards a public television service in Germany. It will be remembered that the Post Office held back from the institution of even an experimental service, so as not to impede technical development, all patents being pooled to accelerate progress.

The new German standard picture, consisting of 180 scanning lines, 40,000 picture points and 25 frames per second, is officially considered sufficient for the opening of regular television transmissions. This will mean the erection of ultra-short-wave transmitters in most large towns. It is impossible to effect this rapidly, but at least Berlin can count on having television broadcast-

Combined cathode-ray television receiver and radio-gramophone shown by Telefunken.

ing by the winter of 1934-35. Broadcasting on a seven metres wave will just be possible, as the modulation frequency, which in this case is just slightly over 500,000 cycles, is still under the permissible maximum for undistorted transmission (one per cent. of the frequency).

Radio Löwe were showing what was the best cathode-ray image of the new standard picture. The size was about half-plate and the colour of the screen, reddish-yellow and black, was as close to white as possible. Herr von Mihaly was showing a new mechanical means of reception based on the following principle: instead of revolving the mirror-drum he revolves the actual picture on to a circle of mirrors and reflects the resulting image on to a ground glass screen. This means of reception, though rather crude, will, no doubt, have its advantages owing to its inexpensiveness.

Manfred von Ardenne, whose work in developing the cathode-ray tube is well known, had on view his new projector tube. I understand that he intends developing television reception with the help of a well-known Berlin firm.

Tekade were also represented with their 90-line mirror screw. The German Post Office laboratories were showing a 180-line cathode-ray receiver.

Telefunken were exhibiting a complete combination radio gramophone and television receiver for ultra-short-waves. Their picture also shows great improvement over last year's, but they do not seem to have reached the same pitch as Radio Löwe. In the Telefunken picture I still noticed that tell-tale cross which was absent at the Löwe stand. Telefunken have also turned their attention to transmission and were showing in operation a complete intermediate film transmitter. This, it will be remembered, was first shown last year by the Fernseh A.G. The object or the scene to be televised is photographed by means of a normal cinema camera, the film then being developed, fixed and televised in the ordinary manner. It is, of course, much simpler to scan a film frame than to scan directly a badly illuminated scene, and the time lag between the actual happening and its transmission by television is under 10 seconds.

The sensation of the Exhibition was on the Fernseh A.G. stand. Next to direct scanning of one person and reception on a mirror screw (only 90 lines) they were showing for the first time a complete television receiver for large halls. It operates on the intermediate film principle described above, only that Dr. Schubert, of the Fernseh A.G., has further developed this invention. To make the whole system more economical he uses a continuous celluloid band which is emulsioned, exposed to the incoming television picture, developed, fixed, washed, and partly dried before being passed through a normal cinema projector. The image is then washed away and the band is ready for use a second time. This all sounds complicated but in reality the whole apparatus can be easily mounted on a normal sized motor car chassis if this should be required. The invention was only completed in May, and it is astonishing that the Fernseh A.G. have been able to produce the complete apparatus within so short a time. The quality of the picture at present is equal to that of a slightly under-exposed home talking film. Dr. Schubert told me that in about two month's time the slight imperfections still contained in the picture would be completely removed. Readers will realise the gigantic effort behind this new receiver if they bear in mind that the incoming television impulses can only impress the film for one millionth of a second.

Argentina

Television in this country is still in its infancy. But the " infant " is gradually approaching a promising degree of strength, owing largely to the enthusiasm of amateur constructors. A group of keen experimenters in Buenos Aires has now formed an association known as " Centro de Television " for the purpose of advancing this new branch of radio. For the moment they have built a quite simple apparatus for the purpose of carrying out demonstrations. This is a good start, for we all know that the first step is the most difficult one and serves as a basis for further development.

News from Abroad

From our Own Correspondents

France

M. DE FRANCE, who is well known in French television circles, has during the last twelve months made considerable progress in developing his television system. The latest developments are the transmission of 144-line images both by line and radio links; and the perfection of an amplifier having a straight characteristic over 20-400,000 cycles and which responds to over 1,000 k.c. with only a slight falling off. Details of the amplifier have been promised and will be published in TELEVISION as early as possible.

M. De France televises both films and living images and in the latter case flood-lighting is employed as opposed to the travelling light spot system. He has found it possible to reproduce a group of people by this means at a distance of 7 yards from the scanner, which is no mean achievement. The scanner is of the Nipkow type having a four-turn spiral geared to the main disc, with a reduction of 4-1 and acting as a shutter. Thus each section of the spiral of holes is exposed in turn to the exclusion of the remaining 3 spirals. The speed of the spiral-holed disc is 3,000 r.p.m. The floodlit images are sharply focussed on to the scanner with a good quality lens. The photo-cell is a small standard vacuum type and only one is employed behind the scanner.

For reception M. Defrance arranges a duplicate of the transmitting scanner and a special flat-plate gas discharge lamp with a lens magnifying system to increase the image size. In another model a lensed disc is used with a Kerr cell light modulator projecting on to a screen. In yet another type of receiver a cathode ray tube is used, but results are not as good as they might be owing to the difficulty of modulating the ray without undue alteration of the spot size.

The ninety-line film scanner used by De France, showing the shielded photo-cell and amplifier.

Development is taking place by combining the variation in the speed of the ray with the intensity modulation; this shows great promise. The photograph illustrates the layout of the film scanner which is now being rebuilt.

M. Defrance has also experimented considerably with 90-line and 60-line scanning. In the case of the 90-line pictures, films have been televised with success, the film passing a gate at a constant speed. Floodlighting is employed for the 60-line images, and a slotted disc system has also been used in this connexion. The images seen by a correspondent of this journal gave almost photographic results, even the smoke from a cigarette being clearly visible.

Madeira

"Wireless telephony is so well known to-day," writes a correspondent in the Funchal newspaper *Diario de Noticias* "that it can only now astonish some poor peasant, whose world is measured by the narrow strip of land which bounds his vision. Nearly all the world has heard the midnight chimes from Big Ben in London, cabaret turns from Paris, concerts from Rome, and gramophone records from Holland—and now we have television.

"Mr. W. L. Wraight, who has ived in Madeira for some years, is now using his third television receiver, on which he receives the transmissions sent out from the London National station. These transmissions are intended for a radius of about 100 miles, and Madeira is approximately 1,500 miles from the transmitting station.

"Mr. Wraight recently showed us the apparatus he is at present using, and presented to us a programme from the studios in London—first a girl singer and then some dances. The vision is not yet absolutely clear, but Mr. Wraight is constructing a new receiver from which he hopes to get much better results. This apparatus has a screen measuring 30 cms. square. In the meantime we are much impressed by the performance we have seen."

News from Abroad

From our Own Correspondents

Italy

The fifth National Radio Show, held in Milan last month, may be said to have marked the beginning of practical television in Italy. The outstanding exhibit was the complete television apparatus designed for the transmission and reception of both studio scenes and films. For this purpose a double scanning device was employed having two separate amplifiers, the transmitter being housed in a nearby studio. The modulated current sent out from the amplifiers is sent to an ultra-short wave transmitting station having a power of about three kilowatts.

The Scanning Device

The scanning device used for the transmission of films incorporates a disc revolving at double the speed of the images in the film. The disc has 45 apertures so that the images are scanned by 90 horizontal lines. The disc may, however, be replaced by others, having either 60 or 90 apertures in order to obtain a 120 or 180 line scan. Behind the disc is the photo-electric cell coupled to two four-stage R.C. amplifiers. The amplifiers are designed to give a flat response curve from 40 to 500,000 cycles per second, this high frequency being necessary for the 180 line scanning.

The transmitting studio is equipped with two transmitters each with a 60 hole disc revolving at a speed of 1,200 r.p.m. and transmitting 20 images per second. Although the scanning apparatus is not arranged to follow the movements of artists about the studio, its range is sufficient to allow considerable movement. Eight photo-electric cells pick up the reflected light and the link between transmitter and receiver is a very carefully constructed "feeder" to avoid high frequency losses. Ultra-short waves are used at the transmitting end, the frequency being 48 megacycles and the wavelength 6.30 metres The oscillator has no tuning chokes, condensers or frequency multipliers, but is based on the Lechler wire principle. It consists of two half wave length aluminium pipes and two 20 watt valves working in push-pull. This has been found to give a very pure and constant output on ultra-short waves. The oscillator does not require a powerful amplifier, and following the 20-watt valves, arranged in push-pull, the power is modulated in a 250-watt triode. The modulated then passes to a .5 kW screened valve, arranged in push-pull, and finally to the output stage which consists of two 1.5 kW power valves. The aerial consists of a half wavelength aluminium tube coupled to the transmitter by a special feeder. The anode current is supplied to the valves by means of rectifiers, and no battery is required since the whole apparatus is mains-driven. The sound is transmitted by a 100 watt station working on a wavelength of 230 metres.

Three types of receivers are exhibited, using cathode ray tubes, mirror screws and Nipkow discs. Public demonstrations of films are given on the 90-line mirror screw receiver and studio scenes are seen on the 60-line disc receiver. In both cases the ultra-short wave receiver consists merely of a common detector stage whose only aerial is the tuning choke. A four-stage R.C. amplifier, giving a 250 watt class "A" output, modulates a high-powered sodium tube giving a strong amber light. A 900 candle-power beam passes to the mirror-screw which is, of course, revolved by a synchronous motor. The picture, which is 5 ins. by 7 ins., is not projected on to a screen but is viewed in the revolving mirror screen. It may be seen clearly, even in a lighted room, by about 30 people at a distance of 3 or 4 yards.

Reception was of a very high order and there was a constant stream of "lookers" throughout the exhibition. Sixty-line pictures of simple scenes were seen exceptionally well and the reception of the films was also remarkably good. Perhaps the most impressive item in the pro-

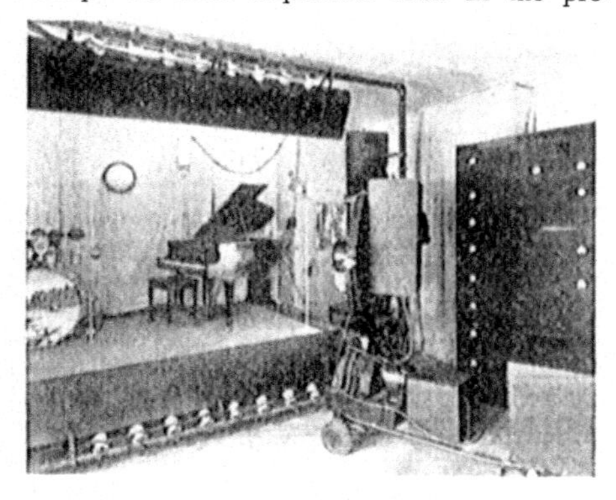

The transmitting studio at Bakersfield, California.

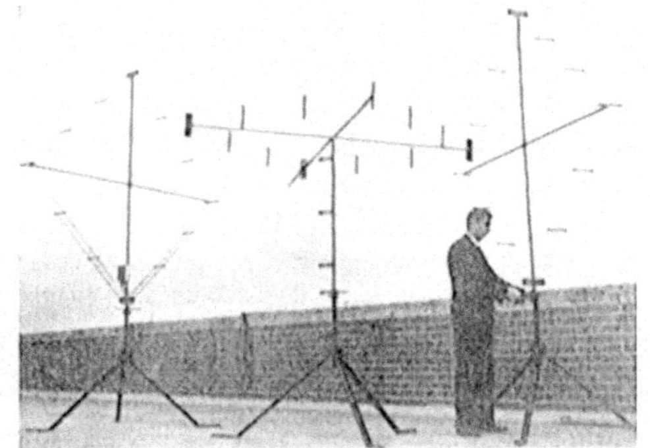

A party of German meteorologists who took part in the International Polar Year visited Tromsö to study the Aurora Borealis, which has an important bearing on radio phenomena. The illustrations show (left) large concave mirror used to condense and reflect the rays from the Aurora Borealis; (right) specially designed aerials erected at Tromsö.

gramme was the transmission of a film showing an ice-hockey match. Every detail was clearly seen as the players raced over the ice, and when a burly player surprised the goal-keeper with a neat, quick shot, I found myself joining in the frantic applause of the spectators who could be seen in the background of the film.

On behalf of TELEVISION, I gained an interview with Signor Banfi, who is in charge of the television activities of the E.I.A.R., the national broadcasting corporation of Italy. Discussing the future, Signor Banfi said that 1934 would see the inauguration of a regular television service, sent out from Milan on ultra-short waves. The aerial will be placed on the Torre Littoria, a metal tower more than 300 feet high. Films will be transmitted on 90, 120 and 180 lines and studio scenes on 60 or 90 lines. Mirror-screw and cathode-ray tube receivers will be established in various parts of the city so that the results may be watched by officials of the E.I.A.R. Radio experts in Milan are determined to make headway with television, and we shall be able to judge the results of their efforts at the Radio Exhibition next year.

United States

A complete vaudeville programme is now being transmitted twice daily from a television studio at Bakersfield, California. Vision is sent out from station W6XAH and sound from station W6XE. The former (1,000 watts) operates on 2,000-2,100 kcs. and the latter (500 watts) on 1,550 kcs., with synchronised sound. The station is owned and operated by the Pioneer Mercantile Company and the television research is in the hands of Mr. Frank Schamblin, Dr. Lee de Forest and Mr. Ralph D. LeMert. The hours of transmission are from 3 to 5 p.m. and from 7 to 9 p.m., and as many as twenty artists appear on the stage at the same time.

A smaller stage is shown in the accompanying photograph. It illustrates the direct pick-up television camera and the main vision amplifier. Fourteen stages of resistance-coupled amplification are used, and the camera is employed for the direct pick-up of outdoor pictures as well as for studio transmissions, where the scene is illuminated by 35 kW of incandescent lighting. Images have been received satisfactorily as far away as 2,000 miles. The system employs 90 lines and the scanning is horizontal with 20 picture frames per second. All the equipment was designed and built by Dr. de Forest, Mr. Schamblin and Mr. LeMert.

"TELEVISION," writes Mr. LeMert, "is the only journal devoted entirely to the science, and its arrival is eagerly awaited each month in the United States."

Television may become an important factor in naval manœuvres when the new U.S. vessels authorised by the National Recovery Act are in service. Thirty-five vessels are included in the plans, and it is definitely known that television equipment will be placed on board at least some of them. The type of apparatus to be used has not been revealed, but these installations will not be complete novelties in U.S. naval circles. Facsimile transmission and reception have been subjected to experimental use for some time past, and it is known that television has been carefully studied in this connection. In fact, actual scenes at sea have been televised in the course of the work. It requires no stretch of the imagination to visualise the value of practical television to the Navy, or for that matter, to any branch of military activities. If apparatus that is not too cumbersome and can be used under field conditions can be developed, it will play an important part in any future conflict.

The actual film used in the television broadcasts from stations W6XS and W6XAO is, of course, seen only by the operators and "lookers" can identify characters only as the televised image of the film appears in their receivers. It was a striking tribute of the detail sometimes obtained when a " looker " reported that one of the young ladies who appeared in a film was wearing a necklace. Casual inspection of the film by the director of the station failed to reveal this fact, and the " looker " was so informed. But a few days later, the director happened to pick up a section of the film in broad daylight and, much to his surprise, detected a fine line round the actress's neck. It was a fine gold chain without even a pendant. The " looker " was immediately notified and congratulated upon the detail he had obtained. The receiver was 10½ miles from the transmitter.

The same two stations have inaugurated a new television service in which full-length films are broadcast. Since the Federal Radio Commission permits only experimental broadcasts in television, each programme is prefaced with a notice to that effect. It is said, however, that prominent film producers are co-operating with the stations and that they are showing great interest in these activities.

" Chain " Television.

The opinion has been expressed by Mr. Hollis Baird, chief engineer of Shortwave and Television Laboratories, Boston, that " chain " television, on the same lines as the chain sound broadcasts in the United States, will be essential for a complete sight and sound service in the future. It is clear, however, that new methods of linking the stations will be required for short wave work. As Mr. Baird points out, the present chain sound broadcasting system uses telephone wires to link the stations, but the sound requirements are only one four-hundredths those of television, and no known telephone circuits, nor any that appears practicable in the near future will be able to carry the television currents. A relay system, therefore, is the solution. At the farthest visible point on the horizon from the main station, a receiver will pick up the television signals and relay them to another similar station. This point-to-point transmission will be directional in character and will be repeated until the desired point has been reached. When the signal is relayed to a city where broadcasting is desired, it will be used to actuate a non-directional antenna and the programme will be broadcast over a circle of about 30 miles radius.

The main reason for the relay system, apart from the fact that some method other than wire transmission must be employed, is that the beam or relay stations can be operated at low power with ample efficiency for the purpose, while a broadcasting station to cover the radius mentioned will have to be as powerful as the largest present-day sound broadcasters. Thus power will be conserved and not wasted needlessly for the relaying of the signal.

Germany

The German Post Office has issued the following statement: " At a conference held in the German Ministry of Posts at which representatives of the Ministry of Propaganda, of the R.R.G. and the German television industry took part, the chairman, Ministerialdirektor Giess, explained the plans of the German Post Office regarding television. The present picture quality of 90 lines is to be increased to 180 lines; on the other hand, the present number of frames per second (25), is to be retained. The German Post Office will adapt the existing high-power ultra-short wave transmitter at Berlin to the new standard. This work will take two or three months to complete. In the meantime the television industry will have had time to produce a number of television receivers for these new high-quality television broadcasts."

The German Post Office has also ordered a second ultra-short wave transmitter so as to be able to transmit sound as well as vision on seven metres wavelength. This transmitter will probably be finished by April next. From that time onwards talkie films will be broadcast together with sound instead of only vision as before. At the same time the German Post Office has ordered a direct scanning apparatus for transmission of persons and scenes as well as films. It is proposed to increase the number of television transmissions. Very probably a programme will be broadcast three times daily on several days in the week.

While it awaits the adaptation of the present ultra-short wave transmitter and the opening of the second one, the German Post Office will occupy itself with experiments as regards transmission and reception of ultra-short waves and with the transmission of these very high frequencies along cables, and will start a special series of experiments with decimetre and dentimetre waves. The moment intermediate film transmitters and receivers have reached the commercial stage, the German Post Office will use this new system for the broadcasting of outdoor events.

News *from* Abroad

By OUR SPECIAL CORRESPONDENTS

Italy - Rome

NEW TRANSMISSIONS SHORTLY.

Ente Italiano par le Audiziono Radiofoniche informs us that at the moment transmissions of television on 25.4 metres have been discontinued. New tests are expected to commence shortly, but a definite date has not been given.

France - Paris

THE BARTHELEMY SYSTEM.

The well-known television research worker, M. Barthélemy, recently gave a demonstration of his television system at the laboratories of the P.T.T. to M. Laurent-Eynac, the new Minister. M. Laurent-Eynac expressed his satisfaction as to the results achieved, and we understand he issued instructions for regular television transmissions to be commenced in the very near future with the co-operation of the administration of the P.T.T.

There is a general increase of interest in television in France.

United States of America

TWENTY-EIGHT TRANSMITTERS.

Great interest is being taken in television development in the U.S.A., and all leading laboratories are conducting intensive research to establish television as a commercial proposition. Licences for experimental broadcasts have been renewed, including N.B.C. transmissions from New York on 2,750-2,850 kilocycles on a power of 5 kilowatts. Sparkes Withington, of Jackson, Michigan and the Western Television Research Corporation, had their licences renewed by the Federal Radio Commission.

At the moment there are twenty-eight televison transmitters in the United States, transmitting on the following frequency bands :—1,600-1,700 megacycles, 2,000-2,300 megacycles, 2,750-2,850 megacycles, 43-46 megacycles, 60-80 megacycles.

NEW AMERICAN TELEVISION SERVICE.

The Don Lee Television Broadcasting stations W6XS and W6XAO, Los Angeles, announce that from now on there will be transmissions of full-length feature broadcasts of cinema films. This is now part of the regular schedule, and is in addition to the regular transmissions of news-reels and close-ups.

To lessen sideband cutting which has so far been experienced, the Don Lee station W6XS changed on November 1 from 2,150 kilocycles to 2,800 kilocycles.

Dr. B. C. Goldmark, who was formerly in charge of the television research department of Pye Radio, Cambridge, has been appointed head of the research division of the Ray-O-Television Manufacturing Corporation in Long Island City.

ULTRA-SHORT WAVES AND FOG.

In Boston, Mass., U.S.A. it has been found from experience that 56-megacycle transmissions are greatly affected by fog and mist. Whereas in normal weather no appreciable increase of signal strength has been observed with the increase of power, in foggy weather quite an appreciable increase has been observed.

STANDARDS AT WHICH TO AIM.

Mr. W. Hoyt Peck discussed in a recent essay what constitutes perfect detail in television. In his opinion, the television image on 180 lines 18 in. deep will contain all details the human eye can see when this picture is viewed from a distance of 9 ft.

Mr. Peck says : "Television may well copy the use of an illusion of detail, of which the motion picture makes use. When a character is seen in a group of a long shot, the features are more or less formless lights and shades, detail being supplied principally from memory, and from very general impressions." Mr. Peck concludes : "Look for this the next time you go to the movies."

In this connection it is interesting to note that the same opinion is expressed in a recent article published by the research engineer of the Leningrad research department concerning television. It is said that it is useless to attempt to reproduce a perfect image as the human eye will indicate perfection in a picture, when actually it is far from being so.

Television Research in Russia

Television research is extensively carried out in Russia by the Leningrad Electrophysical Institute, and also the Institute of Telemechanics. Experimental transmissions are being carried out from transmitters in Leningrad and Moscow.

Germany

180-LINE TRANSMISSIONS.

It is reported that research workers in Germany are concentrating their efforts on producing a commercial system of television, utilising 180 lines. It is stated that for transmissions on 90 lines all difficulties surrounding technical requirements have been eliminated, but 180 lines have not yet emerged from the laboratory stage. The greatest difficulty at the moment is the modulation of ultra-short waves with 180-line television signal. It is expected that when this difficulty is overcome, the first regular television transmitter will be installed in Berlin.

It is stated that 180 lines constitute the transmission limit on a wavelength of 7 metres. In this connection it is interesting to note that other research workers consider 180 lines to be the maximum number of lines required for satisfactory television service.

It is urged that the authorities concerned concentrate on the type of programmes the public wants, as, so far, this side of television has scarcely been touched upon. Opinion in some quarters is that television programmes will be nearly as important as the technical development of the science.

Hungary

A PROPOSED SERVICE.

The broadcasting authorities had a number of television experiments conducted, and sent their television engineer to Berlin to gather the latest information as to the possibilities of television broadcasting.

Regular transmissions do not yet take place, and no definite dates can be given, but great interest is attached to television, and it is hoped to start transmissions in the near future.

The Baird Company's Great Achievement

SUCCESS ON ULTRA-SHORT WAVES

(Photo—Courtesy of Daily Sketch)

An actual photograph of the received picture at Wardour Street.

IMAGINE a picture of a size approximately 16 in. by 18 in. perfectly steady, and with such detail that it will bear quite close inspection without revealing more than a slight line appearance. This is the achievement of the Baird Company—*and the transmission was on the ultra short waves from the Crystal Palace*, the receiver being at Film House, Wardour Street, in the heart of London.

Those who have been closely in touch with television development have been aware that such pictures could be produced as a laboratory experiment with a short line between transmitter and receiver, but the most optimistic have doubted the immediate possibility of broadcasting these pictures and receiving them free of interference. This, however, the Baird Company have accomplished, for during a programme of about an hour's duration the only visible signs of interference were three flashes of light which passed across the screen and did not last for more than a fraction of a second.

That the system is of remarkably wide scope was amply proved by the fact that three classes of transmission were shown—living subjects which included a musician, a conjurer, a mannequin and lecturer; a transmission of a film with a large amount of detail in it, and ordinary illustrations. The reproduced pictures were of a pleasant light sepia shade of ample brilliancy.

The scheme of the new Baird system is shown in a simple manner by the accompanying diagram. It will be seen that a disc scanner is used at the transmitting end and a cathode-ray tube for reception. The actual diameter of the tube is twelve inches and the image on this is magnified by means of a lens placed in front, though in the cabinet model of the receiver which is shown on this page no lens is used, and therefore the picture is approximately twelve inches square. It will be seen that a feature of this model is the simple control, there being only two knobs.

180-line scanning is employed with a picture frequency of twenty-five per second, so flicker, of course, is entirely absent and, as mentioned before, there is only a suggestion of the scanning visible, this being horizontal. Vision is transmitted on a wavelength of 6 metres and sound on 6.2 metres.

It will be appreciated that perhaps the most remarkable feature of the system is the entire elimination of interference. There in the very centre of London with thousands of motor cars, lifts, signs and other electrical devices there was hardly a trace of trouble from these sources. The service range is about thirty miles.

The cabinet model of the Baird cathode-ray receiver: Mr. Baird is at the controls and Capt. West is on the left.

Regarding this demonstration, Sir Harry Greer, the chairman of the Company, in the course of a speech which was made from the Crystal Palace and received in London said:

"It is assumed in some quarters and challenged in others that the B.B.C. is the authorised medium for television transmissions. I am not at this moment prepared to join in this issue. I can, however, most definitely tell you that your Company are in a position to transmit a programme from the Crystal Palace."

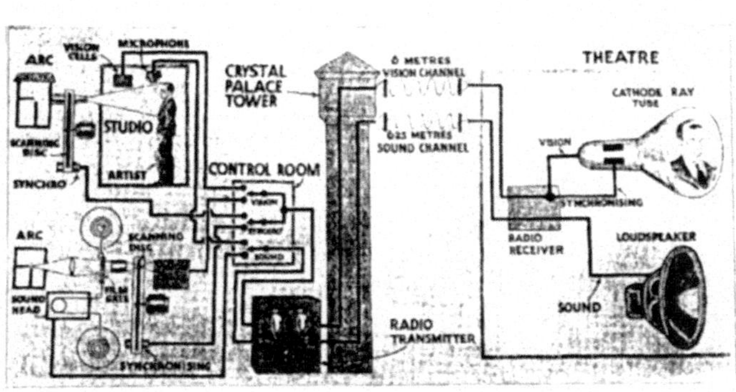

This simple schematic diagram shows the general principle of the Baird system.

REVIEWS OF THE PROGRAMMES AND RECEPTION REPORTS

THE radio receiver used in the transmitter room behind the studio is now mains synchronised, and results during the transmission of *Carmen,* when it was first used, show that it provides a reliable check. A separate radio check was added to a line check when the studio was transferred to Portland Place and, since March, pictures on two screens have been visible from the control desk. It is essential for engineers at the transmitting end to know beyond possibility of doubt what sort of image is being broadcast, and this result is now achieved. Previously one receiver was radio synchronised and, owing to severe mains fluctuations, the picture was unsteady. These fluctuations in the mains are a local condition caused by the heavy and varying loads imposed by the electrical equipment at Broadcasting House nearby and one curious effect was that engineers were receiving better pictures in their homes some miles away than could be got by radio on the spot.

The six-valve receiver in use was designed by D. C. Birkinshaw, research engineer, and consists of an H.F. amplifier (in which great precautions have been taken to prevent sideband cutting), a diode rectifier, a special amplifying stage developed by the B.B.C. research engineers, and three standard resistance-coupled amplifying stages. The job is designed to give the best possible results on a radio picture, and now that automatic mains synchronisation is employed, a better picture is obtained than I have seen before on the screen behind the scanner.

The scenery used during the past month has not been an unqualified success. Though the backcloth looks crude when viewed in the studio, the drawings in black and white become natural when seen by television, I am all in favour of painting the scene in this way. The effect is to clothe the studio, normally rather bare, and I like to see as many changes as possible in the course of a programme. This may be a matter of taste, but there is a positive advantage from the use of scenery in the sense of speed obtained by seeing artists moving in front of a stationary object or design. There is also no doubt that " props " are an aid to perspective in the picture.

A scene from the ballet "The Gods Go A-begging" with Lydia Sokolova and Stanislas Idzikowski.

Susie Salaman has been responsible for the scenic effects in several recent programmes: *Cleopatra, The Gods Go A-Begging* ballet, and *Carmen.* Her bold style suits the medium and her settings for *Cleopatra* and the opera made splendid pictures on the screen. In *The Gods Go A-Begging,* the scene was overcharged and the figures seemed to me to lose definition against a backcloth laden with paint. One looker complained to me that his images were blurred, which may have been due to his apparatus. But the producer is wise enough to consider the effect of his productions in visors which are crude, judged by 1934 standards, and the backcloth will be less elaborate in future.

Nine big photo-electric cells are now in use and, as each new cell is four times as powerful as the old, thirty-six small cells would be needed to produce the same result. Equipment at Broadcasting House comprised only sixteen small cells, and pictures are certainly better lighted by the new arrangement.

A third microphone of the condenser type is now fitted in a stand round the angle of the studio wall for use by the announcer. In this position he will be out of sight where the beam from the scanner cannot reach him and the dim light of his reading lamp will not affect the transmission. It is intended to keep the microphone permanently in this place so that it can be discovered when needed in the darkness. The orchestral microphone is more or less a fixture behind the black curtain which shields the music lamps used by the band, but the artists' microphone must always be mobile to suit the action of the programmes. In the twilight of the studio it is enough to have one length of cable trailing across the floor.

If correspondence is any guide there is less looking, as there is less listening, during the summer months, and the producer for July reports that he is receiving comparatively few letters about programmes. So a telegram of congratulation from Mr. O. H. Relly, of Eastbourne, was all the more welcome after the second Cocktail Club programme. This correspondent considered that the transmission was the best technically that he had seen and the show had amused him more than any other. His reception was so good that he could see the rims of glasses standing on the bar. How tantalising!

Roy Royston was broadcasting for the first time in this programme and is now engaged for his original part in the radio production of *The Girl Friend*. Another case of an artist introduced to the microphone in the television studio and afterwards booked for a broadcasting show. This Cocktail Club is a bright and spontaneous effort, calculated to raise the spirits in the forenoon.

Charles Heslop was a guest that morning and so was Signor Vittorio Podrecca, whose extended season at the Fortune Theatre enabled him to appear again during the month. Marionettes usually fail to move me, but these little people have wit and charm. Lack of detail in the picture hides their imperfections, and was possible to forget that they were dolls. Nearly every emotion is within their range.

In transmission they occupied the whole screen so lookers were not conscious of their size. Before the scanner their diminutive stature gives them an advantage over human artists because they are able to move and dance close to the lens in a focus which would taken only the head and shoulders of an actor. There were moments when their movements became uncannily perfect.

The dolls are operated by long strings, and scaffolding with planks at the top was raised to the ceiling to enable the operators to work from above. Though the ceiling is lofty, there was barely room.

Carmen of all operas needs to be seen as well as heard and was, therefore, a good choice for an experiment in television. Drastic cuts were necessary in the score which had to be compressed to forty minutes and still it was easy to follow the plot. It was a realistic scene when Carmen (Sarah Fischer) was stabbed by Don Jose (Heddle Nash) and behind the curtain there was the usual rush to get changed in time, which was not always achieved. Heddle Nash was delighted with the role and resolved to sing it on the operatic stage.

One of Signor Podrecca's marionettes; the marionettes were "visiting members" in the Cocktail Club programme, and the picture shows a marionette imitating Josephine Baker.

Another exquisite performance was given by Sokolova, this time with Idzikowski in *The Gods Go A-Begging*. Pupils from her school took part in the ballet and it is good for the new art that ballerinas of the future should be trained from youth in the technique and routine of television.

Future programmes for the diary:

Friday, July 27th (in the morning):

First appearance of Wendy Toye, the young dancer from *The Golden Toy* and *Ballerina*. I was charmed by her performance at audition and recommend this transmission, which also includes our friends, Gavin Gordon and Leonie Zifado.

July 31st (at night):

George Sanders, returning to the studio after his success in *Conversation Piece*; and Hilda Mareno, who will sing in Spanish to blend with Reuben Garcia, a newcomer in dances with castenets.

August 3rd (at 11 a.m.):

Leslie French in songs, and Nini Theilade in dances, from *A Midsummer Night's Dream*; both are playing in the open air theatre.

TELEVISION IN JAPAN

THE increasing importance of Japan in commercial fields of the world has been a subject of topical interest recently. It is, therefore, interesting to visualise progressive activity in technical circles.

With the latter idea in mind, our representative set out to interview Dr. Kawarada, Professor of the Engineering College in Waseda University, near Tokio, during a recent visit to this country.

Questioned about the work being done in Japan at the present time, we were interested to receive some details of the layout at the Waseda Laboratory for televising baseball matches.

The apparatus is located at a corner of the baseball pitch, behind a plate-glass screen, and it is possible to follow two or three of the players with a telephoto lens. A scanner on the Nipkow disc principle is employed, using sixty lines and framing eighteen pictures per second. The wavelength used is 159 metres with a power of about twenty watts in the aerial. The definition leaves something to be desired, but satisfactory results are obtained in conjunction with a running commentary.

Six stages of amplification have been employed in conjunction with a Kerr cell and mirror-wheel projection apparatus. A new mirror-wheel equipment is in course of construction, and actually two wheels will be used simultaneously, one each for vertical and horizontal scanning respectively. The wheels will run at 6,000 revolutions per minute, and an increase to 120-line scanning will be made. The light will be produced by a 10 kW arc.

It is interesting to note that the Waseda University Laboratory specialises on Kerr cell reception, whilst the cathode-ray principle is being explored at the Technical College at Hamamatsu, about 100 miles from Tokio.

The other experimental centre is at the Electro-technical Laboratory of the Ministry of Communications, Tokio, where neon, sodium, mercury lamps, etc., form the subject of investigation for television purposes.

It appears that there is little to choose to date between the various methods of reception, at least as far as definition is concerned, although, of course, there is the difference between a projected and subjective picture. Dr. Karawada is of opinion that the more important trend of progress for the near future will lie in the large projected picture for public functions.

The three television authorities mentioned above are subsidised by the Broadcasting Corporation of Japan, which holds a monopoly. Thus the joint research workers pool their findings for the benefit of their national broadcasting authority.

Professor Karawada stated that he was especially impressed with the German transmissions and reception, utilising the intermediate film with 180 lines on a wavelength of 7 metres, and so far as this country is concerned with the potentialities in the latest investigations of the Marconi Company.

A TELEVISION SERVICE FOR GERMANY

FIRST AUTHENTIC DETAILS—EXCLUSIVE TO THIS JOURNAL

The ultra-short wave aerials on top of the Berlin Funkturm tower.

IN view of a number of conflicting and exaggerated statements in the daily Press the German Post Office has issued an official communiqué regarding the present television position in Germany.

The Post Office laboratories, under Dr. Banneitz, have made experiments to determine the range of the Berlin ultra short-wave sight and sound transmissions. It has been possible to receive these at sufficient strength on the summit of the Brocken in the Harz mountains, that is to say, a distance of about 140 miles (200 kilometres), from Berlin. Reception on the Brocken is sufficient to provide modulation for a second ultra short-wave transmitter which could be erected there. Taking the height of the position into consideration, an ultra short-wave transmitter on the Brocken would have a range of from 100 to 150 kilometres and could thus provide important towns such as Hanover, Braunschweig, Magdeburg, Kassel and Erfurt with television reception.

The Post Office has ordered a transportable television transmitter which will first be put into operation on the summit of the mountain next summer.

Line Transmission

Experiments with an entirely new type of cable are about to be made. A short length of cable is at present being laid in Berlin for practical tests. If these prove successful it would be possible to connect television transmitters by cable. It would also be possible to provide two-way television communication between two cities such as Munich and Berlin. As the cost of such a cable would at present be 8 million marks there is no possibility of such a service being opened for some years to come.

So far the official communiqué. Our German correspondent has been able to obtain independent information as regards the position in Berlin.

He interviewed Dr. Kirschstein, the head of the television department of the German Broadcasting Company. Dr. Kirschstein, who had just received the visit of the members of the British Television Committee, who were in Berlin at the time, told him that the morning and afternoon sight and sound television broadcasts would continue to be provided by the Post-Office as they were purely experimental and mainly intended for the laboratories.

A period of one hour in the evening from 9 to 10 p.m. is to be reserved for an experimental service intended for a larger public. It will be necessary for programmes with definite entertainment value to be provided for this. The R.R.G. intend supplying these as, in all probability, the future development of television will be closely connected to broadcasting and therefore it is expedient that the broadcasting company should take up television.

Dr. Kirschstein pointed out that at the present moment it was difficult to see exactly when the evening television programmes would commence. There were a certain number of technical and organisation difficulties to be overcome. He briefly outlined the programme policy for the experimental service. The one hour in the evening will be taken up by 10 minutes special broadcast, then broadcasts of excerpts from the news reels supplied by film companies and of a shortened version of a popular film. At the present moment negotiations between the film companies and the R.R.G. are not quite complete, but it is hoped to reach some mutually satisfactory arrangement within a short time.

The 10 minutes special programme will be filmed by the broadcasters. A small news reel car of the ordinary sound film type will be specially built for the R.R.G. and a second car will be provided for the necessary artificial light apparatus. This team of two cars will accompany the "Echo of the Day" car on its daily round and what will be called a "Mirror of the Day" will be produced for the television programme. A film taken in the normal manner will be processed in the normal manner and broadcast or rather televised in the evening.

Big events such as Hitler speeches will be handled by the special television van first shown at the Berlin Exhibition. In this case television will take place within a few *seconds* of the actual event.

Studio in Berlin

At the Berlin Broadcasting House a studio is being provided where the intermediate-film television transmitter contained in the big van can be installed for the televising of short variety or dramatic scenes. But this possibility is very much of the future, as for the present it will be less expensive and equally entertaining to draw upon existing films as far as the entertainment part of the programme is concerned.

To recapitulate then: Berlin will shortly have an experimental sight and sound, high-definition, ultra-short-wave television service. Technical development will continue and

(Continued at foot of page 545.)

"A Television Service for Germany"

(*Continued from page* 542.)

next summer a second ultra-short-wave transmitter will be erected on the Brocken mountain which will receive its modulation by wireless link from Berlin. The technical development remains in the hands or rather under the guidance of the Post-Office whereas the R.R.G. will now begin to provide programmes (out of ordinary broadcast listeners' money as long as the service remains experimental). The Post-Office have decided to install public looking-in posts at various points in Berlin. The first of these, at the Reichspostmuseum in the centre of the town, will shortly be opened. The exact date of the opening of the television service is not fixed at the time of writing, but it certainly will be before the end of the winter.

Radio Löewe is at present in a position to supply cathode-ray tube receivers at the remarkably low price of RM. 600; these receivers will be available to the general public the moment the experimental programme service commences.

CHAPTER VIII

THE TELEVISION MACHINE

The Secret of the Cabinet—Cabinet Lighting Device—Making a Dimmer—The Cabinet Scenery—Behind the Cabinet—Other Tricks with this Cabinet

We know that television is going to be popular soon. It won't be long before we can tune in the living pictures of our radio stars as easily as we tune in their voices now. But in spite of some public presentations, television is not widely used as yet. The experimenters are still perfecting it.

Invite your friends to a demonstration of a remarkable new television machine and your promise may be regarded with some doubts, but it is pretty certain to rouse keen interest. The guests arrive and are ushered into a room where the machine stands in a curtained doorway. It is a cabinet about the size of a large cabinet radio; in fact, if you wish, it may be housed in a regular radio cabinet bought secondhand from some radio dealer. There are tuning dials and the usual radio gadgets, but the interesting part is the lower half of the cabinet which has a front like a small puppet theatre. There is no curtain, and the inside of the box is in dead blackness.

With a careful eye on your watch, you announce that you will show your audience Alexander and Mose in person. It might be any other suitable radio sketch, of course, but they will serve for an example. Then the lights in the room are extinguished.

There comes the slight hum of the radio—it is the household radio as a matter of fact, tuned in from another room. The audience hears the chimes and the station announcement and the theme music of the act.

Then comes a faint flicker of light from the cabinet, a gradual glowing, and they are staring at a small stage set. It looks pretty much like a cardboard stage set, which is just what it is, except that the lighting is a little dim. Just as the spectators are wondering what happens next, it happens—Alexander and Mose are in the picture, and not puppets but very evidently live human beings. Perhaps they look a little like two boy friends of yours with their faces blacked up, but there is no denying they are in a box which can't possibly be big enough to hold them in the flesh. As the real Alexander and Mose go through the evening sketch in sound, the "television" actors go through it in dumb show, and if they are clever at adapting their actions to the radio material it isn't hard to imagine that this really is television.

THE SECRET OF THE CABINET

Of course the machine is a hoax. It is based on the principle of an old stage illusion which has produced ghosts and mysteries of magic for many generations of magicians. I do not recall ever having seen the present adaptation of the idea, and I am passing it along to you as something new and timely. To build the cabinet is not difficult, and the material needed will not prove very expensive.

The actors in the sketch just described were in a room adjoining the one where the audience sat, but the scenery was in the cabinet itself, a small set painted on cardboard.

To make the actors seem a part of the cabinet picture, you will need a good plate-glass mirror and a sheet of clear, highly polished, thin glass. These and a few tiny flashlight

The Secret of the Cabinet.

The actor stands under a brilliant light against a black curtain. His image is reflected by mirror in the box on to the miniature stage, while radio brings in the broadcast programme.

incandescent bulbs, one special incandescent bulb, and an electric dry-cell battery make up about all the materials aside from the cabinet itself.

The cabinet I fixed up stands 30 in. high, 21 in. wide and 19 in. deep, and is made of plywood. It was a part of the packing in which an electric refrigerator was shipped, and if you are looking for such a cabinet it is not a bad idea to talk to the local agent for one of these or similar machines and see if he won't be glad to help you out.

In the lower half of the front is the gilt cardboard mask of a theatre front, with an opening cut out that measures 10 in. wide by 9 in. high. I found that size frame just about perfect for the size image the illusion would produce.

The upper and blank part of this cabinet front has several old radio dials attached to give it a weighty and scientific aspect. Needless to say, the dials have nothing to do with it. If you should be able to get hold of a large, old-fashioned radio cabinet as I suggested, the dials will be there already and your show can be housed very handsomely.

The mirror will prove the most expensive item. Good-quality mirror glass is needed, and the glazier must cut you a mirror just 18½ by 20 in. for such a cabinet as I used. That is, the mirror must completely fill the upper half of the box when placed there at an angle of 45 degrees. If there is an old sideboard about the house, or you know where you can get hold of a remnant of large broken mirror of that size, take that to your glazier to be cut and you will save yourself money. If bought new, such a piece of glass should cost about five shillings.

For the clear glass, which should measure the same size as the mirror, I suggest you go to the nearest car cemetery and try striking a bargain for the plate-glass window out of an old closed car. Or buy the thin, clear, high-grade glass like that in picture frames. The thinner and more highly polished the glass is the better.

Now the cabinet will need a little carpentry. A hole the size of the opening in the theatre front must be sawed out of the front wall and in the middle of the lower half. To each side wall fix slats that will support the mirror and the clear glass. Notice by the drawing that the mirror slopes at an angle of 45 degrees, its top toward the opening in the rear of the box, and the clear glass which occupies the lower half of the box slopes exactly parallel to the mirror.

THE TELEVISION MACHINE.

Sketch shows one side removed.

A—the mirror which faces downward. B—Clear glass. C—Cardboard scenery. The dials and knobs on the front are for show only.

The opening in the rear wall should be exactly the dimension of the upper half. The mirror must have a clear view so that it will reflect the picture against the clear glass.

When you have placed the slats and experimented with the two glasses until you are sure they fit securely and snugly, remove them from the box and paint the entire inside of the cabinet a dull, solid black. The best paint for this is lampblack mixed with turpentine. No detail, not even a pin point, must be overlooked by your paint brush.

CABINET LIGHTING DEVICE

Lights for the cabinet are the next consideration. Five of the small bulbs sold for flashlamps are ample, or if you use more battery cells two of the small motor-car headlights which burn on an 8-volt circuit are enough. The tiny flash-lamp bulbs can be lighted from a single 4½-volt dry cell.

In a board base, bore five holes just big enough to fit the metal sockets of the little lamps. Across the middle of these five holes saw a notch to the same depth as the holes. Stretch a small copper wire along this notch and make it fast at either end to a screw set in the side of the base. Now put another screw at one end of the base and make the end of a second length of wire fast to it. Just where the wire reaches the first hole make a snug turn of it around the base of a bulb. Fasten the turn by tacking the wire to the board just where the bends end. This makes a socket into which you can screw the bulb until its lower end touches the copper wire in the slot. In this way you have the bulb attached to two wires which can be connected to the two terminals of a dry-cell battery. The other lights are wired in the same way, the same wire serving for all.

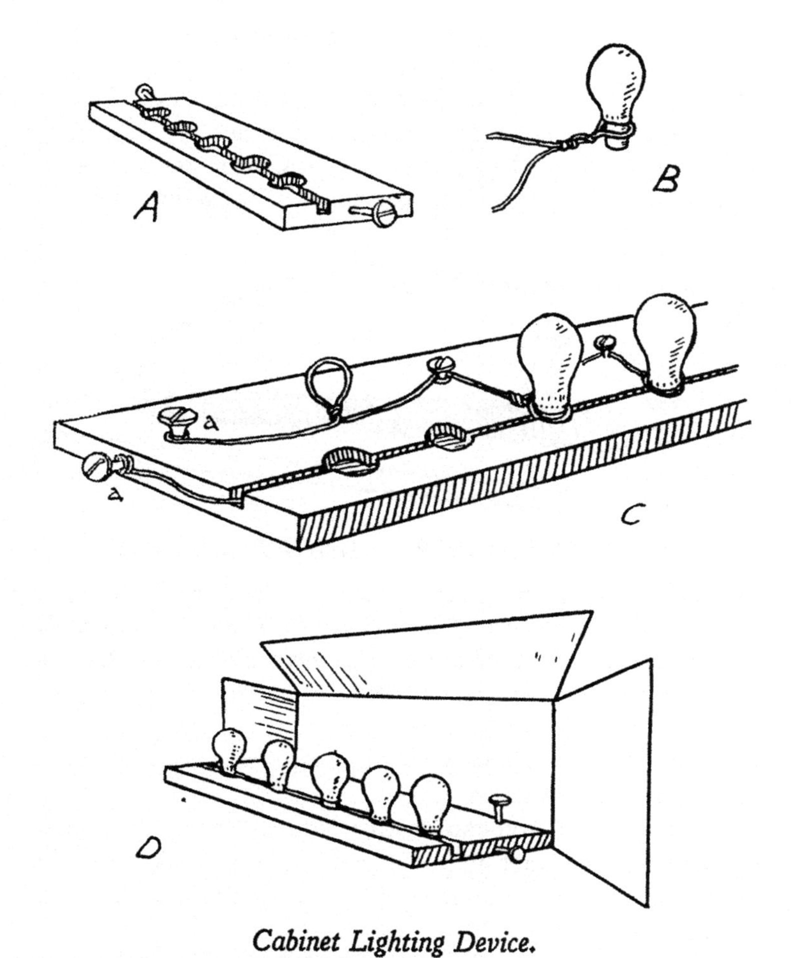

Cabinet Lighting Device.

A. The base-board. B. Wire lamp-holder. C. The circuit; points aa lead to battery. D. The complete job.

MAKING A DIMMER

The lighting of your show is a most important feature. Your aim is to show the audience a stage set lighted by half-light, and, on top of that picture, to superimpose the reflections of living actors moving about under a brilliant

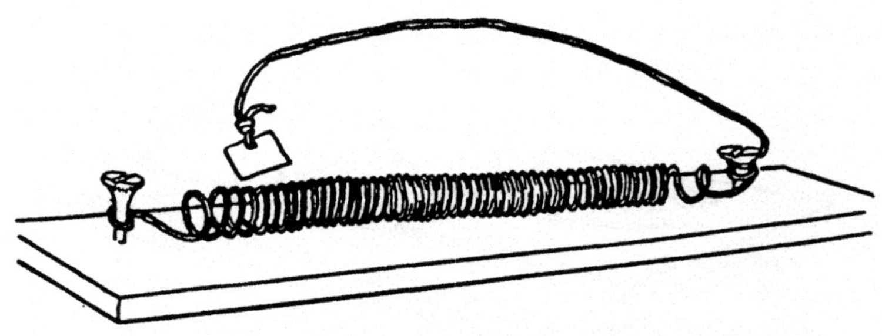

A DIMMER FOR CABINET LIGHTS.

Use the coil spring from a blind roller. A piece of tin or a copper wire controls the amount of current used. The end screws enable the dimmer to be inserted into the lamp circuit as explained in text.

light so that their images completely obscure the detail of scenery behind them.

To make sure you can do this, you will need a dimming device for the cabinet lights. The dimmer is an old stand-by of the theatre where lighting effects contribute so much to the illusion.

A good dimmer for five little flashlamp bulbs can be made out of the coiled inner spring of an old blind roller. Take out the spring and mount it on a suitable narrow wood base, making each end fast to a screw as shown in the drawing. In order to bend the hardened steel wire you will have to hold a short length of either end over the furnace fire or over an open gas flame until it has become white-hot. Let it cool slowly and it will prove pliable.

Now cut a piece of copper wire a little longer than the coil spring and fasten one end firmly to one of the screws. To the other end of the wire attach a small square of copper or tin. The dimmer is ready for use as soon as you have connected the lamps to the dry cell in the way already described, to make sure that they all light properly.

To wire the dimmer into the circuit, disconnect one of the wires from the dry cell and make it fast to one of the dimmer screws. Insert a new piece of copper wire leading from the second dimmer screw to the vacated terminal on the cell. Now, when the tin square is moved along the coil spring, the lights will brighten from a dim red spark to a brilliant white. The tin may be stuck in the spring at the point where the desired effect is obtained.

Fasten the dimmer to the rear of the cabinet top. Fasten the row of lights, which should have a small cardboard shade and reflector, into the front of the cabinet, just under the clear glass at the bottom. The rear of the reflector, which faces the audience, and every detail of the baseboard must be painted the same dead black as the inside of the cabinet.

THE CABINET SCENERY

What scene you use will depend upon which broadcast you mean to present by "television". Or perhaps you will want to make a different use of the cabinet.

If you wish, you can use a full back-drop that just fits into the rear bottom half of the cabinet and add some cut-out flats in front of it. Or you can use a few cut-out pieces against the black back wall of the cabinet. Avoid any dead white in this scenery as that will tend to show up under any light and spoil the illusion. Better paint in browns and blues and dark red or heavy grey. Any solid blacks are apt to vanish completely under this unusual lighting.

BEHIND THE CABINET

Behind the cabinet is the very important "back-stage" of this pseudo-scientific show. It is here that the living actors perform.

The cabinet itself is set in a doorway, and around it a curtain should be draped so that the hole in the back of the cabinet is left unobstructed but no ray of light can reach the audience in front.

Whatever of this rear room is visible in the reflection—and that you can determine only by experimenting—must be covered with cloth of dead black. For my machine, which presents living persons in half-length portrait only, 4 yards of black muslin tacked to the rear wall of the room was plenty. Remember no furniture or any detail of the room must be in range of the cabinet opening, nothing but dead-black wall. It is the actors only who are reflected in the clear glass, and all around them, *through* the glass, the audience sees the cardboard scenery.

The actors therefore must have a brilliant lighting. The simplest way to make this light is to purchase one of the special lamps used for making indoor moving pictures. The life of these lamps is only an hour or two; but as their cost is low and they will not be used long at a time, they are about the best way to light your show. If such a light is not used, the actors can be spotlighted by a very brilliant incandescent bulb fastened into a shaded bridge lamp and so adjusted that they get the full benefit of the beam. Or the brilliant beam from an amateur moving-picture projector or a good magic lantern could be utilized. The important thing is to light your actors with all the brilliance that can be arranged. In principle this illusion is not unlike the method by which moving-picture actors are photographed into scenes taken a thousand miles from the studio. Though the cabinet does not make a photograph, it "double-prints", superimposing one image on top of another ; and unless the added image is stronger, more brilliantly lighted, the result is a double exposure or ghost.

OTHER TRICKS WITH THIS CABINET

Speaking of ghosts, do you know you can produce an excellent ghost illusion with this same cabinet ?

By lighting the scenery inside the cabinet a little more brilliantly than the living actor behind it, a phantom figure seems to move about inside ; and if the figure wears an appropriate white sheet, the effect is rather eerie.

Another use for such a cabinet is the transformation illusion. A living actor who occupies a fixed position can be transformed into a grinning skeleton. The skeleton must be a toy and the same size as the image of the living person, which you can determine only by experiment. The puppet can be made easily by painting white bones on black cardboard of just the right size. If the lights are managed properly the skeleton will be invisible when the cabinet lights are out but will fade in when the lights at the rear are dimmed and the cabinet lighting increased in brilliance. Perhaps a more cheerful demonstration of the magic of the lights would be to fade the image of some boy you know into a little cardboard ballet-dancer which dances inside the cabinet.

Another pleasing and mystifying use for the cabinet would be to give a show performed by living actors in a setting of cardboard scenery. Unless the room behind the cabinet is very large, it will be impossible to get full-length reflections of your living actors. This might be overcome if the mirror in the cabinet were one of those curved reducing mirrors, which reflect a much smaller image than flat glass. Unfortunately for most of us showmen, such mirrors are rather expensive.

However, I am very sure that ingenious young experimenters who read this and build such a cabinet will find some excellent new uses for it.